PROCEEDINGS OF THE FIFTH INTERNATIONAL SCHOOL

ADVANCED COURSES OF MATHEMATICAL ANALYSIS V

EDITORS

JUAN CARLOS NAVARRO PASCUAL
EL AMÍN KAIDI
Universidad de Almería, Spain

PROCEEDINGS OF THE FIFTH INTERNATIONAL SCHOOL

ADVANCED COURSES OF MATHEMATICAL ANALYSIS V

Universidad de Almería, Almería, Spain 12 – 16 September 2011

World Scientific

NEW JERSEY • LONDON • SINGAPORE • BEIJING • SHANGHAI • HONG KONG • TAIPEI • CHENNAI • TOKYO

Published by

World Scientific Publishing Co. Pte. Ltd.
5 Toh Tuck Link, Singapore 596224
USA office: 27 Warren Street, Suite 401-402, Hackensack, NJ 07601
UK office: 57 Shelton Street, Covent Garden, London WC2H 9HE

British Library Cataloguing-in-Publication Data
A catalogue record for this book is available from the British Library.

ADVANCED COURSES OF MATHEMATICAL ANALYSIS V
Proceedings of the Fifth International School

ISBN 978-981-4699-68-6

Printed in Singapore

PREFACE

The courses of Mathematical Analysis in Andalusia started out at the University of Cádiz in September 2002. The initiative was launched by our dear colleague, Professor Antonio Aizpuru, who had passed on early. His proposal was supported by several groups of Mathematical Analysis from different Andalusian universities.

The objectives that inspired this project can be shown in three fundamental ideas:

- Providing a meeting place and collaboration between different groups of Mathematical Analysis in Andalusia.
- Contributing to the strengthening in the training of young researchers in this area, providing easier access to new trends, results, problems, and advanced techniques.
- Putting at the disposal of the international scientific community the course contents through a publication of the widest possible dissemination.

After the already mentioned edition of the event held in 2002, other editions took place in Granada (2004), Huelva (2007) and Cádiz (2009).

This volume is dedicated to the V International Course of Mathematical Analysis in Andalusia, which was carried out at the University of Almería from 12 to 16 September 2011.

For the organization of the event we counted on the financial support of the Spanish national, Andalusian regional and Almería local institutions. Likewise, we received economic assistance from the i-MATH project, the University of Almería and the cooperative group Cajamar.

Four interventions (minicourses) for three hours each and five hourly plenary lectures were scheduled. Moreover, the interested participants presented their most recent contributions as short communications or posters.

Taking into account the high scientific level, depth and current relevance of exposed material, as well as the carefulness of the experts in its preparation, we think the publication can be very useful for graduate students and researchers from multiple domains more or less close to Mathematical Analysis.

This book contains an updated and extended version of the minicourses and plenary lectures taught by Professors Bernardo Cascales, Fernando Cobos, Luboš Pick, Bernard Russo (minicourses) and Javier Duoandikoetxea, José María Martell, Julio D. Rossi, Enrique A. Sánchez Pérez and Charles Swartz (plenary lectures).

In his section, Bernardo Cascales discusses the notions of lower semicontinuity, upper semicontinuity and measurability of multifunctions. The author summarizes his contributions in this field with an agile and entertaining style, including interesting applications of analytical or topological nature while presenting the current interest of the notions and results.

Fernando Cobos provides a brief and elegant introduction to the theory of interpolation of operators, a branch of functional analysis with important applications to numerous areas of mathematics. Following the presentation of classical theorems on interpolation of operators between L_p-spaces, he reviews the main interpolation methods and studies the behaviour under interpolation of compactness and weak compactness of operators.

For his part, Luboš Pick presents the recent developments in the theory of function spaces and their applications to sharp Sobolev embeddings. He blends to perfection the depth of the contents with an instructive and motivating speech. The author shows particularly the suitability in this context of the so-called rearrangement-invariant spaces.

The extensive and illuminating contribution by Bernard Russo is an introduction to nonassociative algebra, including triple systems as well as algebras, continuous cohomology, and the interplay between Jordan theory and operator space theory. The starting point of the whole discussion is the product rule for differentiation. The author highlights the role of contractive projections in the structure theory of Jordan triples and quantum functional analysis.

Javier Duoandikoetxea presents a simple proof of the extrapolation theorem of J. L. Rubio de Francia in the context of Muckenhoupt weights. The modified forms of factorization of weights proposed by the author allow to obtain the sharp form of the extrapolation theorem in the case of the classical Ap weights.

José María Martell's section contains a joint work with David Cruz-Uribe and Carlos Pérez showing that the Muckenhoupt-Wheeden conjectures are true for off-diagonal inequalities. The authors also prove an analogous result for the Haar shift operators with the Hardy-Littlewood maximal operator replaced by the dyadic maximal operator.

With a didactic and motivating style, Julio D. Rossi introduces the reader to some techniques and relevant results for the theory of second order elliptic partial differential equations and their connections with the game theory.

The goal of the contribution by Enrique A. Sánchez Pérez is to show the factorization version of the Radon-Nikodym theorem for vector measures and its applications in general functional analysis and harmonic analysis.

Charles Swartz's exposition is devoted to the Orlicz-Pettis theorem, one of the first and more influential abstract results in functional analysis. The author describes a generalization of this theorem to multiplier convergent series.

It only remains to express our deep appreciation to each and every one of the authors mentioned by their dedication in preparing manuscripts and their friendly cooperation in all stages of this initiative. We also thank the support of the various members of committees involved in the course, the enthusiasm of all participants and the assistance of the government institutions and collaborating entities.

The editors

J. C. Navarro Pascual — Almería, Spain
El Amín Kaidi — Spring 2015

SPONSORS

OFFICIAL PLANS AND GOVERNMENT INSTITUTIONS

Ministerio de Ciencia e Innovación (Acción Complementaria MTM2011-14555-E)
Consolider Ingenio Mathematica (MIGS-CEXT-0465)
Plan Propio de Investigación de la Universidad de Almería
Facultad de Ciencias Experimentales
Departamento de Álgebra y Análisis Matemático de la Universidad de Almería
Ayuntamiento de Almería
Diputación de Almería
Ayuntamiento de Roquetas de Mar

FOUNDATIONS

Grupo Cooperativo Cajamar

ANDALUSIAN UNIVERSITIES

Universidad de Almería
Universidad de Cádiz
Universidad de Granada
Universidad de Huelva
Universidad de Jaén
Universidad de Málaga
Universidad de Sevilla
Universidad Pablo de Olavide

ORGANIZING COMMITTEES

SCIENTIFIC/ORGANIZING COMMITTEE

Tomás Domínguez Benavides – Universidad de Sevilla
Antonio Fernández Carrión – Universidad de Sevilla
Daniel Girela Álvarez – Universidad de Málaga
El Amín Kaidi – Universidad de Almería
Fernando León Saavedra – Universidad de Cádiz
Miguel Marano Calzolari – Universidad de Jaén
Francisco Javier Martín Reyes – Universidad de Málaga
Juan Francisco Mena Jurado – Universidad de Granada
Juan Carlos Navarro Pascual – Universidad de Almería
Francisco Javier Pérez Fernández – Universidad de Cádiz
Rafael Payá Albert – Universidad de Granada
Carlos Pérez Moreno – Universidad de Sevilla
Cándido Piñeiro Gómez – Universidad de Huelva
Francisco Roca Rodríguez – Universidad de Jaén
Ramón Jaime Rodríguez Álvarez – Universidad de Huelva
Ángel Rodríguez Palacios – Universidad de Granada
Luis Rodríguez Piazza – Universidad de Sevilla
María Victoria Velasco Collado – Universidad de Granada
Antonio Villar Notario – Universidad Pablo de Olavide

LOCAL ORGANIZING COMMITTEE

Enrique de Amo Artero – Universidad de Almería
José Carmona Tapia – Universidad de Almería
Antonio Jiménez Vargas – Universidad de Almería
Antonio Morales Campoy – Universidad de Almería
Miguel Ángel Navarro Pascual – Universidad de Almería
María Isabel Ramírez Álvarez – Universidad de Almería
José Juan Rodríguez Cano – Universidad de Almería
Agripina Rubio Flores – Universidad de Almería
Mª Gracia Sánchez-Lirola Ortega – Universidad de Almería
Moisés Villegas Vallecillos – Universidad de Almería

CONTENTS

Measurability and semi-continuity of multifunctions

Bernardo Cascales

Departamento de Matemáticas, Universidad de Murcia,
30.100 Espinardo. Murcia, Spain
E-mail: beca@um.es

The following pages contain details of a mini-course of three lectures given at the *V International Course of Mathematical Analysis of Andalucía (CIDAMA), Almeria, September 12-17, 2011.* When I was invited to give this mini-course and thought about possible topics for it, I decided to talk about multifunctions because they have always been present in my research on fields theoretically apart from each other as topology and integration theory. Therefore you will find here my biased views regarding part of the research that I have done over the years. The proofs for this material have been published elsewhere by me or by some other authors. This mini-survey is written attending to the invitation of the editors of this book with the sole purpose of witnessing the given mini-course and with the aim of providing the reader with connections and ideas that usually are not written in research papers. I thank the organizers of CIDAMA V as well as the editors of the book for their kind invitation to give the lecture and write this mini-survey.

In these notes we shall deal with multifunctions (or set-valued maps). Multifunctions naturally appear in analysis and topology, for instance via inequalities, performing unions or intersections with sets indexed in another set, considering the set of points minimizing an expression, etc. First, we will present some results about semi-continuity of multifunctions, namely, lower semi-continuity and an application of Michael's selection theorem. Then we will deal with upper semi-continuity of multifunctions and an application to the generation of K-analytic structures with consequences in topology and functional analysis. We will finish by showing a few results about measurability for multifunctions related to the Kuratowski-Ryll-Narzesdky selection theorem and their implications to integrability of multifunctions for non separable Banach spaces.

Keywords: Set-valued map; multifunction; lower semi-continuous; upper semi-continuous; measurable; compactness; metrizability; Lindelöf property; K-analytic space; Pettis integrability; Effros measurability.

1. Settings, first definitions and introduction

Our notation and terminology is standard and it is either explained when needed or can be found in our references for Banach spaces[1,2], topology[3,4] and vector measures and integration[5].

By capital letters $D, E, S, X, Y, \Omega, \ldots$ we denote sets. Sometimes these sets are endowed with a topology, *i.e.*, they are topological spaces. In particular by $(E, \|\cdot\|)$ we denote a real Banach space (or simply E if $\|\cdot\|$ is tacitly assumed): B_E stands for the closed unit ball in E, S_E for the unit sphere, E^* for the dual space of E and E^{**} for the bidual space of E; w is the weak topology and w^* is the weak* topology in the dual. Throughout this paper (Ω, Σ, μ) is a complete finite measure space.

Definition 1.1. A multifunction (set-valued map) is a map ψ from a set X into the family of subsets 2^Y of another set Y, *i.e.*, for each $x \in X$ the image $\psi(x)$ is a subset of Y.

Example 1.1.

(1) The map $\log : \mathbb{C} \setminus \{0\} \to 2^{\mathbb{C}}$ that sends every $z \in \mathbb{C} \setminus \{0\}$ to the set $\log(z)$ of all logarithms of z is a multifunction, see[6] p. 39.
(2) If $g, G : X \to \mathbb{R}$ are two given functions with $g(t) \leq G(t)$ for every $t \in X$, then $\psi(t) := [g(t), G(t)]$ defines a multifunction $\psi : X \to 2^{\mathbb{R}}$, see figure 1.

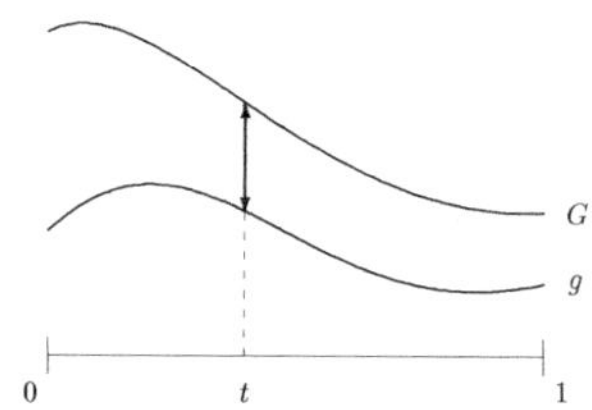

Fig. 1. Example of multifuncion

(3) If $f : Y \to X$ is an onto map, then $\psi(x) := f^{-1}(x)$, $x \in X$, defines multifunction $\psi : X \to 2^Y$.
(4) If K is a Hausdorff compact space the map $\psi : C(K) \to 2^K$ given by

$$\psi(f) := \{x \in K : |f(x)| = \sup_{t \in K} |f(t)| =: \|f\|_\infty\}$$

is a multifunction defined in the Banach space of scalar-valued continuous functions $C(K)$.
(5) If E is a Banach space the duality mapping $J : E \to 2^{B_{E^*}}$ given by

$$J(x) := \{x^* \in B_{E^*} : \|x\| = x^*(x)\}$$

is a multifunction, see[2] p. 343.

(6) If E is a Banach space and $Y \subset E$ is a closed proximinal subspace, then the metric projection $P_Y : E \to 2^Y$ given by

$$P_Y(x) := \{y \in Y : \|x - y\| = \inf_{z \in Y} \|x - z\| =: d(x, Y)\}$$

is a multifunction (recall that by definition Y being proximinal means $P_Y(x) \neq \emptyset$ for every $x \in E$, see[7] §5).

(7) If E is a Frchet space, see[8] §18.2, and $U_1 \supset U_2 \supset \cdots \supset U_n \supset \cdots$ is a basis of neighborhoods of 0 then $\psi : \mathbb{N}^{\mathbb{N}} \to 2^E$ given by

$$\psi(\alpha) := \bigcap_{k=1}^{\infty} n_k U_k, \text{ with } \alpha = (n_k)_k,$$

is a multifunction with $\psi(\mathbb{N}^{\mathbb{N}}) = E$, $\psi(\alpha) \subset \psi(\beta)$ if $\alpha \leq \beta$ (coordinatewise) in $\mathbb{N}^{\mathbb{N}}$ and $\{\psi(\alpha) : \alpha \in \mathbb{N}^{\mathbb{N}}\}$ is a fundamental family of bounded sets of E.

(8) If $E = \lim\limits_{\to} E_n$ is an (LF) space, see[8] §19.5, and

$$U_1^m \supset U_2^m \supset \cdots \supset U_n^m \supset \cdots$$

is a basis of neighborhoods of 0 in E_m then $\psi : \mathbb{N}^{\mathbb{N}} \to 2^{E'}$ given by

$$\psi(\alpha) := \overline{\operatorname{aco}\Big(\bigcup_{k=1}^{\infty} U_{n_k}^k\Big)}^{\circ}, \text{ with } \alpha = (n_k)_k,$$

is a multifunction with $\psi(\mathbb{N}^{\mathbb{N}}) = E'$, $\psi(\alpha) \subset \psi(\beta)$ if $\alpha \leq \beta$ (coordinatewise) in $\mathbb{N}^{\mathbb{N}}$ and $\{\psi(\alpha) : \alpha \in \mathbb{N}^{\mathbb{N}}\}$ is a fundamental family of equicontinuous subsets of E' (polars A° are taken in the dual pair $\langle E, E' \rangle$, see[8] §20.8).

(9) If E is a Banach space, $f : \Omega \to E$ and $r : \Omega \to [0, \infty)$ are functions then $F : \Omega \to 2^E$ given by

$$F(\omega) := f(\omega) + r(\omega) B_E, \, \omega \in \Omega,$$

is a multifunction.

(10) If $\{f_i : \Omega \to E\}_{i \in I}$ is a family of functions we can consider the multifunction $F : \Omega \to 2^E$ defined by

$$F(\omega) := \overline{\operatorname{co}\{f_i(\omega) : \, i \in I\}}.$$

The three intimately connected notions below are the ones that we shall deal with in these notes.

Definition 1.2. Let X and Y be topological spaces and let $\psi : X \to 2^Y$ be a multifunction. We say that ψ is lower semi-continuous (l.s.c.) if the set

$$\{x \in X : \psi(x) \cap O \neq \emptyset\}$$

is open for every open subset O of Y, see[9] §43 and[10] Ch. 7.

Definition 1.3. Let X and Y be topological spaces and let $\psi : X \to 2^Y$ be a multifunction. We say that ψ is upper semi-continuous (u.s.c.) if the set

$$\{x \in X : \psi(x) \cap F \neq \emptyset\}$$

is closed for every closed subset F of Y, see[9] §43 and[10] Ch. 7.

It is easy to check that ψ as above is u.s.c. if, and only if, for every $x_0 \in X$ and every open set $V \supset \psi(x_0)$ in Y, there is an open neighborhood $U \subset X$ of x_0 such that $\psi(x) \subset V$ for every $x \in U$, see figure 2.

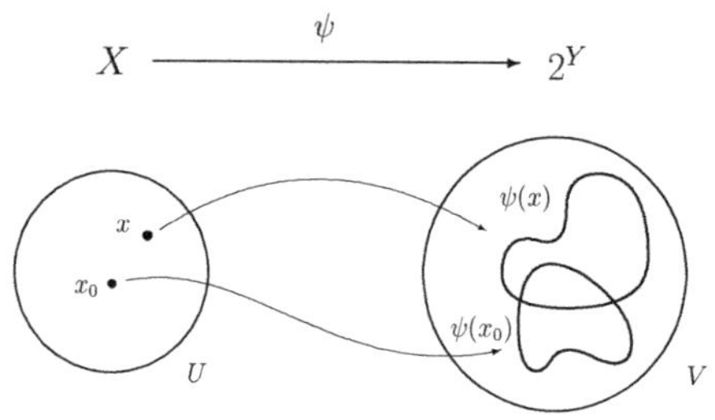

Fig. 2. Upper semi-continuity

Definition 1.4. Let (Ω, Σ) be a measurable space and let E be a Banach space. A multifunction $F : \Omega \to 2^E$ is said to be Effros measurable if

$$\{t \in \Omega : \ F(t) \cap O \neq \emptyset\} \in \Sigma \text{ for each open set } O \subset E. \qquad \text{(E)}$$

More general notions of measurability can be found in the literature: we remark that the notion above makes sense for any topological space in the range, see[10–12].

Natural examples illustrating the above notions are easy to provide. Beyond those spread out in the literature we isolate, for the purposes of these notes, the following ones.

Example 1.2.

(1) *(A lower semi-continuous multifunction)* If X is a topological space and we assume that in example 1.1.(2) $g : X \to \mathbb{R}$ is upper semi-continuous and $G : X \to \mathbb{R}$ is lower semi-continuos, then it is easily checked that $\psi(t) := [g(t), G(t)]$ defines a lower semi-continuous multifunction $\psi : X \to 2^{\mathbb{R}}$.

(2) *(An upper semi-continuous multifunction)* Let us consider $\mathbb{N}$ endowed with its discrete topology and $\mathbb{N}^{\mathbb{N}}$ with its product topology. The multifunction ψ defined in example 1.1.(7) is upper semi-continuous whenever E is Frchet-Montel, see[8] §27.2 for the definition, and the basis $U_1 \supset U_2 \supset \cdots \supset U_n \supset \cdots$ of neighborhoods of 0 is made up of closed sets.

(3) *(A measurable multifunction)* Assume here that E is a separable Banach space. When dealing with Effros measurability for Borel σ-algebras, the first examples that come to mind are l.s.c. multifunctions (and u.s.c. multifunctions if they take compact values, see[11] Cor. III.3). A quite remarkable example regarding measurability of multifunctions is the one provided by example 1.1.(10) when $I = \mathbb{N}$ and each $f_n : \Omega \to E$ is measurable. A celebrated result by Castaing-Valadier says that all Effros measurable multifunctions $\psi : \Omega \to 2^E$ with closed values are of the form described in example 1.1.(10) with $I = \mathbb{N}$ and each f_n measurable, see[11] Th. III.9.

Definition 1.5. Given a multifunction $\psi : X \to 2^Y$ a *selector* (*selection*) for ψ is a single-valued function $f : X \to Y$ such that

$$f(x) \in \psi(x),$$

for every $x \in X$, see figure 3.

In our views the leading role of multifunctions in many aspects of mathematical analysis and topology is due to their proliferation and the strong consequences that can be obtained from their study. In the rest of these notes we shall present some results connected with our research that repeatedly go once and again to one of the ideas below:

(a) when dealing with multifunctions defined between topological spaces $\psi : X \to 2^Y$ semi-continuity properties of ψ can be used many times:
   - to transfer properties from X to Y;
   - to find "good" selectors for ψ (from a topological point of view);

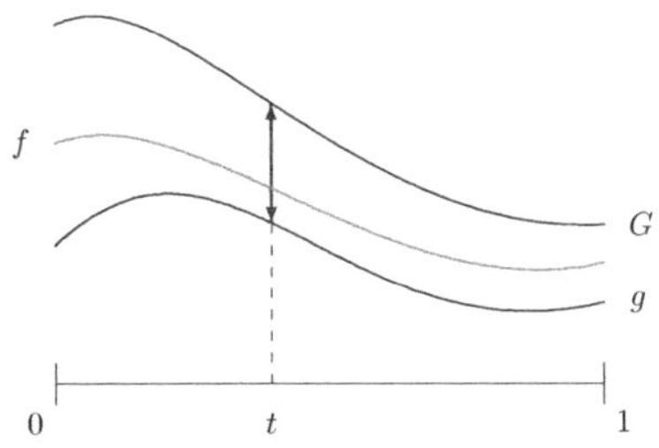

Fig. 3. Selection

(b) when dealing with multifunctions $F : \Omega \to 2^E$ their measurability can be used many times:

- to find "good" selectors for F (from a measurability point of view);
- to study properties of integrability for F.

For the study of questions as in (a) many names come to our minds, a few of which are: Argyros, Arkhangel'skiĭ, Jayne, Kuratowski, Mercourakis, Michael, Negrepontis, Talagrand, Rogers, etc. For the study of questions as in (b) authors like Aumann, Debreu, Hess, Kuratowsky, Ryll Nardzewski, etc. made very important contributions. Many other authors have made quite important contributions too to topics related to (a) and (b) above. Since it is imposible to name all of them we cut our list short without diminishing the importance of contributions of those that we cannot name. Let us stress though that very in particular, Debreu[13] and Aumann[14] established very important results in mathematics and in some models in economy when dealing with multifunctions (notice that Debreu and Aumann received the Nobel prize in economy, 1983 and 2055 respectively).

We finish this introduction collecting three superb selection results.

Theorem 1.1 (Michael,[15]). *Assume that X is a paracompact space, that E is a Banach space and that $\psi : X \to 2^E$ is a l.s.c. multifunction such that $\psi(x)$ is closed, convex and nonempty for every $x \in X$. Then ψ has a continuos selector,* i.e., *there is a continuous function $f : X \to E$ such that $f(x) \in \psi(x)$ for every $x \in X$.*

Amongst the many applications of Michael's theorem we can mention Bartle-Graves' theorem (if $F \subset E$ is a closed subspace of E, the quotient map $\pi : E \to E/F$ has a positive homogeneous lifting, see[16] Prop. 1.19) and Borsuk-Kakutani-Didunji's theorem (if K is compact and $H \subset K$ is closed and metrizable, there is a simultaneous extension continuous operator $T : C(H) \to C(K)$ such that $\|T\| = 1$ and $T1 = 1$, see[16] Prop. 1.21).

Theorem 1.2 (Jayne-Rogers,[17] Th. 5.4). *Let E be an Banach space. The following statements are equivalent:*

(i) E is Asplund, i.e., *every separable subspace has separable dual;*
(ii) the duality mapping $J : E \to 2^{B_{E^*}}$ has a Baire-1 selector, i.e., *there is a sequence of norm-to-norm continuous maps $f_n : E \to E^*$ such that for every $x \in E$ there exists $\lim_n f_n(x) \in J(x)$.*

We should note that the implication $(i) \Rightarrow (ii)$ is based on the fact that the duality mapping J is norm-to-w^* upper semi-continuos and that whenever E is an Asplund space then (B_{E^*}, w^*) is norm-fragmented, see[18,19]. The above result can be found in[20] Th. 5.2, Rem. 5.11. Such a remarkable selection result has played a fundamental role in renorming theory and in the study of boundaries in Banach spaces,[17,20–22].

Theorem 1.3 (Kuratowski-Ryll Nardzewski,[23]). *Let (Ω, Σ, μ) be a complete probability space and $F : \Omega \to 2^E$ a multifunction with closed non empty values of E. If E is separable and F is Effros measurable, then F admits a measurable selector f,* i.e., *there is a $f : \Omega \to E$ such that $f^{-1}(O) \in \Sigma$ for every open set $O \subset E$ and $f(\omega) \in F(\omega)$ for every $\omega \in \Omega$.*

A proof for the above Kuratowski-Ryll Nardzewski's theorem can be found in[11] Th. III.6 and[10] Th. 14.2.1. Over the years Kuratowski-Ryll Nardzewski's theorem has been the milestone result to build up several theories of multifunction integration that henceforth have been presented only for separable Banach spaces as range spaces.

2. Lower semi-continuity for multifunctions, an application

This section is the witness of how lower-semicontinuity and Michael selection theorem ignited the appearance of tools that allowed to rewrite most of the known results about pointwise and weak compactness in C_p-theory and functional analysis from a quantitative point of view.

A straightforward application of Michael's selection theorem 1.1 is the following result.

Theorem 2.1 ([16] Pro. 1.18). *Let X be a paracompact space and let $f_1 \leq f_2$ be two real functions on X such that f_1 is upper semi-continuous and f_2 is lower semi-continuous. Then, there exists a continuous function $h \in C(X)$ such that $f_1(x) \leq h(x) \leq f_2(x)$ for all $x \in X$.*

Proof. It is easily proved that the multifunction $\psi : X \to 2^{\mathbb{R}}$ given by $\psi(x) := [f_1(x), f_2(x)]$, $x \in X$, is l.s.c. and therefore theorem 1.1 can be used to conclude the existence of the continuous selection h, see figure 4. □

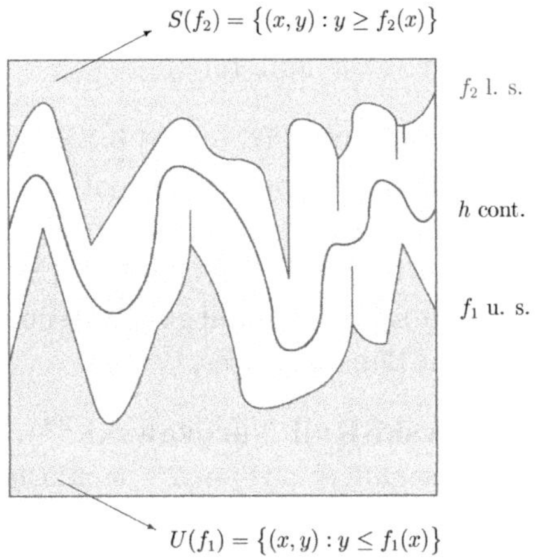

Fig. 4. A sandwich result

As a consequence of the above result we have.

Theorem 2.2 ([16] Pro. 1.19). *Let X be a paracompact space. For a given bounded function $f \in \mathbb{R}^X$ the distance of f to the subspace of bounded and continuous functions on X is given by*

$$d(f, C_b(X)) = \frac{1}{2}\operatorname{osc}(f)$$

where

$$\operatorname{osc}(f) = \sup_{x \in X} \operatorname{osc}(f, x) = \sup_{x \in X} \inf\{\operatorname{diam} f(U) : U \subset X \text{ open}, x \in U\}.$$

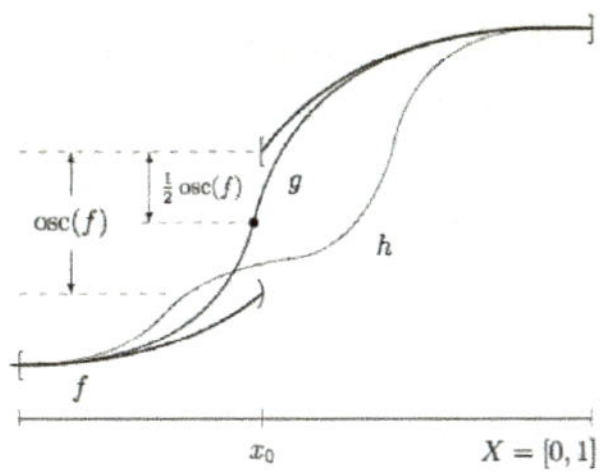

Fig. 5. Distance to continuous

Proof. As in[16] Pro. 1.18 put $\delta = \frac{1}{2}\operatorname{osc}(f)$. It is clear that the distance is at least δ. To prove the other direction, define

$$f_1(x) := \inf_{U \in \mathcal{V}_x} \sup_{z \in U} f(z) - \delta$$

$$f_2(x) := \sup_{U \in \mathcal{V}_x} \inf_{z \in U} f(z) + \delta$$

Then $f_1 \leq f_2$. It is easy to check that f_1 is upper semi-continuous and f_2 is lower semi-continuous. By theorem 2.1, there is a continuous function $h \in C(X)$ such that

$$f_1(x) \leq h(x) \leq f_2(x)$$

for every $x \in X$. On the other hand, for every $x \in X$ we have

$$f_2(x) - \delta \leq f(x) \leq f_1(x) + \delta$$

and therefore

$$h(x) - \delta \leq f_2(x) - \delta \leq f(x) \leq f_1(x) + \delta \leq h(x) + \delta.$$

So $d(f,h) \leq \delta = \frac{1}{2}\operatorname{osc}(f)$ and this finishes the proof. □

When X is only a normal space and the functions are not necessarily bounded a proof for the above result can be found in[24].

Theorem 2.2 has been the key and inspiration to prove the four results that follow.

Theorem 2.3 ([25,26])**.** *Let K be a compact space and let H be a uniformly bounded subset of $C(K)$. We have*

$$\operatorname{ck}(H) \leq \hat{d}(\overline{H}^{\mathbb{R}^K}, C(K)) \leq \gamma_K(H) \leq 2\operatorname{ck}(H).$$

Theorem 2.4 ([26]). *Let K be a compact topological space and let H be a uniformly bounded subset of $\mathbb{R}^K$. Then*

$$\gamma_K(H) = \gamma_K(co(H))$$

and as a consequence for $H \subset C(K)$ we obtain that

$$\hat{d}(\overline{co(H)}^{\mathbb{R}^K}, C(K)) \leq 2\hat{d}(\overline{H}^{\mathbb{R}^K}, C(K))$$

and if $H \subset \mathbb{R}^K$ is uniformly bounded then

$$\hat{d}(\overline{co(H)}^{\mathbb{R}^K}, C(K)) \leq 5\hat{d}(\overline{H}^{\mathbb{R}^K}, C(K)).$$

Theorem 2.5 ([26]). *Let E be a Banach space and let B_{E^*} be the closed unit ball in the dual E^* endowed with the w^*-topology. Let $i : E \to E^{**}$ and $j : E^{**} \to \ell_\infty(B_{E^*})$ be the canonical embeddings. Then, for every $x^{**} \in E^{**}$ we have*

$$d(x^{**}, i(E)) = d(j(x^{**}), C(B_{E^*}))\,.$$

Theorem 2.6 ([26,27]). *Let H be a bounded subset of a Banach space E. Then*

$$\mathrm{ck}(H) \leq \mathrm{k}(H) \leq \gamma(H) \leq 2\,\mathrm{ck}(H) \leq 2\,\mathrm{k}(H) \leq 2\omega(H) \tag{1}$$

$$\gamma(H) = \gamma(\mathrm{co}(H)) \quad \text{and} \quad \omega(H) = \omega(\mathrm{co}(H).$$

For any $x^{**} \in \overline{H}^{w^*}$, there is a sequence $(x_n)_n$ in H such that

$$\|x^{**} - y^{**}\| \leq \gamma(H)$$

for any cluster point y^{**} of $(x_n)_n$ in E^{**}. Furthermore, H is relatively compact in (E, w) if, and only if, it is zero one (equivalently all) of the numbers $\mathrm{ck}(H), \mathrm{k}(H), \gamma(H)$ and $\omega(H)$.

The notation used is the following:

(1) The distance d in $\mathbb{R}^K$ or $C(K)$ always refers to the supremum distance.
(2) If T be a topological space and A subset of T, then $A^{\mathbb{N}}$ is considered as the set of all sequences in A. The set of all cluster points in T of a sequence $\varphi \in A^{\mathbb{N}}$ is denoted by $\mathrm{clust}_T(\varphi)$.
(3) If H be a subset $\mathbb{R}^K$ we define:

$$\mathrm{ck}(H) := \sup_{\varphi \in H^{\mathbb{N}}} d(\mathrm{clust}_{\mathbb{R}^K}(\varphi), C(K)),$$

$$\hat{d}(H, C(K)) := \sup_{g \in H} d(g, C(K)),$$

and

$$\gamma_K(H) := \sup\{\big|\lim_n \lim_m f_m(x_n), \lim_m \lim_n f_m(x_n)\big| : (f_m) \subset H, (x_n) \subset K\},$$

assuming that the involved limits exist.

(4) If E is a Banach space and $H \subset E$ is a bounded set, then $\overline{H}^{w^*}$ stands for its w^*-closure in E^{**} and

$$k(H) = \hat{d}(\overline{H}^{w^*}, E) = \sup_{y \in \overline{H}^{w^*}} \inf_{x \in E} \|y - x\|,$$

$$\gamma(H) := \sup\{|\lim_n \lim_m f_m(x_n) - \lim_m \lim_n f_m(x_n)| : (f_m) \subset B_{E^*}, (x_n) \subset H\},$$

assuming the involved limits exist,

$$\text{ck}(H) := \sup_{\varphi \in H^{\mathbb{N}}} d(\text{clust}_{E^{**}, w^*}(\varphi), E)$$

and $\omega(H) := \inf\{\varepsilon > 0 : H \subset K_\varepsilon + \varepsilon B_E \text{ and } K_\varepsilon \subset X \text{ is } w\text{-compact}\}$.

For obvious reasons the quantities that appear in theorem 2.6 are called measures of weak noncompactness, see[28,29]. Measures of noncompactness or weak noncompactness have been successfully applied to the study of compactness, operator theory, differential equations and integral equations, see for instance[26,27,29–39]. Theorem 2.6 tells us that *all classical approaches used so far to study weak compactness in Banach spaces (Tychonoff's theorem, Eberlein-Šmulian's theorem, Eberlein-Grothendieck double-limit criterion) are qualitatively and quantitatively equivalent.* Quantitative versions of James compactness theorem can be found in[33]. Surveys about these questions are[24,40,41].

3. Upper semi-continuity for multifunctions, applications

This section explains how one can exploit the use of multifunctions $\psi : X \to 2^Y$ between topological spaces from two different but connected angles: (a) transferring properties of X to properties of Y when ψ is upper semi-continuous; (b) ensuring how to automatically produce upper-semicontinuity from descriptive properties.

The two results that follow, theorems 3.1 and 3.2, have been during our years of research the most useful ones that we have ever found. The first one is related to property (a) above and the second one to property (b). The ideas behind them can be traced back to references[42–45].

Recall that the *weight* $w(X)$ of a topological space X is the minimal cardinality of a basis for the topology of X. By the *density* $d(X)$ we mean the minimal cardinality of a dense subset of X. The *Lindelöf number* $l(X)$ of X is the smallest infinite cardinal number $\mathfrak{m}$ such that every open cover of X has a subcover of cardinality $\leq \mathfrak{m}$.

Theorem 3.1 ([46] **Pro. 2.1**). *Let X and Y be topological spaces and let $\psi : X \to 2^Y$ be an upper semi-continuous compact-valued map such that the set $Y = \bigcup\{\psi(x) : x \in X\}$. Assume that $w(X)$ is infinite. Then,*

(1) the Lindelöf number $l(Y^n) \leq w(X)$, for every $n = 1, 2, \ldots$;
(2) if Y is moreover assumed to be metric then $d(Y) \leq w(X)$.

Proof. The proof below is the one that was published in [46] Pro. 2.1 and it is included in order that the reader can get the flavour of the techniques needed.

To prove (1) we observe first that for every $n = 1, 2, \ldots$ the multi-valued map $\psi^n : X^n \to 2^{Y^n}$ given by

$$\psi^n(x_1, x_2, \ldots, x_n) := \psi(x_1) \times \psi(x_2) \times \cdots \times \psi(x_n)$$

is compact-valued, upper semi-continuous and

$$Y^n = \bigcup\{\psi^n(x_1, x_2, \ldots, x_n) : (x_1, x_2, \ldots, x_n) \in X^n\}.$$

Since $w(X)$ is infinite we have that $w(X^n) = w(X)$ and therefore we only need to prove (1) for $n = 1$. Take $(G_i)_{i \in I}$ any open cover of Y. For each $x \in X$ the compact set $\psi(x)$ is covered by the family $(G_i)_{i \in I}$ and therefore we can choose a finite subset $I(x)$ of I such that

$$\psi(x) \subset \bigcup_{i \in I(x)} G_i.$$

By upper semi-continuity, for each x in X we can take an open set O_x of X such that $x \in O_x$ and

$$\psi(O_x) \subset \bigcup_{i \in I(x)} G_i.$$

The family $(O_x)_{x \in X}$ is an open cover of X and therefore there is a set $F \subset X$ such that $|F| \leq w(X)$ and $X = \bigcup_{x \in F} O_x$, see [3] Theorem 1.1.14. Then

$$Y = \psi(X) = \bigcup_{x \in F} \psi(O_x) = \bigcup_{x \in F} \bigcup_{i \in I(x)} G_i.$$

Hence $(G_i)_{i \in I}$ has a subcover of at most $w(X)$ elements.

For the proof of (2) we refer to [3] Theorem 4.1.15. □

Theorem 3.2 ([46] Th. 2.3, [45]). *Let X be a first-countable topological space, Y a topological space in which the relatively countably compact subsets are relatively compact and let $\phi : X \to 2^Y$ be a multifunction satisfying the property*

$$\bigcup_{n \in \mathbb{N}} \phi(x_n) \text{ is relatively compact for each convergent sequence } (x_n)_n \text{ in } X. \tag{2}$$

If for each x in X we define

$$\psi(x) := \bigcap \{\overline{\phi(V)} : V \text{ neighborhood of } x \text{ in } X\}, \tag{3}$$

then the multifunction so defined $\psi : X \to 2^Y$ is upper semi-continuous, compact-valued and satisfies $\phi(x) \subset \psi(x)$ for every x in X.

We recall that a topological space Y is said to be *K-analytic* if there is a usco map $T : \mathbb{N}^{\mathbb{N}} \to 2^Y$ such that $T(\mathbb{N}^{\mathbb{N}}) := \bigcup\{T(\alpha) : \alpha \in \mathbb{N}^{\mathbb{N}}\} = Y$, [47]. Recall also that a regular topological space T is *angelic* if every relatively countably compact subset A of T is relatively compact and its closure $\overline{A}$ is made up of the limits of sequences from A. In angelic spaces the different concepts of compactness and relative compactness coincide: the (relatively) countably compact, (relatively) compact and (relatively) sequentially compact subsets are the same, as seen in [7]. Examples of angelic spaces include metric spaces, spaces $C_p(K)$, when K is a countably compact space, see [48,49] and all Banach spaces in their weak topologies.

Corollary 3.1 ([43] Corollary 1.1). *Let Y be an angelic space. Assume that there is a family of subsets $\{A_\alpha : \alpha \in \mathbb{N}^{\mathbb{N}}\}$ of Y with the properties:*

(α) *A_α is compact for every $\alpha \in \mathbb{N}^{\mathbb{N}}$;*
(β) *$A_\alpha \subset A_\beta$ if $\alpha \leq \beta$;*
(γ) *$Y = \bigcup\{A_\alpha : \alpha \in \mathbb{N}^{\mathbb{N}}\}$.*

Then,

(1) Y is K analytic;
(2) if moreover Y metrizable, then Y is separable.

Proof. Let us prove (1). To do so we will use theorem 3.2. We define $\phi(\alpha) := A_\alpha$, $\alpha \in \mathbb{N}^\mathbb{N}$. We check that ϕ satisfies the assumptions (2). Indeed, let $\pi_j : \mathbb{N}^\mathbb{N} \to \mathbb{N}$ be the j-th projection onto $\mathbb{N}$ and if $\alpha_n \to \alpha$ in $\mathbb{N}^\mathbb{N}$ we define, for every $j \in \mathbb{N}$,

$$m_j := \max\{\pi_j(\alpha_n) : n \in \mathbb{N}\},$$

and we write $\beta := (m_j)$. Note that $\alpha_n \leq \beta$ for every $n \in \mathbb{N}$ and then condition (β) ensures that $A_{\alpha_n} \subset A_\beta$ for every $n \in \mathbb{N}$. Thus

$$\bigcup_{n\in\mathbb{N}} \phi(\alpha_n) = \bigcup_{n\in\mathbb{N}} A_{\alpha_n} \subset A_\beta.$$

and since condition (α) guarantees that A_β is compact, we conclude that requirement (2) is fulfilled. Therefore we can use theorem 3.2 and produce the usco map $\psi : \mathbb{N}^\mathbb{N} \to 2^Y$ with the property $\phi(\alpha) \subset \psi(\alpha)$ for every $\alpha \in \mathbb{N}^\mathbb{N}$. Now, condition ($\gamma$) applies to conclude that $Y = \bigcup\{\psi(\alpha) : \alpha \in \mathbb{N}^\mathbb{N}\}$ and therefore Y is K-analytic.

Statement (2) straightforwardly follows from statement (1) in combination with (2) in theorem 3.1, if we bear in mind that $\mathbb{N}^\mathbb{N}$ is second countable, *i.e.*, the weight $w(\mathbb{N}^\mathbb{N})$ is countable. □

Theorem 3.3 ([40] Theorem 2.6). *Let K be a compact space and let Δ be the diagonal of $K \times K$. The following statements are equivalent:*

(1) K is metrizable;
(2) $(C(K), \|\cdot\|_\infty)$ is separable;
(3) Δ is a G_δ subset of $K \times K$;
(4) $\Delta = \bigcap_n G_n$ with each G_n open in $K \times K$ and $\{G_n : n \in \mathbb{N}\}$ being a basis of open neighbourhoods of Δ;
(5) $(K \times K) \setminus \Delta = \bigcup_n F_n$, with $\{F_n : n \in \mathbb{N}\}$ an increasing family of compact subsets in $(K \times K) \setminus \Delta$;
(6) $(K \times K) \setminus \Delta = \bigcup_n F_n$, with $\{F_n : n \in \mathbb{N}\}$ an increasing family of compact sets that swallows all the compact subsets in $(K \times K) \setminus \Delta$;
(7) $(K \times K) \setminus \Delta = \bigcup\{A_\alpha : \alpha \in \mathbb{N}^\mathbb{N}\}$ with $\{A_\alpha : \alpha \in \mathbb{N}^\mathbb{N}\}$ a family of compact sets that swallows all the compact subsets in $(K \times K) \setminus \Delta$ such that $A_\alpha \subset A_\beta$ whenever $\alpha \leq \beta$;
(8) $(K \times K) \setminus \Delta$ is Lindelöf.

Proof. We refer to the proof of this theorem to[40] Theorem 2.6. We reproduce here only the implication (7) ⇒ (2). Assume that (7) holds and let us define $O_\alpha := (K \times K) \setminus A_\alpha$, $\alpha \in \mathbb{N}$. The family $\mathcal{O} := \{O_\alpha : \alpha \in \mathbb{N}^\mathbb{N}\}$ is a basis of open neighbourhoods of Δ that satisfies the decreasing condition

$$O_\beta \subset O_\alpha, \text{ if } \alpha \leq \beta \text{ in } \mathbb{N}^{\mathbb{N}}. \tag{4}$$

Given $\alpha = (n_k) \in \mathbb{N}^{\mathbb{N}}$ and any $m \in \mathbb{N}$ we write

$$\alpha|^m := (n_m, n_{m+1}, n_{m+2}, \dots)$$

and define

$$B_\alpha := \Big\{ f \in n_1 B_{C(K)} : (m \in \mathbb{N}, \text{ and } (x,y) \in O_{\alpha|^m}) \Rightarrow |f(x) - f(y)| \leq \frac{1}{m} \Big\}.$$

Note that each B_α is $\|\cdot\|_\infty$-bounded, closed and equicontinuous as a family of functions defined on K. Therefore, Ascoli's theorem, see[4] p. 234, implies that B_α is compact in $(C(K), \|\cdot\|_\infty)$. The decreasing property (4) implies that $B_\alpha \subset B_\beta$ if $\alpha \leq \beta$ in $\mathbb{N}^{\mathbb{N}}$. We claim that $C(K) = \bigcup\{B_\alpha : \alpha \in \mathbb{N}\}$. To see this, given $f \in C(K)$ take $M > 0$ such that $\|f\|_\infty \leq M$. On the other hand since $\mathcal{O}$ is a basis of neighborhoods of Δ, there exists a sequence $\big(\alpha_m = (n_k^m)\big)$ in $\mathbb{N}^{\mathbb{N}}$ such that

$$|f(x) - f(y)| \leq \frac{1}{m} \text{ for every } (x,y) \in O_{\alpha_m}.$$

If we define now $n_1 := \max\{n_1^1, M\}$ and $n_k := \max\{n_k^1, n_{k-1}^2, \dots, n_1^k\}$, $k = 2, 3, \dots$, then for the sequence $\alpha = (n_k) \in \mathbb{N}^{\mathbb{N}}$ we have that $f \in B_\alpha$. The family $\{B_\alpha : \alpha \in \mathbb{N}\}$ of subsets of $(C(K), \|\cdot\|_\infty)$ satisfies the hypothesis of corollary 3.1 and we conclude that $(C(K), \|\cdot\|_\infty)$ is separable. This finishes the proof of (7) $\Rightarrow$ (2) . □

Do not be misled by the purely topological aspect of the above theorem. Our contribution there, that is, implication (7) $\Rightarrow$ (2), was first stated also in a topological setting (apparently different) in[44] Theorem 1 as kind of lemma to establish metrizability results for compact sets in locally convex spaces. On the light of this result we introduced the class $\mathfrak{G}$ of locally convex spaces:

Definition 3.1 ([44]). *A locally convex space E belongs to the class $\mathfrak{G}$ if there is a family $\{A_\alpha : \alpha \in \mathbb{N}^{\mathbb{N}}\}$ of subsets of E' satisfying the properties:*

(a) for any $\alpha \in \mathbb{N}^{\mathbb{N}}$ the countable subsets of A_α are equicontinuous;
(b) $A_\alpha \subset A_\beta$ if $\alpha \leq \beta$;
(c) $X = \bigcup\{A_\alpha : \alpha \in \mathbb{N}^{\mathbb{N}}\}$.

The class $\mathfrak{G}$ is a very wide class of locally convex spaces and it is stable under the usual operations in functional analysis of countable type (completions, closed subspaces, quotients, direct sums, products, etc.) that contains metrizable locally convex spaces and their duals and for which (7) $\Rightarrow$ (2) collected in theorem 3.3 implies:

Theorem 3.4 ([44]). *If E is a locally convex space in class $\mathfrak{G}$, then its compact (even its precompact) subsets are metrizable.*

Proof. See[44] Theorem 2. □

We should mention that theorem 3.4 solved a number of open question in those times, for instance one posed by Floret in[50], in which he asked about the sequential behaviour of compact subsets of (LM)-spaces, *i.e.*, inductive limits of metrizable locally convex spaces. Note that since $\mathfrak{G}$ contains metric spaces and their duals theorem 3.4 provides metrizability of compact subsets for (LM)-spaces as well as many other classes of spaces for which these properties were unknown by then. Since the appearance of[44] a number of authors coming from topology and functional analysis have been working on topics connected with those developed there: we refer to the recent book[51] for more references, applications and consequences of these ideas. Another possible survey reference is[40].

Next we isolate the result below to show the simplicity, beauty and power of the techniques involving K-analytic structures.

Theorem 3.5 (Dieudonné, Theorem §.2.(5)[8]). *Every Fréchet-Montel space E is separable (in particular, for any open set $\Omega \subset \mathbb{C}$ the space of holomorphic functions $(\mathcal{H}(\Omega), \tau_k)$ with its compact-open topology is separable).*

Proof. Fix $U_1 \supset U_2 \supset \cdots \supset U_n \ldots$ a basis of absolutely convex closed neighborhoods of 0. Given $\alpha = (n_k) \in \mathbb{N}^{\mathbb{N}}$, let us define

$$A_\alpha := \bigcap_{k=1}^{\infty} n_k U_k.$$

The family $\{A_\alpha : \alpha \in \mathbb{N}^{\mathbb{N}}\}$ is made up of closed bounded sets, covers E and satisfies $A_\alpha \subset A_\beta$ if $\alpha \leq \beta$. Since E is Montel, each A_α is compact and since E is Fréchet it is metrizable and therefore corollary 3.1 applies to say that E is separable. □

We finish pointing out some some recent developments that got started with[52,53] where the following definition can be found.

Definition 3.2 ([52,53]). *Given topological spaces M and Y, an M-ordered compact cover of a space Y is a family $\mathcal{F} = \{F_K : K \in \mathcal{K}(M)\} \subset \mathcal{K}(Y)$ such that*

$$\bigcup \mathcal{F} = Y \text{ and } K \subset L \text{ implies } F_K \subset F_L \text{ for any } K, L \in \mathcal{K}(M).$$

Y is said to be dominated *(resp.* strongly dominated*) by the space M if there exists an M-ordered compact cover $\mathcal{F}$ (resp. that moreover swallows all compact subsets of Y, in the sense that for any compact $C \subset Y$ there is $F \in \mathcal{F}$ such that $C \subset F$) of the space Y.*

It can be proved[53], that condition (7) in theorem 3.3 is equivalent to $(K \times K)\setminus\Delta$ to be strongly dominated by a Polish space (Polish space means topological space that is metrizable, separable and complete for some metric given the topology).

Proposition 3.1 ([54]). *Let K be a compact space and $\mathfrak{m}$ a cardinal number. The following statements are equivalent:*

(1) $w(K) \leq \mathfrak{m}$;
(2) There exists a metric space M with $w(M) \leq \mathfrak{m}$ and a family $\mathcal{O} = \{O_L : L \in \mathcal{K}(M)\}$ of open subsets in $K \times K$ that is basis of the neighborhoods of Δ such that $O_{L_1} \subset O_{L_2}$ whenever $L_2 \subset L_1$ in $\mathcal{K}(M)$;
(3) $(K \times K)\setminus\Delta$ is strongly dominated by a metric space M with $w(M) \leq \mathfrak{m}$.

The following questions is to the best of our knowledge still unanswered in full generality.

Question 3.1. if K is a compact space such that $(K \times K)\setminus\Delta$ is dominated by a Polish space, is K metrizable?

In the presence of some extra set theoretical axiom the answer is positive.

Theorem 3.6 ([53]). *Under $MA(\omega_1)$, if K is a compact space such that $(K \times K) \setminus \Delta$ is dominated by a Polish space then K is metrizable.*

4. Measurability for multifunctions, an application

In this section we will briefly present why we have been interested about measurable selectors for multifunctions and how we came across what we called property (P) that has shown to be useful for the existence of these measurable selectors.

Our interest about measurable selectors for measurable multifunctions goes back to our interest about integration for multifunctions. As said already, integration for multifunctions has its origin in the papers by Aumann[14] and Debreu[13]: good references on measurable selections and integration of multifunctions are the monographs[10,11] and the survey[12]; a common thing in all these studies that deal with multifunctions whose values are subsets of a Banach space E is that E was always assumed to be *separable*. The main reason for this limitation on E relies on the fact that an integrable multifunction should have integrable (measurable) selectors and the tool to find these measurable selectors has always been the well-known selection theorem of Kuratowski and Ryll-Nardzewski[23] that only works when the range space is separable. Therefore if one wishes to find measurable selectors outside the universe of the E's being separable a different approach should be done. With this in mind the following definition was introduced.

Definition 4.1 ([55] Definition 2.1). *A multifunction $F : \Omega \to 2^E$ is said to satisfy property (P) if for each $\varepsilon > 0$ and each $A \in \Sigma^+$ there exist $B \in \Sigma^+_A$ and $D \subset E$ with* $\operatorname{diam}(D) < \varepsilon$ *such that*

$$F(t) \cap D \neq \emptyset \text{ for every } t \in B,$$

see figure 6.

Here the notation that we use is the following: starting with our complete probability space (Ω, Σ, μ) we write Σ^+ to denote the family of all $A \in \Sigma$ with $\mu(A) > 0$; given $A \in \Sigma^+$, the collection of all subsets of A belonging to Σ^+ is denoted by Σ^+_A. We should mention here that our property (P) above is inspired in the topological notion of *fragmentability* for multifunctions that can be found in[56].

It can be proved, see[55], that for a multifunction $F : \Omega \to 2^E$ we have the following properties:

(i) *If there exists a multifunction $G : \Omega \to 2^E$ satisfying property (P) such that $G(t) \subset F(t)$ for μ-a.e. $t \in \Omega$, then F satisfies property (P) as well.*
(ii) *If there exists a strongly measurable function $f : \Omega \to E$ such that $F(t) = \{f(t)\}$ for μ-a.e. $t \in \Omega$, then F satisfies property (P).*
(iii) *If F admits strongly measurable selectors, then F satisfies property (P).*

Recall that $f : \Omega \to E$ is said to be strongly measurable if it is the μ-a.e. limit of a sequence of measurable simple functions. It is easy to observe

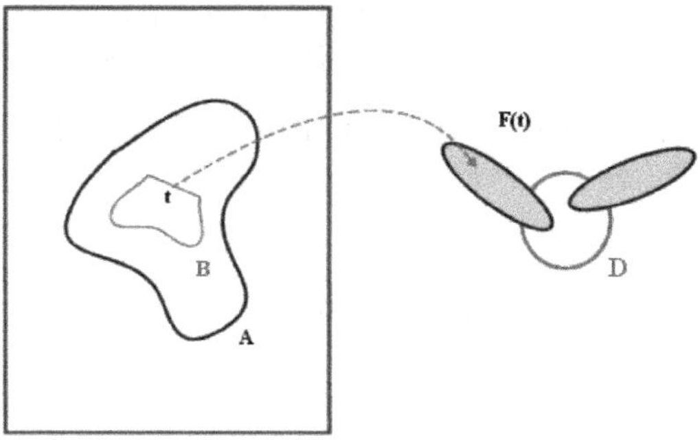

Fig. 6. Property (P)

that property (P) for a multifunction helps to isolate the ideas behind the classical Kuratowski and Ryll-Nardzewski theorem that we present below: proposition 4.1, lemma 4.1 and the proof of theorem 4.2 are co-authored with V. Kadets and J. Rodríguez. These results were written in some preliminary version of[55] but we finally took them out from the version that was sent off for publication.

Proposition 4.1 (Cascales, Kadets and Rodríguez). *Suppose E is separable. Let $F : \Omega \to 2^E$ be a Effros measurable multifunction. Then F satisfies property (P).*

Proof. Fix $\varepsilon > 0$ and $A \in \Sigma^+$. Since E is separable, we can write $E = \bigcup_{n\in\mathbb{N}} C_n$, where each C_n is an open ball with $\operatorname{diam}(C_n) \leq \varepsilon$. By hypothesis, all the sets $B_n := \{t \in \Omega : F(t) \cap C_n \neq \emptyset\}$ belong to Σ and, moreover, $\Omega = \bigcup_{n\in\mathbb{N}} B_n$. Since $\mu(A) > 0$, there is $n \in \mathbb{N}$ such that $B := A \cap B_n \in \Sigma_A^+$. Now, the set $D := C_n$ intersects $F(t)$ for all $t \in B$. □

Lemma 4.1 (Cascales, Kadets and Rodríguez). *Let $F : \Omega \to 2^E$ be a multifunction satisfying property (P). Then for each $\varepsilon > 0$ there exists a strongly measurable countably-valued function $f : \Omega \to E$ such that $F(t) \cap \overline{B(f(t), \varepsilon)} \neq \emptyset$ for μ-a.e. $t \in \Omega$.*

Proof. Property (P) and a standard exhaustion argument allow us to find a sequence (A_n) of pairwise disjoint measurable subsets of Ω with $\mu(\Omega \setminus \bigcup_{n\in\mathbb{N}} A_n) = 0$ and a sequence (D_n) of subsets of X with diameter less than or equal to ε such that, for each $n \in \mathbb{N}$, we have $F(t) \cap D_n \neq \emptyset$ for every $t \in A_n$. Take $x_n \in D_n$ for all $n \in \mathbb{N}$ and define $f : \Omega \to E$ by $f := \sum_{n\in\mathbb{N}} x_n \chi_{A_n}$. This function satisfies the desired property: for each

$n \in \mathbb{N}$ and each $t \in A_n$ there is some $y \in F(t) \cap \overline{D_n}$ and, bearing in mind that $\text{diam}(D_n) \leq \varepsilon$, we get $y \in F(t) \cap \overline{B(f(t), \varepsilon)}$. The proof is over. □

Proposition 4.2 (Kuratowski and Ryll-Nardzewski). *Suppose E is separable. Let $F : \Omega \to 2^E$ be an Effros measurable multifunction having norm closed values. Then F admits strongly measurable selectors.*

Proof. [by Cascales, Kadets and Rodríguez]. Let (ε_m) be a decreasing sequence of positive real numbers converging to 0.

By Proposition 4.1, F satisfies property (P) and therefore Lemma 4.1 ensures the existence of a strongly measurable countably-valued function $f_1 : \Omega \to E$ such that $F(t) \cap \overline{B(f_1(t), \varepsilon_1)} \neq \emptyset$ for all $t \in A_1$, where $A_1 \in \Sigma$ and $\mu(\Omega \setminus A_1) = 0$.

Define $F_1(t) := F(t) \cap \overline{B(f_1(t), \varepsilon)}$ if $t \in A_1$, $F_1(t) = \{0\}$ otherwise. It is easily checked that the multifunction F_1 is Effros measurable. Again, proposition 4.1 and lemma 4.1 allow us to find a strongly measurable countably-valued function $f_2 : \Omega \to E$ such that $F_1(t) \cap \overline{B(f_2(t), \varepsilon_2)} \neq \emptyset$ for all $t \in A_2$, where $A_2 \in \Sigma$ and $\mu(\Omega \setminus A_2) = 0$. In this way, we can construct a sequence $f_m : \Omega \to E$ of strongly measurable countably-valued functions and a sequence (A_m) in Σ with $\mu(\Omega \setminus A_m) = 0$ such that

$$F(t) \cap \left(\bigcap_{m=1}^{p} \overline{B(f_m(t), \varepsilon_m)} \right) \neq \emptyset \quad \text{for all } p \in \mathbb{N} \tag{5}$$

whenever $t \in A := \bigcap_{m \in \mathbb{N}} A_m$.

Fix $t \in A$. We claim that the sequence $(f_m(t))$ converges in norm to some point in $F(t)$. Indeed, given $j \geq i$ we can use (5) to find

$$x_j \in F(t) \cap \left(\bigcap_{m=1}^{j} \overline{B(f_m(t), \varepsilon_m)} \right),$$

so that $\|f_i(t) - x_j\| \leq \varepsilon_i$ and $\|f_j(t) - x_j\| \leq \varepsilon_j$, hence $\|f_i(t) - f_j(t)\| \leq \varepsilon_i + \varepsilon_j \leq 2\varepsilon_i$. This shows that $(f_m(t))$ is Cauchy and so it converges in norm. Since x_j belongs to $F(t)$ and $\|f_j(t) - x_j\| \leq \varepsilon_j$ for all $j \in \mathbb{N}$, the limit of $(f_m(t))$ also belongs to $F(t)$, as claimed.

Let $f : \Omega \to E$ be a function such that $f(t) = \lim_{m \to \infty} f_m(t)$ whenever $t \in A$ and an arbitrary $f(t) \in F(t)$ whenever $t \in \Omega \setminus A$. Clearly, f is a selector of F. Since each f_m is strongly measurable and $\mu(\Omega \setminus A) = 0$, it follows that f is strongly measurable and the proof is finished. □

The good thing regarding property (P) above is that beyond giving a new insight for the classical proof of Kuratowski and Ryll-Nardzewski's theorem it allowed us to characterize when a given multifunction does have measurable selectors.

In what follows the symbol $wk(E)$ (resp. $cwk(E)$) stands for the collection of all weakly compact (resp. convex weakly compact) non-empty subsets of the Banach space E.

Theorem 4.1 ([55] Theorem 2.5). *For a multifunction $F : \Omega \to wk(E)$ the following statements are equivalent:*

(i) F admits a strongly measurable selector.
(ii) F satisfies property (P).
(iii) There exist a set of measure zero $\Omega_0 \in \Sigma$, a separable subspace $Y \subset E$ and a multifunction $G : \Omega \setminus \Omega_0 \to wk(Y)$ that is Effros measurable and such that $G(t) \subset F(t)$ for every $t \in \Omega \setminus \Omega_0$.

We write $\delta^*(x^*, C) := \sup\{x^*(x) : x \in C\}$ for any set $C \subset E$ and any $x^* \in E^*$. A multifunction $F : \Omega \to 2^E$ is said to be scalarly measurable if for each $x^* \in E^*$ the function $t \mapsto \delta^*(x^*, F(t))$ is measurable. In particular a single valued function $f : \Omega \to E$ is scalarly measurable if the composition $x^* \circ f$ is measurable for every $x^* \in E^*$. Note that every Effros measurable multifunction F is scalarly measurable.

Here is second result about scalar measurability for multifunctions that seems that has had some impact in integration for multifunctions.

Theorem 4.2 ([55] Theorem 3.8). *Every scalarly measurable multifunction $F : \Omega \to wk(E)$ admits a scalarly measurable selector.*

To finish let us mention the impact of measurable selections on multifunction integration. A multifunction $F : \Omega \to cwk(E)$ is said to be *Pettis integrable* if

- $\delta^*(x^*, F)$ is integrable for each $x^* \in E^*$;
- for each $A \in \Sigma$, there is $\int_A F\, d\mu \in cwk(E)$ such that

$$\delta^*\Big(x^*, \int_A F\, d\mu\Big) = \int_A \delta^*(x^*, F)\, d\mu \quad \text{for every } x^* \in E^*.$$

For the notion of Pettis ingegrability for single valued functions we refer to[5].

Theorem 4.3 ([57] Theorem 2.5). *Let $F : \Omega \to cwk(E)$ be a Pettis integrable multifunction. Then F admits a Pettis integrable selector.*

Theorem 4.4 ([57] Theorem 2.6). *Let $F : \Omega \to cwk(E)$ be a Pettis integrable multifunction. Then F admits a collection $\{f_\alpha\}_{\alpha<\mathrm{dens}(E^*,w^*)}$ of Pettis integrable selectors such that*

$$F(\omega) = \overline{\{f_\alpha(\omega) : \alpha < \mathrm{dens}(E^*, w^*)\}} \quad \textit{for every } \omega \in \Omega.$$

Moreover,

$$\int_A F \, d\mu = \overline{\left\{ \int_A f \, d\mu : f \textit{ is a Pettis integrable selector of } F \right\}}$$

for every $A \in \Sigma$.

References

1. J. B. Conway, *A course in functional analysis*, Graduate Texts in Mathematics, Vol. 96 (Springer-Verlag, New York, 1985).
2. M. Fabian, P. Habala, P. Hájek, V. Montesinos and V. Zizler, *Banach space theory* (CMS Books in Mathematics / Ouvrages de Mathématiques de la SMC, Springer, New York, 2011).
3. R. Engelking, *General topology* (PWN—Polish Scientific Publishers, Warsaw, 1977), Translated from the Polish by the author, Monografie Matematyczne, Tom 60. [Mathematical Monographs, Vol. 60].
4. J. L. Kelley, *General topology* (Springer-Verlag, New York-Berlin, 1975), Reprint of the 1955 edition [Van Nostrand, Toronto, Ont.], Graduate Texts in Mathematics, No. 27.
5. J. Diestel and J. J. Uhl, Jr., *Vector measures* (American Mathematical Society, Providence, R.I., 1977), With a foreword by B. J. Pettis, Mathematical Surveys, No. 15.
6. J. B. Conway, *Functions of one complex variable*, Graduate Texts in Mathematics, Vol. 11, second edn. (Springer-Verlag, New York-Berlin, 1978).
7. K. Floret, *Weakly compact sets*, Lecture Notes in Mathematics, Vol. 801 (Springer, Berlin, 1980), Lectures held at S.U.N.Y., Buffalo, in Spring 1978.
8. G. Köthe, *Topological vector spaces. I* (Springer-Verlag New York Inc., New York, 1969).
9. K. Kuratowski, *Topology. Vol. II* (Academic Press, New York-London; Państwowe Wydawnictwo Naukowe Polish Scientific Publishers, Warsaw, 1968).

10. E. Klein and A. C. Thompson, *Theory of correspondences* (Canadian Mathematical Society Series of Monographs and Advanced Texts, John Wiley & Sons, Inc., New York, 1984).
11. C. Castaing and M. Valadier, *Convex analysis and measurable multifunctions* (Lecture Notes in Mathematics, Vol. 580, Springer-Verlag, Berlin-New York, 1977).
12. C. Hess, Set-valued integration and set-valued probability theory: an overview, in *Handbook of measure theory, Vol. I, II*, ed. E. Pap (North-Holland, Amsterdam, 2002) pp. 617–673.
13. G. Debreu, Integration of correspondences, in *Proc. Fifth Berkeley Sympos. Math. Statist. and Probability (Berkeley, Calif., 1965/66), Vol. II: Contributions to Probability Theory, Part 1*, eds. L. M. Lecam and J. Neyman (Univ. California Press, Berkeley, Calif., 1967) pp. 351–372.
14. R. J. Aumann, Integrals of set-valued functions, *J. Math. Anal. Appl.* **12**, 1 (1965).
15. E. Michael, Continuous selections. I, *Ann. of Math.* **63**, 361 (1956).
16. Y. Benyamini and J. Lindenstrauss, *Geometric nonlinear functional analysis. Vol. 1*, American Mathematical Society Colloquium Publications, Vol. 48 (American Mathematical Society, Providence, RI, 2000).
17. J. E. Jayne and C. A. Rogers, *Selectors* (Princeton University Press, Princeton, NJ, 2002).
18. I. Namioka and R. R. Phelps, Banach spaces which are Asplund spaces, *Duke Math. J.* **42**, 735 (1975).
19. C. Stegall, The duality between Asplund spaces and spaces with the Radon-Nikodým property, *Israel J. Math.* **29**, 408 (1978).
20. R. Deville, G. Godefroy and V. Zizler, *Smoothness and renormings in Banach spaces*, Pitman Monographs and Surveys in Pure and Applied Mathematics, Vol. 64 (Longman Scientific & Technical, Harlow; copublished in the United States with John Wiley & Sons, Inc., New York, 1993).
21. B. Cascales, V. P. Fonf, J. Orihuela and S. Troyanski, Boundaries of Asplund spaces, *J. Funct. Anal.* **259**, 1346 (2010).
22. A. Moltó, J. Orihuela, S. Troyanski and M. Valdivia, *A nonlinear transfer technique for renorming*, Lecture Notes in Mathematics, Vol. 1951 (Springer-Verlag, Berlin, 2009).
23. K. Kuratowski and C. Ryll-Nardzewski, A general theorem on selectors, *Bull. Acad. Polon. Sci. Sér. Sci. Math. Astronom. Phys.* **13**, 397 (1965).
24. C. Angosto and B. Cascales, A new look at compactness via distances

to function spaces, in *Advanced courses of mathematical analysis III*, eds. J. M. Delgado Sánchez and T. Domínguez Benavides (World Sci. Publ., Hackensack, NJ, 2008) pp. 49–66.

25. C. Angosto and B. Cascales, The quantitative difference between countable compactness and compactness, *J. Math. Anal. Appl.* **343**, 479 (2008).
26. B. Cascales, W. Marciszewski and M. Raja, Distance to spaces of continuous functions, *Topology Appl.* **153**, 2303 (2006).
27. C. Angosto and B. Cascales, Measures of weak noncompactness in Banach spaces, *Topology Appl.* **156**, 1412 (2009).
28. J. Banaś and A. Martinón, Measures of weak noncompactness in Banach sequence spaces, *Portugal. Math.* **52**, 131 (1995).
29. A. Kryczka, S. Prus and M. Szczepanik, Measure of weak noncompactness and real interpolation of operators, *Bull. Austral. Math. Soc.* **62**, 389 (2000).
30. C. Angosto, *Distance to function spaces* (PhD thesis, Universidad de Murcia, 2007).
31. K. Astala and H.-O. Tylli, Seminorms related to weak compactness and to Tauberian operators, *Math. Proc. Cambridge Philos. Soc.* **107**, 367 (1990).
32. F. S. De Blasi, On a property of the unit sphere in a banach space, *Colloq. Math* **65**, 333 (1992).
33. B. Cascales, O. F. K. Kalenda and J. Spurný, A quantitative version of James's compactness theorem, *Proc. Edinb. Math. Soc.* **55**, 369 (2012).
34. M. Fabian, P. Hájek, V. Montesinos and V. Zizler, A quantitative version of Krein's theorem, *Rev. Mat. Iberoamericana* **21**, 237 (2005).
35. A. S. Granero, P. Hájek and V. M. Santalucía, Convexity and w^*-compactness in Banach spaces, *Math. Ann.* **328**, 625 (2004).
36. A. S. Granero, An extension of the Krein-Šmulian theorem, *Rev. Mat. Iberoam.* **22**, 93 (2006).
37. A. S. Granero and M. Sánchez, Convexity, compactness and distances, in *Methods in Banach space theory*, eds. J. M. F. Castillo and W. B. Johnson (London Math. Soc. Lecture Note Ser., Cambridge Univ. Press, Cambridge, 2006) pp. 215–237.
38. A. Kryczka, Quantitative approach to weak noncompactness in the polygon interpolation method, *Bull. Austral. Math. Soc.* **69**, 49 (2004).
39. A. Kryczka and S. Prus, Measure of weak noncompactness under complex interpolation, *Studia Math.* **147**, 89 (2001).
40. B. Cascales and J. Orihuela, A biased view of topology as a tool in

functional analysis, in *Recent progress in general topology III*, eds. K. P. Hart, J. van Mill and P. Simon (Atlantis Press, Paris, 2014) pp. 93–164.

41. B. Cascales, J. Orihuela and M. Ruiz Galán, Compactness, optimality, and risk, in *Computational and analytical mathematics*, eds. D. H. Bailey, H. H. Bauschke, P. Borwein, F. Garvan, M. Théra, J. D. Vanderwerff and H. Wolkowicz, Springer Proc. Math. Stat., Vol. 50 (Springer, New York, 2013) pp. 161–218. In honor of Jonathan Borwein's 60th birthday, Papers from the workshop (JonFest) held at Simon Fraser University, Burnaby, BC, May 16–20, 2011.
42. B. Cascales and J. Orihuela, Metrizability of precompact subsets in (LF)-spaces, *Proc. Roy. Soc. Edinburgh Sect. A* **103**, 293 (1986).
43. B. Cascales, On K-analytic locally convex spaces, *Arch. Math. (Basel)* **49**, 232 (1987).
44. B. Cascales and J. Orihuela, On compactness in locally convex spaces, *Math. Z.* **195**, 365 (1987).
45. B. Cascales and J. Orihuela, A sequential property of set-valued maps, *J. Math. Anal. Appl.* **156**, 86 (1991).
46. B. Cascales, J. Kąkol and S. A. Saxon, Weight of precompact subsets and tightness, *J. Math. Anal. Appl.* **269**, 500 (2002).
47. M. Talagrand, Espaces de Banach faiblement $\mathcal{K}$-analytiques, *Ann. of Math.* **110**, 407 (1979).
48. A. Grothendieck, Critères de compacité dans les espaces fonctionnels généraux, *Amer. J. Math.* **74**, 168 (1952).
49. J. L. Kelley and I. Namioka, *Linear topological spaces* (Springer-Verlag, New York-Heidelberg, 1976), With the collaboration of W. F. Donoghue, Jr., Kenneth R. Lucas, B. J. Pettis, Ebbe Thue Poulsen, G. Baley Price, Wendy Robertson, W. R. Scott, and Kennan T. Smith, Second corrected printing, Graduate Texts in Mathematics, No. 36.
50. K. Floret, Some aspects of the theory of locally convex inductive limits, in *Functional analysis: surveys and recent results II (Proc. Second Conf. Functional Anal., Univ. Paderborn, Paderborn, 1979)*, eds. K. D. Bierstedt and B. Fuchssteiner, North-Holland Math. Stud., Vol. 38 (North-Holland, Amsterdam-New York, 1980) pp. 205–237.
51. J. Kąkol, W. Kubiś and M. López-Pellicer, *Descriptive topology in selected topics of functional analysis*, Developments in Mathematics, Vol. 24 (Springer, New York, 2011).
52. V. V. Tkachuk, A space $C_p(X)$ is dominated by irrationals if and only if it is K-analytic, *Acta Math. Hungar.* **107**, 253 (2005).
53. B. Cascales, J. Orihuela and V. V. Tkachuk, Domination by second

countable spaces and Lindelöf Σ-property, *Topology Appl.* **158**, 204 (2011).
54. B. Cascales, M. Muñoz and J. Orihuela, The number of K-determination of topological spaces, *Rev. R. Acad. Cienc. Exactas Fís. Nat. Ser. A Math. RACSAM* **106**, 341 (2012).
55. B. Cascales, V. Kadets and J. Rodríguez, Measurability and selections of multi-functions in Banach spaces, *J. Convex Anal.* **17**, 229 (2010).
56. J. E. Jayne, J. Orihuela, A. J. Pallarés and G. Vera, σ-fragmentability of multivalued maps and selection theorems, *J. Funct. Anal.* **117**, 243 (1993).
57. B. Cascales, V. Kadets and J. Rodríguez, Measurable selectors and set-valued Pettis integral in non-separable Banach spaces, *J. Funct. Anal.* **256**, 673 (2009).

Introduction to interpolation theory

Fernando Cobos

Departamento de Análisis Matemático, Facultad de Matemáticas,
Universidad Complutense de Madrid,
Plaza de Ciencias 3, 28040 Madrid, Spain
E-mail: cobos@mat.ucm.es

We describe the main interpolation methods for Banach couples, their roots and some of their properties and applications. We consider also factorization of weakly compact operators and interpolation of couples of Banach algebras.

Keywords: L_p-spaces; real interpolation; complex interpolation; compact operators; weakly compact operators; interpolation of Banach algebras.

1. Introduction

The theory of interpolation of operators is a branch of functional analysis with important applications to harmonic analysis, partial differential equations, approximation theory, function spaces, operator theory and other more distant areas of mathematics (see, for example, the monographs by Butzer and Berens[1], Bergh and Löfström[2], Triebel[3], Kreĭn, Petunin and Semenov[4], Bennett and Sharpley[5], Brudnyĭ and Krugljak[6], Connes[7] or Amrein, Boutet de Monvel and Georgescu[8]). In this course we give a brief introduction to interpolation theory and some of its application.

We start by describing the roots of interpolation theory: The classical theorems due to Riesz and Thorin and to Marcinkiewicz on interpolation of operators between L_p-spaces. This is done in Section 2 where we also derive the Hausdorff-Young inequality and Young's inequality for convolutions. Then, in Section 3, we review the main interpolation methods: The real method of Lions and Peetre and the complex method of Calderón. We also show the spaces that come out applying those methods to certain couples of Banach spaces and Paley's inequality for Fourier coefficients.

The author has been supported in part by the Spanish Ministerio de Economía y Competitividad (MTM2010-15814).
2000 Mathematics Subject Classification: 46B70, 46E30, 47B10, 47B06.

Besides boundedness, compactness and weak compactness are other important properties that an operator may have. In the final Section 4 we describe the behaviour under interpolation of these properties. We also discuss the factorization theorem for weakly compact operators due to Davis, Figiel, Johnson and Pelczyński and the interpolation of couples of Banach algebras.

2. Classical theorems on interpolation of operators between L_p-spaces

Let (U,μ) and (V,ν) be σ-finite measure spaces. We denote by $L_p = L_p(U) = L_p(U,d\mu)$ the Lebesgue space, formed by all (equivalent classes of) scalar-valued μ-measurable functions f on U which have a finite norm

$$\|f\|_{L_p} = \Big(\int_U |f(x)|^p d\mu\Big)^{1/p}.$$

Here $1 \leq p < \infty$. The space $L_\infty = L_\infty(U) = L_\infty(U,d\mu)$ consists of all μ-essentially bounded scalar-valued functions, normed by

$$\|f\|_{L_\infty} = \text{ess sup}\{|f(x)| : x \in U\}.$$

By now we consider $\mathbb{C}$ as our scalar field $\mathbb{K}$, but later on it will not be important if $\mathbb{K}$ is $\mathbb{R}$ or $\mathbb{C}$.

Theorem 2.1 (Riesz-Thorin). *Assume that $1 \leq p_0, p_1, q_0, q_1 \leq \infty$ and let T be a linear operator from $L_{p_0}(\mu) + L_{p_1}(\mu)$ into $L_{q_0}(\nu) + L_{q_1}(\nu)$ which maps continuously*

$$L_{p_0}(U,d\mu) \longrightarrow L_{q_0}(V,d\nu) \quad \textit{with norm} \quad M_0$$

and

$$L_{p_1}(U,d\mu) \longrightarrow L_{q_1}(V,d\nu) \quad \textit{with norm} \quad M_1\,.$$

Take any $0 < \theta < 1$ and put $1/p = (1-\theta)/p_0 + \theta/p_1$, $1/q = (1-\theta)/q_0 + \theta/q_1$. Then T maps continuously

$$L_p(U,d\mu) \longrightarrow L_q(V,d\nu)$$

with norm $M \leq M_0^{1-\theta} M_1^{\theta}$.

This result was established by Marcel Riesz in 1926 with the restriction $p \leq q$ (which is essential when scalars are real). Later, in 1938, Thorin gave a new proof removing the condition $p \leq q$.

Notice that the estimate for the norms is logarithmically convex

$$\log M \leq (1-\theta)\log M_0 + \theta \log M_1 .$$

Besides, we have

$$(1/p\,,\,1/q) = (1-\theta)(1/p_0\,,\,1/q_0) + \theta(1/p_1\,,\,1/q_1).$$

Before giving the proof, let us illustrate the interest of this result with an application. Take the interval $[0, 2\pi]$ with the Lebesgue measure and for any $f \in L_p([0, 2\pi])$ let

$$\hat{f}(m) = \frac{1}{2\pi}\int_0^{2\pi} f(x)e^{-imx}dx \quad , \quad m \in \mathbb{Z},$$

be the Fourier coefficients of f.

We know that the functions $((2\pi)^{-1/2}e^{imx})_{m\in\mathbb{Z}}$ are an orthonormal basis in the Hilbert space $L_2([0, 2\pi])$. Hence, given any $f \in L_2([0, 2\pi])$, Parseval's equality yields that the sequence $(\hat{f}(m))$ belongs to ℓ_2 with

$$\|(\hat{f}(m))\|_{\ell_2} = \frac{1}{\sqrt{2\pi}}\|f\|_{L_2} .$$

Consider now the classical problem of describing the relationship between integrability properties of a function and decay of its Fourier coefficients. That is, to determine in which sequence space are the Fourier coefficients of a function f of $L_p([0, 2\pi])$ for $p \neq 2$.

Besides Parseval's equality, there is another easy estimate that we can do: If $f \in L_1([0, 2\pi])$, we can estimate directly the ℓ_∞-norm of $(\hat{f}(m))$. Namely,

$$|\hat{f}(m)| = |\frac{1}{2\pi}\int_0^{2\pi} f(x)e^{-imx}dx| \leq \frac{1}{2\pi}\int_0^{2\pi} |f(x)|dx = \frac{1}{2\pi}\|f\|_{L_1} .$$

Call T to the operator associating to any function f the sequence of its Fourier coefficients, $T(f) = (\hat{f}(m))$, then we can formulate our estimates as follows

$$T : L_1([0, 2\pi]) \longrightarrow \ell_\infty \quad \text{is bounded with norm} \quad M_0 \leq \frac{1}{2\pi}$$

and

$$T : L_2([0, 2\pi]) \longrightarrow \ell_2 \quad \text{is bounded with norm} \quad M_1 = \frac{1}{\sqrt{2\pi}} .$$

The problem is to show the behaviour of T on $L_p([0,2\pi])$ for the other values of p.

Consider first the case $1 < p < 2$. We can choose $0 < \theta < 1$ such that $1/p = (1-\theta)/1 + \theta/2$. Let p' be the conjugate index of p, $1/p + 1/p' = 1$. Clearly, $1/p' = \theta/2 = (1-\theta)/\infty + \theta/2$. Hence, applying the Riesz-Thorin theorem, we derive that $T : L_p([0,2\pi]) \longrightarrow \ell_{p'}$ is bounded with norm

$$M \leq \left(\frac{1}{2\pi}\right)^{1-\theta}\left(\frac{1}{\sqrt{2\pi}}\right)^{\theta} = \left(\frac{1}{2\pi}\right)^{1/p}.$$

In other words, we have derived the following classical result.

Corollary 2.2 (The Hausdorff-Young inequality). *If* $1 < p < 2$ *and* $f \in L_p([0,2\pi])$ *then* $(\hat{f}(m)) \in \ell_{p'}$ *and*

$$\Big(\sum_{m\in\mathbb{Z}} |\hat{f}(m)|^{p'}\Big)^{1/p'} \leq \Big(\frac{1}{2\pi}\int_0^{2\pi} |f(x)|^p dx\Big)^{1/p}.$$

The remaining case $2 < p \leq \infty$ is easy because $L_p([0,2\pi]) \hookrightarrow L_2([0,2\pi])$. So, if $f \in L_p([0,2\pi])$ we have that $(\hat{f}(m)) \in \ell_2$, and this is the best that one can say because it is known that if $r < 2$ there are continuous functions whose Fourier coefficients belong to ℓ_2 but they are not in ℓ_r (see[9]).

This application shows the interest of Riesz-Thorin theorem. In fact, often in Analysis one deals with operators which can act boundedly between different couples of Banach spaces. For this reason, it is important to have results which give relationships between properties of a given operator considered in different spaces. For example, this is of special interest working with integral operators.

For the proof of Riesz-Thorin theorem we are going to need the following result from complex analysis.

Lemma 2.3 (The three line lemma). *Let* $F(z)$ *be a bounded continuous function on the strip* $0 \leq Rez \leq 1$ *which is analytic on the interior of the strip* $0 < Rez < 1$. *If* $|F(z)| \leq M_0$ *for* $Rez = 0$ *and* $|F(z)| \leq M_1$ *for* $Rez = 1$, *then* $|F(z)| \leq M_0^{1-\theta} M_1^{\theta}$ *for* $Rez = \theta,\ \ 0 < \theta < 1$.

Proof of Riesz-Thorin theorem. We start with the case $p_0 = p_1 = p$.

Since $1=(1-\theta)q/q_0+\theta q/q_1$, using Hölder's inequality we obtain

$$\begin{aligned}\|Tf\|_{L_q} &= \Big(\int_V |Tf(y)|^{(1-\theta)q}|Tf(y)|^{\theta q}d\nu\Big)^{1/q}\\ &\le \Big(\int_V |Tf(y)|^{q_0}d\nu\Big)^{(1-\theta)/q_0}\Big(\int_V |Tf(y)|^{q_1}d\nu\Big)^{\theta/q_1}\\ &= \|Tf\|_{L_{q_0}}^{1-\theta}\|Tf\|_{L_{q_1}}^{\theta} \le M_0^{1-\theta}\|f\|_{L_{p_0}}^{1-\theta}M_1^{\theta}\|f\|_{L_{p_1}}^{\theta} = M_0^{1-\theta}M_1^{\theta}\|f\|_{L_p}.\end{aligned}$$

Suppose now $p_0 \neq p_1$. Since $1/p=(1-\theta)/p_0+\theta/p_1$, we have $1<p<\infty$. Let S_U [resp. S_V] be the space of all simple functions on U [resp. V] which vanish outside sets of finite measure. Then $S_U \subset L_p(U)$ and $S_V \subset L_q(V)$. Moreover, S_U is dense in $L_p(U)$ because $p<\infty$. We should show that

$$M=\sup\{\|Tf\|_{L_q} : \|f\|_{L_p}=1\} \le M_0^{1-\theta}M_1^{\theta}.$$

First observe that

$$\|Tf\|_{L_q}=\sup\Big\{|\int_V (Tf)gd\nu| : g\in S_V \text{ and } \|g\|_{L_{q'}}=1\Big\}.$$

Since S_U is dense in $L_p(U)$, it suffices to prove then that for all $f\in S_U$ with $\|f\|_{L_p}=1$ and for all $g\in S_V$ with $\|g\|_{L_{q'}}=1$ we have

$$|\int_V (Tf)gd\nu| \le M_0^{1-\theta}M_1^{\theta}.$$

Write

$$<Tf,g>=\int_V (Tf)gd\nu.$$

In order to use the three line lemma, put

$$\begin{aligned}\frac{1}{p(z)} &= \frac{1-z}{p_0}+\frac{z}{p_1},\\ \frac{1}{q'(z)} &= \frac{1-z}{q_0'}+\frac{z}{q_1'},\\ \varphi_z(x) &= |f(x)|^{p/p(z)}\frac{f(x)}{|f(x)|},\ x\in U,\\ \psi_z(y) &= |g(y)|^{q'/q'(z)}\frac{g(y)}{|g(y)|},\ y\in V,\\ F(z) &= <T\varphi_z,\psi_z>.\end{aligned}$$

It is not hard to check that $F(z)$ is analytic on the open strip $0<Rez<1$ and bounded and continuous on the closed strip $0\le Rez\le 1$. Moreover, since $1/p(it)=(1-it)/p_0+it/p_1$, we have

$$\|\varphi_{it}\|_{L_{p_0}}=\||f|^{p/p_0}\|_{L_{p_0}}=\|f\|_{L_p}^{p/p_0}=1.$$

Similarly

$$\|\varphi_{1+it}\|_{L_{p_1}} = 1 = \|\psi_{it}\|_{L_{q_0'}} = \|\psi_{1+it}\|_{L_{q_1'}} .$$

Now we can estimate $|F(z)|$ in the lines which are the boundary of the strip. We have for $-\infty < t < \infty$

$$\begin{aligned} |F(it)| &= |< T\varphi_{it}, \psi_{it} >| \leq \|T\varphi_{it}\|_{L_{q_0}} \|\psi_{it}\|_{L_{q_0'}} \\ &\leq M_0 \|\varphi_{it}\|_{L_{p_0}} \|\psi_{it}\|_{L_{q_0'}} = M_0 . \end{aligned}$$

Analogously, $|F(1+it)| \leq M_1$. Consequently, the three line lemma implies that

$$|F(\theta)| \leq M_0^{1-\theta} M_1^{\theta} .$$

But $1/p(\theta) = (1-\theta)/p_0 + \theta/p_1 = 1/p$ and so $\varphi_\theta = f$. We also have that $\psi_\theta = g$. Therefore

$$|\int_V (Tf) g d\nu| = |< Tf, g >| = |< T\varphi_\theta, \psi_\theta >| = |F(\theta)|$$

and the result follows. □

Next we show another application of Riesz-Thorin theorem. Take $U = V = \mathbb{R}^n$ and $d\mu = d\nu = dx$ the Lebesgue measure. Recall that the *convolution* of two functions f and g is defined by

$$f * g(x) = \int f(x-y) g(y) dy$$

provided the integral exists.

Theorem 2.4 (Young's inequality). *If $f \in L_p$ and $g \in L_q$, then $f * g \in L_r$ with $1/p + 1/q = 1/r + 1$ and*

$$\|f * g\|_{L_r} \leq \|f\|_{L_p} \|g\|_{L_q} .$$

Proof. Fix any $f \in L_p$. Given any $g \in L_1$ by Minkowski's inequality for integrals we have

$$\begin{aligned} \|f * g\|_{L_p} &= \Big(\int | \int f(x-y) g(y) dy|^p dx \Big)^{1/p} \\ &\leq \int \Big(\int |f(x-y) g(y)|^p dx \Big)^{1/p} dy = \|f\|_{L_p} \|g\|_{L_1} . \end{aligned}$$

On the other hand, for $g \in L_{p'}$ with $1/p + 1/p' = 1$, using Hölder's inequality

$$\begin{aligned} \int |f(x-y) g(y)| dy &\leq \Big(\int |f(x-y)|^p dy \Big)^{1/p} \Big(\int |g(y)|^{p'} dy \Big)^{1/p'} \\ &= \|f\|_{L_p} \|g\|_{L_{p'}} . \end{aligned}$$

That is to say, $\|f * g\|_{L_\infty} \le \|f\|_{L_p}\|g\|_{L_{p'}}$.

Consider now the convolution operator $Tg = f * g$. Clearly T is linear and, as we have just shown, T acts boundedly

$$T : L_1 \longrightarrow L_p \quad \text{with norm} \quad M_0 \le \|f\|_{L_p}$$

and

$$T : L_{p'} \longrightarrow L_\infty \quad \text{with norm} \quad M_1 \le \|f\|_{L_p}.$$

Choose $0 < \theta < 1$ so that $1/q = (1-\theta)/1 + \theta/p'$. Assumption on parameters yields that this choice is possible. It follows that $1/r = (1-\theta)/p$. Whence, using the Riesz-Thorin theorem, we derive that

$$T : L_q \longrightarrow L_r \quad \text{is bounded with norm} \quad M \le \|f\|_{L_p}^{1-\theta}\|f\|_{L_p}^{\theta} = \|f\|_{L_p}.$$

That is to say

$$\|f * g\|_{L_r} \le \|f\|_{L_p}\|g\|_{L_q}.$$

$\square$

Subsequently, the scalar field $\mathbb{K}$ can be either $\mathbb{R}$ or $\mathbb{C}$.

The other classical interpolation theorem is due to Marcinkiewicz. In order to formulate it, we need the concept of "weak L_p-space". Let $f : U \longrightarrow \mathbb{K}$ be a μ-measurable function which is finite almost everywhere. The *non-increasing rearrangement* of f on $(0, \infty)$ is given by

$$f^*(t) = \inf\{\delta > 0 : \mu(\{x \in U : |f(x)| > \delta\}) \le t\}.$$

This is a non-negative, non-increasing function which is continuous to the right and, among others properties, satisfies that

$$\|f\|_{L_p(U)} = \left(\int_0^\infty f^*(t)^p dt\right)^{1/p} = \|f^*\|_{L_p((0,\infty))}$$

(see [1,2,5]).

For $1 < p < \infty$ the weak L_p-space is defined by

$$L_{p,\infty}(U) = \left\{f : \|f\|_{p,\infty} = \sup_{t>0}\left\{t^{(1/p)-1}\int_0^t f^*(s)ds\right\} < \infty\right\}.$$

The functional $\|\cdot\|_{p,\infty}$ is a norm equivalent to the quasi-norm

$$|||f|||_{p,\infty} = \sup_{t>0}\{t^{1/p}f^*(t)\}.$$

Indeed, we have

$$\begin{aligned}|||f|||_{p,\infty} &\leq \sup_{t>0}\left\{t^{(1/p)-1}f^*(t)\int_0^t ds\right\}\\ &\leq \sup_{t>0}\left\{t^{(1/p)-1}\int_0^t f^*(s)ds\right\}\\ &= \|f\|_{p,\infty}\,.\end{aligned}$$

On the other hand

$$\|f\|_{p,\infty} \leq \sup_{t>0}\left\{t^{(1/p)-1}\int_0^t |||f|||_{p,\infty}s^{-1/p}ds\right\} \leq \frac{1}{1-(1/p)}|||f|||_{p,\infty}\,.$$

The space $L_{p,\infty}$ is larger than L_p because for any $t>0$ we have

$$\begin{aligned}\|f\|_p = \Big(\int_0^\infty f^*(s)^p ds\Big)^{1/p} &\geq \Big(\int_0^t f^*(s)^p ds\Big)^{1/p}\\ &\geq \Big(f^*(t)^p\int_0^t ds\Big)^{1/p} = t^{1/p}f^*(t).\end{aligned}$$

Hence

$$\|f\|_{p,\infty} \leq \frac{1}{1-(1/p)}|||f|||_{p,\infty} \leq \frac{1}{1-(1/p)}\|f\|_p\,.$$

But still we can recover L_p-estimates from weak L_p-estimates as the following result shows.

Theorem 2.5 (The Marcinkiewicz interpolation theorem). *Assume that* $1\leq p_0, p_1\leq\infty$, $1<q_0\neq q_1<\infty$ *and let* T *be a linear operator which maps continuously*

$$L_{p_0}(U,d\mu)\longrightarrow L_{q_0,\infty}(V,d\nu)\quad \textit{with norm}\quad M_0$$

and

$$L_{p_1}(U,d\mu)\longrightarrow L_{q_1,\infty}(V,d\nu)\quad \textit{with norm}\quad M_1\,.$$

Let $0<\theta<1$, *put* $1/p=(1-\theta)/p_0+\theta/p_1\,, 1/q=(1-\theta)/q_0+\theta/q_1$ *and assume that* $p\leq q$. *Then* T *maps continuously*

$$L_p(U,d\mu)\longrightarrow L_q(V,d\nu)$$

with norm $M\leq CM_0^{1-\theta}M_1^{\theta}$ *(where the constant* C *does not depend on* T*).*

This result appeared without proof in a paper by Marcinkiewicz in 1939. In 1956, Zygmund gave a proof and also applications that cannot be obtained by the Riesz-Thorin theorem.

At the beginning of the sixties, Lions, Peetre, Calderón, Gagliardo, Krein, Triebel and other authors investigated the validity of these results for general couples of Banach spaces, developing the abstract interpolation theory. Two main interpolation methods were developed: The complex method related to ideas of Riesz-Thorin theorem and the real method associated to the Marcinkiewicz theorem. In the next section we describe them and we also give a proof for the Marcinkiewicz interpolation theorem.

3. Main interpolation methods. Examples

By a *Banach couple* $\bar{A} = (A_0, A_1)$ we mean two Banach spaces A_j, $j = 0, 1$, which are continuously embedded in some Hausdorff topological vector space $\mathcal{A}$. Then we can form their sum

$$A_0 + A_1 = \{a \in \mathcal{A} : a = a_0 + a_1, a_j \in A_j\,, j = 0, 1\}$$

and their intersection

$$A_0 \cap A_1 = \{a \in \mathcal{A} : a \in A_0, a \in A_1\}\,.$$

These spaces become Banach spaces when endowed with the norms

$$\|a\|_{A_0+A_1} = \inf\{\|a_0\|_{A_0} + \|a_1\|_{A_1} : a = a_0 + a_1, a_j \in A_j\}$$

and

$$\|a\|_{A_0\cap A_1} = \max\{\|a\|_{A_0}, \|a\|_{A_1}\}\,.$$

We want to find procedures for assigning to each couple $\bar{A}$ an *intermediate space*, that is, a Banach space in the position

$$A_0 \cap A_1 \hookrightarrow \bigstar \hookrightarrow A_0 + A_1\,.$$

Here $\hookrightarrow$ means continuous embedding. We also want that an abstract version of the Riesz-Thorin theorem holds.

To define the real method we need the *Peetre's K-functional* which is the function that we get when we replace in the definition of $\|\cdot\|_{A_0+A_1}$ the norm of A_1 by a multiple

$$K(t,a) = K(t,a;\bar{A}) = \inf\{\|a_0\|_{A_0}+t\|a_1\|_{A_1} : a = a_0+a_1,\ a_j \in A_j\}, \quad t > 0.$$

For $0 < \theta < 1$ and $1 \le q \le \infty$, the *real interpolation space* $(A_0, A_1)_{\theta,q}$ is formed by all those $a \in A_0 + A_1$ which have a finite norm

$$\|a\|_{\theta,q} = \left(\int_0^\infty \left(t^{-\theta}K(t,a)\right)^q \frac{dt}{t}\right)^{\frac{1}{q}} \quad \text{if} \quad 1 \le q < \infty,$$

$$\|a\|_{\theta,\infty} = \sup_{t>0}\{t^{-\theta}K(t,a)\}$$

(see[10], [1], [2], [3]).

It turns out that $(A_0, A_1)_{\theta,q}$ is a Banach space, which is intermediate with respect to $\bar{A}$

$$A_0 \cap A_1 \hookrightarrow (A_0, A_1)_{\theta,q} \hookrightarrow A_0 + A_1\,,$$

with $A_0 \cap A_1$ dense in $(A_0, A_1)_{\theta,q}$ if $q < \infty$ and with the interpolation property. In order to describe this property, let $\bar{B} = (B_0, B_1)$ be another Banach couple. We put $T \in \mathcal{L}(\bar{A}, \bar{B})$ to mean that T is a linear operator from $A_0 + A_1$ into $B_0 + B_1$, whose restrictions $T : A_j \longrightarrow B_j$ are bounded for $j = 0, 1$.

Theorem 3.1 (Interpolation theorem). *Let $\bar{A} = (A_0, A_1)$, $\bar{B} = (B_0, B_1)$ be Banach couples and let $T \in \mathcal{L}(\bar{A}, \bar{B})$. For any $0 < \theta < 1$ and $1 \leq q \leq \infty$, the restriction of T to $(A_0, A_1)_{\theta,q}$ is a bounded operator $T : (A_0, A_1)_{\theta,q} \longrightarrow (B_0, B_1)_{\theta,q}$ with norm*

$$\|T\|_{(A_0,A_1)_{\theta,q},(B_0,B_1)_{\theta,q}} \leq \|T\|^{1-\theta}_{A_0,B_0}\|T\|^{\theta}_{A_1,B_1}\,.$$

Proof. Let $M_j = \|T\|_{A_j,B_j}$, $j = 0, 1$. Given any decomposition $a = a_0+a_1$ with $a_j \in A_j$, we have $Ta = Ta_0 + Ta_1$ with $Ta_j \in B_j$ and

$$\begin{aligned} K(t, Ta; \bar{B}) &\leq \|Ta_0\|_{B_0} + t\|Ta_1\|_{B_1} \leq M_0\|a_0\|_{A_0} + tM_1\|a_1\|_{A_1} \\ &= M_0(\|a_0\|_{A_0} + \frac{tM_1}{M_0}\|a_1\|_{A_1})\,. \end{aligned}$$

This yields that

$$K(t, Ta; \bar{B}) \leq M_0 K(\frac{tM_1}{M_0}, a; \bar{A})\,.$$

Consequently,

$$\begin{aligned} \|Ta\|_{(B_0,B_1)_{\theta,q}} &\leq M_0\Big(\int_0^\infty (t^{-\theta}K(\frac{tM_1}{M_0}, a; \bar{A}))^q\frac{dt}{t}\Big)^{1/q} \\ &= M_0(\frac{M_1}{M_0})^\theta\Big(\int_0^\infty (s^{-\theta}K(s, a; \bar{A}))^q\frac{ds}{s}\Big)^{1/q} \\ &= M_0^{1-\theta}M_1^\theta\|a\|_{(A_0,A_1)_{\theta,q}}\,. \end{aligned}$$

□

The following example relates the real method with the results that we discussed in the previous section.

Let (U, μ) be a σ-finite measure space. Then $(L_1(U), L_\infty(U))$ is a Banach couple. As containing space $\mathcal{A}$ we take $\mathcal{L}_\mu$, the space of all μ-measurable functions on U which are finite almost everywhere with the convergence in measure on each measurable set in U of finite measure. It turns out that

$$K(t, f; L_1(U), L_\infty(U)) = \int_0^t f^*(s)ds \quad , \quad t > 0 \tag{1}$$

(see[1–3]). Using this formula, we can determine the real interpolation spaces generated by the couple $(L_1(U), L_\infty(U))$. Recall that for $1 < p < \infty$ and $1 \leq q \leq \infty$, the *Lorentz function space* $L_{p,q} = L_{p,q}(U)$ is defined to be the collection of all (equivalence classes of) μ-measurable functions f on U such that the following norm is finite

$$\|f\|_{L_{p,q}} = \Big(\int_0^\infty (t^{(1/p)-1} \int_0^t f^*(s)ds)^q \frac{dt}{t} \Big)^{1/q}$$

(the first integral should be replaced by the supremum if $q = \infty$).

Notice that $L_{p,\infty}$ is the weak L_p-space introduced in the previous section. One can check that $L_{p,p} = L_p$ with equivalence of norms. Moreover, it turns out that

$$L_{p,q_1} \hookrightarrow L_{p,q_2} \quad \text{if} \quad 1 \leq q_1 \leq q_2 \leq \infty . \tag{2}$$

Hence, the Lorentz scale is a refiner scale than the Lebesgue scale.

Let $1 \leq q \leq \infty, 0 < \theta < 1$ and $1/p = 1 - \theta$, then it follows from (1) that

$$(L_1, L_\infty)_{\theta,q} = L_{p,q}$$

with equivalence of norms. In a more general way, for $1 < p_0 \neq p_1 < \infty$, $0 < \theta < 1$ $1/p = (1-\theta)/p_0 + 1/p_1$ and $1 \leq q_0, q_1, q \leq \infty$ we have with equivalence of norms

$$(L_{p_0,q_0}, L_{p_1,q_1})_{\theta,q} = L_{p,q} \tag{3}$$

(see[1–3]). For the case of Lebesgue spaces, one obtains if $1 \leq p_0, p_1 \leq \infty$ and $1/p = (1-\theta)/p_0 + 1/p_1$ that

$$(L_{p_0}, L_{p_1})_{\theta,p} = L_p \quad \text{(equivalent norms)} . \tag{4}$$

Next we go back to the Marcinkiewicz interpolation theorem to prove it.

Proof of Theorem 2.5. Let T be a linear operator mapping continuously

$$L_{p_0}(U) \longrightarrow L_{q_0,\infty}(V) \quad \text{with norm} \quad M_0$$

and

$$L_{p_1}(U) \longrightarrow L_{q_1,\infty}(V) \quad \text{with norm} \quad M_1 .$$

By the interpolation theorem

$$T : (L_{p_0}(U), L_{p_1}(U))_{\theta,p} \longrightarrow (L_{q_0,\infty}(V), L_{q_1,\infty}(V))_{\theta,p}$$

is bounded with norm less than or equal than $M_0^{1-\theta} M_1^{\theta}$. We proceed to determine the interpolation spaces. Using (4), we have $(L_{p_0}(U), L_{p_1}(U))_{\theta,p} = L_p(U)$. By (3) and (2), since $p \leq q$, we derive $(L_{q_0,\infty}(V), L_{q_1,\infty}(V))_{\theta,p} = L_{q,p} \hookrightarrow L_q$. Equalities are with equivalence of norms. Therefore, we conclude that $T : L_p(U) \longrightarrow L_q(V)$ is bounded with norm $M \leq C M_0^{1-\theta} M_1^{\theta}$ where C is independent of T. □

Next take $U = \mathbb{Z}$ and $\mu\{m\} = 1$ for any $m \in \mathbb{Z}$. Spaces $L_{p,q}(\mathbb{Z}) = \ell_{p,q}$ are the *Lorentz sequence spaces*. For $1 \leq p_0 \neq p_1 \leq \infty$, $0 < \theta < 1$ $1/p = (1-\theta)/p_0 + \theta/p_1$ and $1 \leq q \leq \infty$ we have

$$(\ell_{p_0}, \ell_{p_1})_{\theta,q} = \ell_{p,q} \quad \text{(equivalent norms)} . \tag{5}$$

Now we show a more direct description of $\ell_{p,q}$. Given any null sequence $\xi = (\xi_m) \in c_0$, let $\xi^* = (\xi_n^*)$ be the decreasing rearrangement of $(|\xi_m|)$ defined by

$$\begin{aligned} \xi_1^* &= \max\{|\xi_m| : m \in \mathbb{Z}\} = |\xi_{m_1}| , \\ \xi_2^* &= \max\{|\xi_m| : m \in \mathbb{Z}\backslash\{m_1\}\} = |\xi_{m_2}| , \\ &\vdots \end{aligned}$$

For $f(m) = \xi_m$, $m \in \mathbb{Z}$, we have

$$f^*(t) = \begin{cases} \xi_1^* & \text{for } t \in (0,1) , \\ \xi_n^* & \text{for } t \in [n-1,n), n = 2,3,\ldots \end{cases}$$

It turns out that the norm of $\ell_{p,q}$ is equivalent to

$$\|\xi\|_{\ell_{p,q}} = \Big(\sum_{n=1}^{\infty} (n^{(1/p)-1} \sum_{j=1}^{n} \xi_j^*)^q n^{-1} \Big)^{1/q} .$$

We have $\ell_{p,q} \hookrightarrow \ell_{r,s}$ provided that $p < r$ or $p = r$ and $q < s$. These spaces are different if the couples (p,q) and (r,s) do not coincide. If $\xi \in \ell_{p,q}$ then its decreasing rearrangement (ξ_n^*) decays as $(n^{-1/p} \log(n)^{-1/q})$. In particular, if $1 < p < 2$ then $\ell_{p',p} \subsetneq \ell_{p'}$. This fact allows us to improve the Hausdorff-Young inequality if we work in the scale of Lorentz sequence spaces.

Corollary 3.2 (Paley's inequality). *Let* $1 < p < 2$ *and* $f \in L_p([0, 2\pi])$. *Then* $(\hat{f}(m)) \in \ell_{p',p}$ *and* $\|(\hat{f}(m))\|_{\ell_{p',p}} \le C_p \|f\|_{L_p([0,2\pi])}$ *where* C_p *depends only on* p.

Proof. Let $T(f) = (\hat{f}(m))$. As we have shown in Section 2, the restrictions

$$T : L_1([0, 2\pi]) \longrightarrow \ell_\infty \quad \text{and} \quad T : L_2([0, 2\pi]) \longrightarrow \ell_2$$

are bounded. Using the interpolation theorem and formulae (4) and (5) we derive that

$$T : L_p([0, 2\pi]) \longrightarrow \ell_{p',p}$$

is bounded. This yields the result. □

Next let H be a Hilbert space. Put

$$S_\infty = S_\infty(H) = \mathcal{C}(H) = \{T \in \mathcal{L}(H, H) : T \text{ is compact}\}.$$

The *singular numbers* of $T \in S_\infty$ are defined by

$$s_n(T) = \inf\{\|T - R\|_{H,H} : R \in \mathcal{L}(H, H), \ \ \operatorname{rank} R < n\}.$$

Singular numbers form a non-increasing sequence of non-negative numbers, which coincide with the eigenvalues of the compact positive operator $[T^*T]^{1/2}$. Moreover they are connected with the eigenvalues of T by means of Weyl inequality

$$\sum_{j=1}^{n} |\lambda_j(T)|^p \le \sum_{j=1}^{n} s_j(T)^p, \quad n \in \mathbb{N}, \quad 0 < p < \infty.$$

Here the sequence of eigenvalues of T is supposed to be ordered in non-increasing way with respect to the absolute values, repeating each eigenvalue as many times as its multiplicity indicates (see[11]).

This is the motivation for introducing *the Schatten classes.* For $1 \le p < \infty$, the Schatten p-class is defined by

$$S_p = \{T \in \mathcal{C}(H) : \|T\|_{S_p} = \Big(\sum_{n=1}^{\infty} s_n(T)^p\Big)^{1/p} < \infty\}, \quad S_\infty = \mathcal{C}(H).$$

In the scale $\{S_p\}_{1 \le p \le \infty}$, besides S_∞, other outstanding spaces are S_2, formed by Hilbert-Schmidt operators, and S_1 which consists of nuclear operators.

In a more general way, for $1 < p < \infty$ and $1 \le q \le \infty$, we put

$$S_{p,q} = \Big\{T \in \mathcal{C}(H) :$$
$$\|T\|_{S_{p,q}} = \|(s_n(T))\|_{\ell_{p,q}} = \Big(\sum_{n=1}^{\infty}(n^{(1/p)-1}\sum_{j=1}^{n} s_j(T))^q n^{-1}\Big)^{1/q} < \infty\Big\}$$

(see[3]). It turns out that $K(t,T;S_1,S_\infty) = K(t,(s_n(T));\ell_1,\ell_\infty)$. Hence,

$$(S_1, S_\infty)_{\theta,q} = S_{p,q}\,, \quad 1/p = 1-\theta \quad \text{(equivalent norms)}.$$

We close this section by reviewing the complex interpolation method. Let $\bar{A} = (A_0, A_1)$ be a Banach couple. We need the space $\mathcal{F}(\bar{A})$ of all functions f from the closed strip $S = \{z \in \mathbb{C} : 0 \le \mathrm{Re} z \le 1\}$ into $A_0 + A_1$ such that

- f is bounded and continuous on S and analytic on the interior of S, and
- the functions $t \longrightarrow f(j+it)$ $(j = 0,1)$ are continuous from $\mathbb{R}$ into A_j and tend to zero as $|t| \to \infty$.

The space $\mathcal{F}(\bar{A})$ becomes a Banach space when normed by

$$\|f\|_{\mathcal{F}(\bar{A})} = \max\{\sup_{t\in\mathbb{R}} \|f(it)\|_{A_0}\ ,\ \sup_{t\in\mathbb{R}} \|f(1+it)\|_{A_1}\}.$$

For $0 < \theta < 1$, the *complex interpolation space* $[A_0, A_1]_\theta$ consists of all $a \in A_0 + A_1$ such that $a = f(\theta)$ for some $f \in \mathcal{F}(\bar{A})$. The norm in $[A_0, A_1]_\theta$ is given by

$$\|a\|_{[\theta]} = \inf\{\|f\|_{\mathcal{F}(\bar{A})} : f(\theta) = a\,,\ f \in \mathcal{F}(\bar{A})\}.$$

(see[12],[2],[3]).

This construction produces intermediate spaces

$$A_0 \cap A_1 \hookrightarrow [A_0, A_1]_\theta \hookrightarrow A_0 + A_1\,,$$

and the interpolation theorem (Theorem 3.1) holds for this method with the same estimate for the norms. Moreover, $A_0 \cap A_1$ is dense in $[A_0, A_1]_\theta$.

For $1/p = 1-\theta$, the following formulae are valid, this time with equal norms,

$$[L_1, L_\infty]_\theta = L_p\ ,\quad [\ell_1, \ell_\infty]_\theta = \ell_p \text{ and } [S_1, S_\infty]_\theta = S_p\,.$$

The real and the complex method are related by the following embeddings

$$(A_0, A_1)_{\theta,1} \hookrightarrow [A_0, A_1]_\theta \hookrightarrow (A_0, A_1)_{\theta,\infty}\,.$$

4. Compactness. Factorization of weakly compact operators

In 1960, Krasnoselskii[13] established the following reinforced version of the Riesz-Thorin theorem.

Theorem 4.1. *Assume that* $1 \leq p_0, p_1, q_0, q_1 \leq \infty$, *let* $0 < \theta < 1$, $1/p = (1-\theta)/p_0 + \theta/p_1$, $1/q = (1-\theta)/q_0 + \theta/q_1$ *and let* T *be a linear operator which maps continuously*

$$T : L_{p_0}(U, d\mu) \longrightarrow L_{q_0}(V, d\nu) \quad \text{and} \quad T : L_{p_1}(U, d\mu) \longrightarrow L_{q_1}(V, d\nu)\,.$$

Suppose, in addition, that

$$q_0 < \infty \text{ and } T : L_{p_0}(U, d\mu) \longrightarrow L_{q_0}(V, d\nu) \text{ is compact.}$$

Then

$$T : L_p(U, d\mu) \longrightarrow L_q(V, d\nu) \quad \text{is compact.}$$

Krasnoselskii's theorem led to the investigation of interpolation properties of compact operators between abstract Banach couples. The first results on this question appeared already in the seminal papers by Lions and Peetre[10] and Calderón[12]. Results of[10] work for the real and the complex method but they require that one of the couples reduces to a single Banach space. Compactness result of[12] refers only to the complex method and applies to non-degenerated couples but it requires that the target couple satisfies a certain approximation condition which corresponds to Krasnoselskii's assumption $q_0 < \infty$. Later on, a number of authors have continue the research on this topic. For the real method, the final result was achieved in 1992 in the papers by Cwikel[14] and by Cobos, Kühn and Schonbek[15]. It reads as follows.

Theorem 4.2. *Let* $\bar{A} = (A_0, A_1)$, $\bar{B} = (B_0, B_1)$ *be Banach couples and let* $T \in \mathcal{L}(\bar{A}, \bar{B})$ *such that* $T : A_j \longrightarrow B_j$ *is compact for* $j = 0$ *or* $j = 1$. *For any* $0 < \theta < 1$ *and* $1 \leq q \leq \infty$, *we have that*

$$T : (A_0, A_1)_{\theta,q} \longrightarrow (B_0, B_1)_{\theta,q} \quad \text{is compact.}$$

Even if a restriction of T fails to be compact, the real method produces spaces sufficiently far away from the edges so that the interpolated operator is still compact.

Recently, limiting cases of the real method where $\theta = 0$ or 1 are being extensively studied. In these cases, resulting spaces can be quite close to the edges and a similar statement as Theorem 4.2 fails. Details can be found in the paper by Cobos, Fernández-Cabrera, Kühn and Ullrich[16].

The proof we gave in [15] for Theorem 4.2 is based on properties of vector-valued sequence spaces related to the construction of the real method using series instead of integrals.

As a direct consequence of Theorem 4.2 and formula (3) we obtain the following result of Krasnoselskii's type for Lorentz spaces.

Corollary 4.3. *Assume that*

$$1 < p_0, p_1, r_0, r_1 < \infty, \ 1 \leq q_0, q_1, s_0, s_1, q \leq \infty,$$

let $0 < \theta < 1$ *and put* $1/p = (1-\theta)/p_0 + \theta/p_1$, $1/r = (1-\theta)/r_0 + \theta/r_1$. *If* T *is a linear operator which maps*

$$T : L_{p_0,q_0}(U, d\mu) \longrightarrow L_{r_0,s_0}(V, d\nu) \quad \textit{compactly}$$

and

$$T : L_{p_1,q_1}(U, d\mu) \longrightarrow L_{r_1,s_1}(V, d\nu) \quad \textit{boundedly},$$

then

$$T : L_{p,q}(U, d\mu) \longrightarrow L_{r,q}(V, d\nu) \quad \textit{is compact}.$$

It is natural to wonder if interpolation ideas are also useful to derive a similar result but for Orlicz spaces. Recall that given a σ-finite measure space (U, μ) and an increasing convex function $N(t)$ on $[0, \infty)$ with $N(0) = 0$, the Orlicz space L_N consists of all measurable functions f on U such that $\int N(|f(x)|/\lambda) d\mu(x) < \infty$ for some $\lambda > 0$. The norm of L_N is given by

$$\|f\|_{L_N} = \inf\{\lambda : \int N(|f(x)|/\lambda) d\mu(x) \leq 1\}.$$

Ideas of [15] also work in a more general context and yield the following compactness result for operators between Orlicz spaces. We put $N^{-1}(t)$ for the inverse function of N.

Corollary 4.4. *Let* $(L_{N_0}, L_{N_1}), (L_{M_0}, L_{M_1})$ *be couples of Orlicz spaces and let* $T \in \mathcal{L}((L_{N_0}, L_{N_1}), (L_{M_0}, L_{M_1}))$ *such that*

$$T : L_{N_0} \longrightarrow L_{M_0} \quad \textit{is compact.}$$

Let $0 < \theta < 1$ *and put*

$$N^{-1}(t) = N_0^{-1}(t)[N_1^{-1}(t)/N_0^{-1}(t)]^{\theta}, \ M^{-1}(t) = M_0^{-1}(t)[M_1^{-1}(t)/M_0^{-1}(t)]^{\theta}.$$

Then

$$T : L_N \longrightarrow L_M \quad \textit{is also compact.}$$

The compactness problem for the complex method is not yet solved. A number of sufficient conditions are known to interpolate compact operators by the complex method, but a result of the generality of Theorem 4.2 is not yet known. We refer to the survey[17] for references and more details.

Consider now the case of weakly compact operators. Recall that an operator $T \in \mathcal{L}(A, B)$ between Banach spaces A and B is said to be weakly compact if T carries the closed unit ball of A onto a relatively weakly compact subset of B. An important example of weakly compact operator is the identity map of a reflexive Banach space. Weakly compact operators form an operator ideal. The following result was established by Beauzamy[18] in 1978 for T equals to the identity operator.

Theorem 4.5. *Let $\bar{A} = (A_0, A_1)$, $\bar{B} = (B_0, B_1)$ be Banach couples, let $0 < \theta < 1$, $1 < q < \infty$ and let $T \in \mathcal{L}(\bar{A}, \bar{B})$. A necessary and sufficient condition for $T : (A_0, A_1)_{\theta,q} \longrightarrow (B_0, B_1)_{\theta,q}$ to be weakly compact is that $T : A_0 \cap A_1 \longrightarrow B_0 + B_1$ is weakly compact.*

In particular, the following holds.

Corollary 4.6. *Let $\bar{A} = (A_0, A_1)$ be a Banach couple. If the embedding $A_0 \cap A_1 \hookrightarrow A_0 + A_1$ is weakly compact, then $(A_0, A_1)_{\theta,q}$ is reflexive for any $0 < \theta < 1$ and $1 < q < \infty$.*

A result of this type does not hold for the complex method. In fact, Mastyło showed in[19] an example of a couple of function spaces (E_0, E_1) such that the embedding $E_0 \cap E_1 \hookrightarrow E_0 + E_1$ is weakly compact but $[E_0, E_1]_\theta$ is not reflexive. A sufficient condition to interpolate weakly compact operators by the complex method is given in the following result.

Theorem 4.7. *Let $\bar{A} = (A_0, A_1)$, $\bar{B} = (B_0, B_1)$ be Banach couples, let $0 < \theta < 1$ and let $T \in \mathcal{L}(\bar{A}, \bar{B})$. If $T : A_j \longrightarrow B_j$ is weakly compact for $j = 0$ or $j = 1$, then*

$$T : [A_0, A_1]_\theta \longrightarrow [B_0, B_1]_\theta$$

is also weakly compact.

Next we use the results on interpolation of weakly compact operators by the real method to derive the classical result established by Davis, Figiel, Johnson and Pelczyński[20] in 1974 on factorization of weakly compact operators.

Theorem 4.8. *Let E, F be Banach spaces and let $T \in \mathcal{L}(E, F)$ be a weakly compact operator. Then there is a reflexive Banach space X and bounded* *linear operators* $R \in \mathcal{L}(E, X)$ *and* $S \in \mathcal{L}(X, F)$ *such that the following diagram commutes*

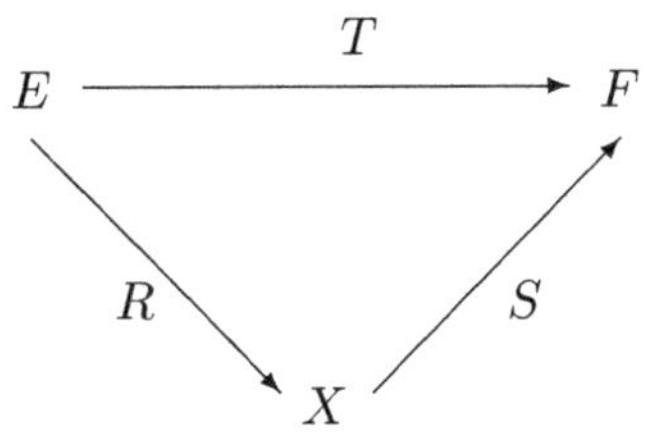

Proof. Let $Ker(T)$ be the kernel of T and let $Y = E/Ker(T)$ be the quotient space. Write $Q(x) = [x]$ for the quotient mapping and let $j([x]) = Tx$. Then the operators

$$Q : E \to Y = E/Ker(T) \quad \text{and} \quad j : Y \to F$$

are linear and bounded. Put

$$A_0 = j(Q(E)) = \{Tx : x \in E\}$$

with

$$\|Tx\|_{A_0} = \|[x]\|_Y = \inf\{\|y\|_E : Tx = Ty\}$$

and let $A_1 = F$. We have

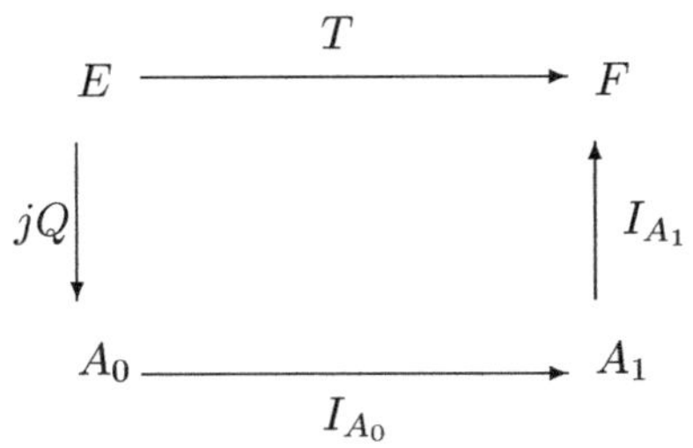

Moreover, using that T is weakly compact, it follows that the embedding $A_0 \hookrightarrow A_1$ is weakly compact. Hence, (A_0, A_1) is a Banach couple such that the embedding $A_0 = A_0 \cap A_1 \hookrightarrow A_0 + A_1 = A_1$ is weakly compact. Take any $0 < \theta < 1$ and $1 < q < \infty$. By the Corollary 4.6, the space $X = (A_0, A_1)_{\theta,q}$ is reflexive, and

$$A_0 \hookrightarrow X \hookrightarrow A_1 .$$

This yields the desired factorization with $R = jQ$ and $S = I_X$. $\square$

We have derived the results of Davis, Figiel, Johnson and Pelczyński from the interpolation properties of weakly compact operators. Looking back to the seventies, the situation was the converse: Since the proof given in[20] had a clear interpolation flavor, it motivated the investigation on the behaviour of weak compactness under interpolation. As one can see in[21], these interpolation techniques are also useful to prove factorization properties for other operator ideals.

In 1992, Galé, Ransford and White[22] asked if the corresponding result holds in the setting of Banach algebras: Whether or not every weakly compact homomorphism between Banach algebras $\Phi : A \longrightarrow B$ factors through a reflexive Banach algebra W with Banach algebra homomorphisms Φ_1, Φ_2 as factors

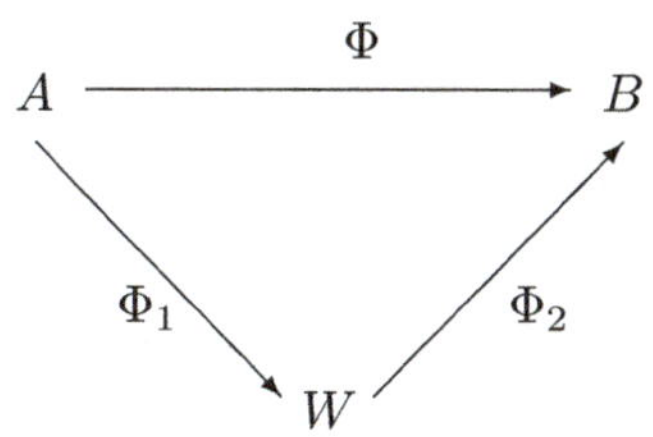

The answer was given in 2004 by Blanco, Kaijser and Ransford[23]. They carried over to the setting of Banach algebras the arguments of Davis, Figiel, Johnson and Pelczyński. However, let us try to established the result of[23] following the previous approach. We need then an interpolation method $\mathfrak{F}$ such that whenever A_0 and A_1 are Banach algebras with $A_0 \hookrightarrow A_1$ and the embedding being weakly compact, then $\mathfrak{F}(A_0, A_1)$ is a reflexive Banach algebra. Let us study it in more detail.

By a *couple of Banach algebras* $\bar{W} = (W_0, W_1)$ we understand a Banach couple formed by Banach algebras W_j , $j = 0, 1$, such that multiplication in W_0 and W_1 coincide in $W_0 \cap W_1$.

For example, working with sequence spaces defined on $\mathbb{Z}$, the couple $\bar{\ell}_1 = (\ell_1, \ell_1(2^{-m}))$ is a couple of Banach algebras if multiplication is defined as convolution

$$\xi * \eta = \Big(\sum_{k=-\infty}^{\infty} \xi_k \eta_{m-k} \Big)_{m \in \mathbb{Z}}, \quad \xi = (\xi_m) \, , \eta = (\eta_m) \, .$$

Subsequently, let $\mathfrak{F}$ be an interpolation method such that, for any Banach couple $\bar{A} = (A_0, A_1)$, $A_0 \cap A_1$ is dense in $\mathfrak{F}(A_0, A_1)$. For example, we can take $\mathfrak{F} = [\cdot,\cdot]_\theta$ or $\mathfrak{F} = (\cdot,\cdot)_{\theta,q}$ with $q < \infty$. We say that $\mathfrak{F}$ *preserve the Banach-algebra structure* if given any couple of Banach algebras $\bar{W} = (W_0, W_1)$ there is a constant $c_{\overline{W}} > 0$ such that for any $x, y \in W_0 \cap W_1$

$$\|xy\|_{\mathfrak{F}(W_0,W_1)} \leq c_{\overline{W}} \|x\|_{\mathfrak{F}(W_0,W_1)} \|y\|_{\mathfrak{F}(W_0,W_1)} \,.$$

Notice that multiplication of x, y is well defined because these vectors belong to the intersection $W_0 \cap W_1$. Observe also that, since $W_0 \cap W_1$ is dense in $\mathfrak{F}(W_0, W_1)$, multiplication can be extended by continuity to the whole of $\mathfrak{F}(W_0, W_1)$, turning $\mathfrak{F}(W_0, W_1)$ into a Banach algebra.

So, to factorize weakly compact homomorphism following the approach described before, we need an interpolation method $\mathfrak{F}$ which preserves the Banach-algebra structure and such that $\mathfrak{F}(A_0, A_1)$ is reflexive if $A_0 \hookrightarrow A_1$ is weakly compact. It follows from a result by Bishop[24] that the first property holds for $\mathfrak{F} = (\cdot,\cdot)_{\theta,1}$. This is the only real method with such property because if $q \neq 1$ we have $(\ell_1, \ell_1(2^{-m}))_{\theta,q} = \ell_q(2^{-\theta m})$ which is not an algebra for convolution. But $\mathfrak{F} = (\cdot,\cdot)_{\theta,1}$ does not satisfy the second property (see[25]). On the other hand, the complex method preserves the Banach-algebra structure (see[12]) but, as we pointed out before, it fails also the second condition.

One can overcome the obstruction by working with *the general real method*, which is defined by replacing the weighted L_q norm of $(\cdot,\cdot)_{\theta,q}$ by a more general lattice norm Γ (see[6]). In the paper by Cobos, Fernández-Cabrera and Martínez[26] it is shown a necessary and sufficient condition on Γ for $\mathfrak{F} = (\cdot,\cdot)_\Gamma$ to preserve the Banach-algebra structure. Doing a suitable choice for Γ we can also obtain for $\mathfrak{F} = (\cdot,\cdot)_\Gamma$ a good behavoiur for weakly compact operators and therefore to complete the proof of the factorization theorem for weakly compact homomorphisms. We refer to[26] for more details.

The problem of interpolation of Banach algebras is clearly related to interpolation of bilinear operators. Results in this direction can be seen in the papers by Astashkin[27] and Cobos and Fernández-Cabrera[28,29].

References

1. P. L. Butzer and H. Berens, *Semi-groups of operators and approximation* (Springer-Verlag New York Inc., New York, 1967).
2. J. Bergh and J. Löfström, *Interpolation spaces. An introduction* (Springer-Verlag, Berlin-New York, 1976).
3. H. Triebel, *Interpolation theory, function spaces, differential operators* (VEB Deutscher Verlag der Wissenschaften, Berlin, 1978).

4. S. G. Kreĭn, Y. Ī. Petunīn and E. M. Semënov, *Interpolation of linear operators* (American Mathematical Society, Providence, R.I., 1982).
5. C. Bennett and R. Sharpley, *Interpolation of operators*, Pure and Applied Mathematics, Vol. 129 (Academic Press, Inc., Boston, MA, 1988).
6. Y. A. Brudnyĭ and N. Y. Krugljak, *Interpolation functors and interpolation spaces. Vol. I*, North-Holland Mathematical Library, Vol. 47 (North-Holland Publishing Co., Amsterdam, 1991).
7. A. Connes, *Noncommutative geometry* (Academic Press, Inc., San Diego, CA, 1994).
8. W. O. Amrein, A. Boutet de Monvel and V. Georgescu, *C_0-groups, commutator methods and spectral theory of N-body Hamiltonians*, Progress in Mathematics, Vol. 135 (Birkhäuser Verlag, Basel, 1996).
9. A. Zygmund, *Trigonometric series: Vols. I, II* (Cambridge University Press, London-New York, 1968).
10. J.-L. Lions and J. Peetre, Sur une classe d'espaces d'interpolation, *Inst. Hautes Études Sci. Publ. Math.* **19**, 5 (1964).
11. A. Pietsch, *Eigenvalues and s-numbers*, Cambridge Studies in Advanced Mathematics, Vol. 13 (Cambridge University Press, Cambridge, 1987).
12. A.-P. Calderón, Intermediate spaces and interpolation, the complex method, *Studia Math.* **24**, 113 (1964).
13. M. A. Krasnoselskii, On a theorem of M. Riesz, *Dokl. Akad. Nauk SSSR* **131**, 246 (1960).
14. M. Cwikel, Real and complex interpolation and extrapolation of compact operators, *Duke Math. J.* **65**, 333 (1992).
15. F. Cobos, T. Kühn and T. Schonbek, One-sided compactness results for Aronszajn-Gagliardo functors, *J. Funct. Anal.* **106**, 274 (1992).
16. F. Cobos, L. M. Fernández-Cabrera, T. Kühn and T. Ullrich, On an extreme class of real interpolation spaces, *J. Funct. Anal.* **256**, 2321 (2009).
17. F. Cobos, Interpolation theory and compactness, in *Function Spaces, Inequalities and Interpolation (Paseky 2009)*, eds. J. Lukeš and L. Pick (Matfyzpress, Prague, 2009) pp. 31–75.
18. B. Beauzamy, *Espaces d'interpolation réels: topologie et géométrie*, Lecture Notes in Mathematics, Vol. 666 (Springer, Berlin, 1978).
19. M. Mastyło, Interpolation spaces not containing ℓ^1, *J. Math. Pures et Appl.* **68**, 153 (1989).
20. W. J. Davis, T. Figiel, W. B. Johnson and A. Pełczyński, Factoring

weakly compact operators, *J. Functional Analysis* **17**, 311 (1974).
21. S. Heinrich, Closed operator ideals and interpolation, *J. Funct. Anal.* **35**, 397 (1980).
22. J. E. Galé, T. J. Ransford and M. C. White, Weakly compact homomorphisms, *Trans. Amer. Math. Soc.* **331**, 815 (1992).
23. A. Blanco, S. Kaijser and T. J. Ransford, Real interpolation of Banach algebras and factorization of weakly compact homomorphisms, *J. Funct. Anal.* **217**, 126 (2004).
24. E. Bishop, Holomorphic completions, analytic continuation, and the interpolation of semi-norms, *Ann. of Math.* **78**, 468 (1963).
25. M. Levy, L'espace d'interpolation réel $(A_0, A_1)_{\theta,p}$ contient l^p, *C. R. Acad. Sci. Paris Sér. A* **289**, 675 (1979).
26. F. Cobos, L. M. Fernández-Cabrera and A. Martínez, On interpolation of Banach algebras and factorization of weakly compact homomorphisms, *Bull. Sci. Math.* **130**, 637 (2006).
27. S. V. Astashkin, Interpolation of bilinear operators by the real method, *Math. Notes* **52**, 641 (1992).
28. F. Cobos and L. M. Fernández-Cabrera, Factoring weakly compact homomorphisms, interpolation of banach algebras and multilinear interpolation, in *Function spaces VIII*, eds. H. Hudzik, J. Musielak, M. Nowak and L. Skrzypczak, Banach Center Publ., Vol. 79 (Polish Acad. Sci. Inst. Math., Warsaw, 2008) pp. 57–69.
29. F. Cobos and L. M. Fernández-Cabrera, On the relationship between interpolation of Banach algebras and interpolation of bilinear operators, *Canad. Math. Bull.* **53**, 51 (2010).

Optimality of function spaces in Sobolev embeddings

Luboš Pick

Department of Mathematical Analysis,
Faculty of Mathematics and Physics, Charles University,
Sokolovská 83, 186 75 Praha 8, Czech Republic
E-mail: pick@karlin.mff.cuni.cz

We study optimality of function spaces that appear in Sobolev embeddings. We focus on rearrangement-invariant Banach function spaces.

Keywords: Sobolev spaces; rearrangement-invariant spaces; optimal range; optimal domain; reduction theorems: compact embeddings; fundamental functions; traces; Gaussian embeddings; isoperimetric function.

1. Introduction

The main purpose of these lectures is to introduce the distinguished audience of the CIDAMA conference to the recent developments in the theory of function spaces and their applications to sharp Sobolev embeddings.

An alternative title of these lecture notes could possibly be "What exactly can one say about a function of several variables given some data about its gradient?" By "some data" we mean certain measurement of its quality, which usually means some information on some of its crucial properties such as the size, continuity or smoothness properties, certain control over the (pointwise or mean) oscillation and so on.

As is well-known from the potential theory, the properties of a function are usually "better" than those of the gradient. For example, when a p-th power of the modulus of the gradient of a function defined on a reasonably smooth bounded domain in $\mathbb{R}^n$ is integrable, then, by the Poincaré inequality, the same is true about the function itself. But, in fact, one can say more about the function – in the particular case given, for example, a considerably higher integrability of the function in question can be established.

2000 Mathematics Subject Classification: 46E35.

A key tool for quantification of such facts is the theory of function spaces. When gradients are involved, the appropriate function spaces are those of Sobolev together with their various modifications and generalizations. It is the embedding properties what makes Sobolev spaces so useful in the theory of partial differential equations and in other applications.

In their most classical form, Sobolev spaces are defined as classes of measurable functions which together with their weak gradient fall into a given Lebesgue space L^p. In such case we say that the Sobolev space is *built upon* the Lebesgue space.

During almost a century of their existence, Lebesgue spaces have constantly played a primary role in analysis. However, it has been known almost from the very beginning that the Lebesgue scale is not sufficiently general to provide a satisfactory description of fine properties of functions required by practical tasks. In the first half of the 20th century, new fine scales of function spaces have been introduced. The efforts of great classics such as Young, Orlicz, Zygmund, Marcinkiewicz, Lorentz, Luxemburg, Hardy, Littlewood, Morrey, Campanato and many others resulted in the development of a powerful and qualitatively new mathematical discipline of function spaces.

In particular, one might be interested in what happens when we build a Sobolev space on a finer function space than just the Lebesgue space. This can be done in a completely analogous way. As a consequence, the rapid development of new finer scales of function spaces has been accompanied by a corresponding increase of sharpness of known Sobolev embeddings, and a natural growth of exactness of the knowledge in applications. A considerable improvement was achieved when Sobolev spaces were built upon Orlicz spaces, and further progress came with the introduction of the non-increasing rearrangement and the related function spaces (Lorentz spaces and their generalizations). Both Orlicz spaces and Lorentz spaces constitute certain refinement of Lebesgue spaces, each of them quite different in nature.

In this text, we concentrate on the question when the Sobolev embeddings are *optimal* (or *sharp*) in a certain sense. This means, we are determined to describe pairs of function spaces, say X and Y, such that the Sobolev embedding

$$W^m(X) \hookrightarrow Y$$

holds and is optimal in the sense that X cannot be replaced by a bigger space and/or Y cannot be replaced by a smaller space within a given "set of rules".

We recall that, for a pair of at least quasinormed linear spaces A and B, we say that A *is embedded* into B, denoted by $A \hookrightarrow B$, if $A \subset B$ and moreover there exists a positive constant C such that $\|x\|_B \leq C\|x\|_A$ for every $x \in A$.

Our results show that neither one of the classes of Lebesgue, Orlicz or Lorentz spaces has the definitive answer, not even any of the known yet finer classes of function spaces such as Lorentz–Zygmund spaces. None of these classes seems to be rich enough in order to provide all the satisfactory ultimate answers as each of them has its own intrinsic restrictions. Instead, we show that the appropriate environment, in which comprehensive answers are available, is that of the so-called *rearrangement-invariant spaces.* Having observed the fact that even this class of function spaces has some limitations and for certain specific tasks one has to settle for function spaces which are not rearrangement invariant (such as weighted spaces, Hölder spaces, Campanato spaces, BMO, etc.), we still find it reasonably wide for our purposes.

The class of rearrangement-invariant spaces, introduced and heavily studied in 1950's by Lorentz, Luxemburg, Zaanen and many others, forms a "common roof" for most of the above-mentioned function spaces, in particular for Lebesgue, Orlicz and Lorentz spaces. Somewhat surprisingly, this general and rather abstract setting seems to be easier to work in than to grapple intrinsic properties of each of the particular classes of function spaces.

2. Sobolev embeddings involving customary function spaces

In this section we want to take a reader to a journey through the history of improvements of Sobolev embeddings dependent on function spaces. We wish to concentrate solely on the function spaces involved and not to be distracted by any other difficulties; therefore we restrict ourselves to embeddings of the first order and only to those happening on a nice domain. we shall return to the question of higher order embeddings and also to the question of bad domains later.

Suppose Ω is a bounded domain in $\mathbb{R}^n$, $n \geq 2$, having a Lipschitz boundary. In its most classical form, the *Sobolev inequality* asserts that, given $1 < p < n$ and setting $p^* = \frac{np}{n-p}$, there exists $C > 0$ such that

$$\left(\int_\Omega |u(x)|^{p^*}\,dx\right)^{\frac{1}{p^*}} \leq C\left(\int_\Omega |(\nabla u)(x)|^p + |u(x)|^p\,dx\right)^{\frac{1}{p}} \quad \text{for all } W^{1,p}(\Omega).$$

(Here and throughout, C denotes a constant independent of important quantities, not necessarily the same at each occurrence.) We can restate this result in the form of a *Sobolev embedding*, namely,

$$W^{1,p}(\Omega) \hookrightarrow L_{p^*}(\Omega), \qquad 1 < p < n, \tag{1}$$

where $W^{1,p}(\Omega)$ is the classical *Sobolev space*, that is, a collection of weakly-differentiable functions on Ω such that $u \in L_p(\Omega)$ and $|\nabla u| \in L_p(\Omega)$, endowed with the norm

$$\|u\|_{W^{1,p}(\Omega)} = \|u\|_{L_p(\Omega)} + \|\nabla u\|_{L_p(\Omega)},$$

and $\|\cdot\|_{L_p(\Omega)}$ is the usual Lebesgue norm.

As mentioned in the Intoduction, we say that the space on the left side of (1) is a *Sobolev space built upon* $L_p(\Omega)$. In this sense, we recognize $L_p(\Omega)$ as the *domain space* of the embedding, and the space $L_{p^*}(\Omega)$ on the right as its *range space*.

We will focus on the question, how *sharp* are the domain space and the range space in the Sobolev embedding.

First thing we note is that this question is dependent on an environment within which it is investigated. For example, the embedding (1) can not be improved within the environment of Lebesgue spaces. This should be understood as follows: if we replace the domain space $L_p(\Omega)$ in (1) by a larger Lebesgue space, say, $L_q(\Omega)$ with $q < p$, then the resulting embedding

$$W^{1,q}(\Omega) \hookrightarrow L_{p^*}(\Omega),$$

can no longer be true. Likewise, if we replace the range space $L_{p^*}(\Omega)$ by a smaller Lebesgue space, say $L_r(\Omega)$, $r > p^*$, then, again, the resulting embedding

$$W^{1,p}(\Omega) \hookrightarrow L_r(\Omega),$$

does not hold any more. In this sense, the embedding (1) is, at least within the environment of Lebesgue spaces, *sharp* (or *optimal*), and it cannot be effectively improved. In other words, if we want to improve the embedding and to get thereby a finer result, we need to use finer classes of function spaces than the Lebesgue scale.

The fact that the Lebesgue scale is simply not delicate enough in order to describe all the interesting details about embeddings, is perhaps best illustrated by the so-called *limiting* or *critical* case of the embedding (1), corresponding to the case $p = n$. When we let p tend to n from the left, then, of course, p^* tends to ∞. However, the limiting embedding

$$W^{1,n}(\Omega) \hookrightarrow L_\infty(\Omega)$$

is not true. It is well known that one can have unbounded functions (typically with logarithmic singularities) in $W^{1,n}(\Omega)$. Therefore, the only information which we can formulate in the Lebesgue spaces environment for the limiting embedding is

$$W^{1,n}(\Omega) \hookrightarrow L_q(\Omega) \qquad \text{for every } q < \infty. \tag{2}$$

Again, this information is optimal within the environment of Lebesgue spaces, where no improvement is available. However, it is quite clear that this result is very unsatisfactory as it does not provide any definite range function space. Such a space can be obtained, but not among Lebesgue spaces. We need a refinement of the Lebesgue scale. One of the most well-known and most widely used such refinements of Lebesgue spaces are Orlicz spaces. We first shortly recall their definition.

Given any *Young function* $A : [0,\infty) \to [0,\infty)$, namely a convex increasing function vanishing at 0, the *Orlicz space* $L_A(\Omega)$ is the r.i. space of all measurable functions u in Ω such that the *Luxemburg norm*

$$\|u\|_{L_A(\Omega)} = \inf\left\{\lambda > 0; \int_\Omega A\left(\frac{|u(x)|}{\lambda}\right)\,dx \le 1\right\}$$

is finite. Of course, if $A(t) = t^p$, then we recover Lebesgue spaces. Other important examples of Orlicz spaces are the *logarithmic Zygmund classes* $L_p(\log L)^\alpha(\Omega)$, generated by the Young function

$$A(t) = t^p\left(\log(e+t)\right)^\alpha, \quad t \in (1,\infty),$$

with $p \in [1,\infty)$ and $\alpha \in \mathbb{R}$, and the *exponential Zygmund classes* $\exp L^\alpha(\Omega)$, generated by the Young function

$$A(t) = \exp\left(t^\alpha\right), \quad t \in (1,\infty), \quad \alpha > 0.$$

Equipped with Orlicz spaces, we can formulate the following limiting case of the Sobolev embedding:

$$W^{1,n}(\Omega) \hookrightarrow \exp L^{n'}(\Omega), \tag{3}$$

where

$$n' = \frac{n}{n-1}.$$

This result is traditionally attributed to Trudinger[1], however, in certain modified form, it appeared earlier in works of Yudovich[2], Peetre[3] and Pokhozhayev[4].

We can now, again, ask how sharp this result is. It turns out, that, remarkably, the range space $\exp L^{n'}(\Omega)$ is sharp within the environment of Orlicz spaces. In other words, it is the smallest possible Orlicz space that still renders this embedding true. This optimality result is due to Hempel, Morris and Trudinger[5].

In this context it might be of interest to ask whether also the classical Sobolev embedding, (1), has the optimal *Orlicz* range space. Of course, as we already know, it is the optimal *Lebesgue* range space, but now we are asking about optimality in a much broader sense, so the question is sensible. The answer is positive, as follows from the result of Cianchi[6].

However, it turns out that nontrivial improvements of both (1) and (3) are still available. To this end, we have to introduce function spaces whose norms involve the so-called nonincreasing rearrangement.

We denote by $\mathcal{M}(\Omega)$ the class of real-valued measurable functions on Ω and by $\mathcal{M}_+(\Omega)$ the class of nonnegative functions in $\mathcal{M}(\Omega)$. Given $f \in \mathcal{M}(\Omega)$, its *non-increasing rearrangement* is defined by

$$f^*(t) = \inf\left\{\lambda > 0;\ |\{x \in \Omega;\ |f(x)| > \lambda\}| \le t\right\}, \quad t \in [0, \infty).$$

We also define the *maximal non-increasing rearrangement* of f by

$$f^{**}(t) = t^{-1} \int_0^t f^*(s)\,ds, \quad t \in [0, \infty).$$

We note that below we shall use the rearrangements defined only on $(0, |\Omega|)$, but it can be as well defined on $[0, \infty)$, extended by zero for $t > |\Omega|$.

We shall work with several classes of function spaces defined with the help of the operation $f \mapsto f^*$. The first such example will be the scale of the two-parameter Lorentz spaces.

Assume that $0 < p, q \le \infty$. The *Lorentz space* $L_{p,q}(\Omega)$ is the collection of all $f \in \mathcal{M}(\Omega)$ such that $\|f\|_{L_{p,q}(\Omega)} < \infty$, where

$$\|f\|_{L_{p,q}(\Omega)} = \|t^{\frac{1}{p}-\frac{1}{q}} f^*(t)\|_{L_q(0,1)}.$$

The Lorentz spaces are nested in the following sense. For every $p \in (0, \infty]$ and $0 < q < r \le \infty$, we have

$$L_{p,q(\Omega)} \hookrightarrow L_{p,r(\Omega)}, \tag{4}$$

and this embedding is strict.

With the help of Lorentz spaces, we have the following refinement of (1):

$$W^{1,p}(\Omega) \hookrightarrow L_{p^*,p}(\Omega), \qquad 1 < p < n. \tag{5}$$

Note that, thanks to (4) and the obvious inequality $p < p^*$, this is nontrivial improvement of the range space in (1). The embedding (5) is due to Peetre[3], and it can be also traced in works of O'Neil[7] and Hunt[8].

A natural question arises, whether a similar Lorentz-type refinement is possible also for the limiting embedding (3). The answer is positive again, but we need to introduce a yet more general function scale.

Let $0 < p, q \leq \infty$ and $\alpha \in \mathbb{R}$. The *Lorentz–Zygmund space* $L_{p,q;\alpha}(\Omega)$ is the collection of all $f \in \mathcal{M}(\Omega)$ such that $\|f\|_{L_{p,q;\alpha}(\Omega)} < \infty$, where

$$\|f\|_{L_{p,q;\alpha}(\Omega)} := \|t^{\frac{1}{p}-\frac{1}{q}} \log^{\alpha}(e/t)\, f^*(t)\|_{L_q(0,1)}.$$

Occasionally we will have to work with a modification of Lorentz–Zygmund spaces in which f^* is replaced by f^{**}. We denote such space by $L_{(p,q;\alpha)}(\Omega)$, hence

$$\|f\|_{L_{(p,q;\alpha)}(\Omega)} := \|t^{\frac{1}{p}-\frac{1}{q}} \log^{\alpha}(e/t)\, f^{**}(t)\|_{L_q(0,1)}.$$

These spaces were introduced and studied by Bennett and Rudnick[9]. If one of the following conditions

$$\begin{cases} 1 < p < \infty, \ 1 \leq q \leq \infty, \ \alpha \in \mathbb{R}; \\ p = 1, \ q = 1, \ \alpha \geq 0; \\ p = \infty, \ q = \infty, \ \alpha \leq 0; \\ p = \infty, \ 1 \leq q < \infty, \ \alpha + \frac{1}{q} < 0, \end{cases} \tag{6}$$

is satisfied, then $L_{p,q;\alpha}(\Omega)$ is a rearrangement-invariant Banach function space.

Equipped with Lorentz–Zygmund spaces, we have the following refinement of the Trudinger embedding (3):

$$W^{1,n}(\Omega) \hookrightarrow L_{\infty,n;-1}(\Omega). \tag{7}$$

This embedding in particular implies that if the gradient of a function u belongs to $L_n(\Omega)$, then its non-increasing rearrangement u^* satisfies

$$\int_0^1 \left(\frac{u^*(t)}{1+\log(\frac{1}{t})} \right)^n \frac{dt}{t} < \infty.$$

The first one to note this fact was Maz'ya who formulated it in a somewhat implicit form involving capacitary estimates, see[10] pp. 105 and 109.

Explicit formulations were given by Hansson[11] and Brézis–Wainger in[12], the result can be also traced in the work of Brudnyi[13]. A more general assertion was proved by Cwikel and Pustylnik[14].

The range space in (7) is a very interesting function space. It is not a Zygmund class of neither logarithmic or exponential type. Moreover, as the relations between Lorentz–Zygmund spaces from[9] show, it satisfies

$$L_{\infty,n;-1}(\Omega) \hookrightarrow \exp L^{n'}(\Omega),$$

and this inclusion is strict. We thus get a nontrivial improvement of (3).

The embedding (7) can be viewed in some sense as the limiting case of (5) when $p \to n+$. Indeed, both these results allow a unified approach, as shown in[15] (again, restricted to functions vanishing on the boundary), where it was noticed that, for $1 < p < n$, we have

$$\int_0^1 t^{\frac{p}{p^*}-1} u^*(t)^p \, dt \le C \int_\Omega |\nabla u(x)|^p \, dx$$

for all $u \in W^{1,p}(\Omega)$, while, in the limiting case, we have

$$\int_0^1 \left(\frac{u^*(t)}{\log\left(\frac{e}{t}\right)} \right)^n \frac{dt}{t} \le C \int_\Omega |\nabla u|^n(x) \, dx$$

for all $u \in W^{1,n}(\Omega)$. Both these results were proved in an elementary way by first establishing a weak version of the Sobolev–Gagliardo–Nirenberg embedding, namely

$$\lambda\left(|\{|u| \ge \lambda\}|\right)^{\frac{1}{n'}} \le C \int_\Omega |\nabla u| \, dx, \qquad u \in W^{1,1}(\Omega),\ \lambda > 0,$$

and then using a truncation argument due to Maz'ya.

In the course of the proof it turned out that yet further improvement of (7) is possible, Namely, it was shown that

$$W^{1,n}(\Omega) \hookrightarrow W_n(\Omega),$$

where , for $0 < p \le \infty$, the space $W_p(\Omega)$ is defined as the family of all measurable functions on Ω for which

$$\|u\|_{W_p(\Omega)} = \begin{cases} \left(\int_0^1 \left(u^*(\frac{t}{2}) - u^*(t)\right)^p \frac{dt}{t} \right)^{\frac{1}{p}} < \infty & \text{when } p < \infty; \\ \sup\limits_{0<t<1} \left(u^*(\frac{t}{2}) - u^*(t)\right) & \text{when } p = \infty. \end{cases}$$

The space $W_p(\Omega)$ was shown to have some interesting properties, such as, for example:

(i) $\|\chi_E\|_{W_p(\Omega)} = (\log 2)^{\frac{1}{p}}$ for every measurable $E \subset \Omega$ and $p \in (0, \infty)$;
(ii) $L_\infty = W_1(\Omega)$;
(iii) for $p \in [1, \infty)$, each integer-valued $u \in W_p(\Omega)$ is bounded;
(iv) for $p \in (1, \infty)$, $W_p(\Omega)$ is not a linear set;
(v) for $p \in (1, \infty)$, $W_p(\Omega) \subsetneqq L_{\infty,p;-1}(\Omega)$;
(vi) $W_p(\Omega) \subsetneqq W_q(\Omega)$ for every $0 < p < q \leq \infty$.

The norm of the space $W_p(\Omega)$ involves the functional $f^*(\frac{t}{2}) - f^*(t)$. In[16], Bastero, Milman and Ruiz showed that it can be equivalently replaced with $f^{**}(t) - f^*(t)$. The quantity $f^{**}(t) - f^*(t)$, which measures, in some sense, the oscillation of f, had been used in function spaces theory before. Function spaces involving this functional have been particularly popular since 1981 when Bennett, De Vore and Sharpley[17] introduced the "weak L_∞", the rearrangement-invariant space of functions for which $f^{**}(t) - f^*(t)$ is bounded.

The problem of optimality of function spaces in Sobolev embeddings can be also viewed from a reversed angle. So far we have focused solely on the question of optimality of the range space in various contexts. However, one can also ask whether the *domain* space is optimal. For example, it is clear that (1) and (5) have the best possible Lebesgue domain spaces. We can however ask whether these domain spaces are also optimal as *Orlicz* spaces. The answer is interesting and perhaps even surprising. While, in the non-limiting embedding (1), the space $L_p(\Omega)$ is indeed the optimal Orlicz range for $L_{p^*}(\Omega)$ ([18] Corollary 4.9), the situation in the limiting case is quite different. Not only that $L_n(\Omega)$ *is not* the largest Orlicz space for which the Trudinger inequality (3) holds, but, oddly enough, there is no such optimal Orlicz space at all. This should be understood as follows: for every Orlicz space $L_A(\Omega)$ such that

$$W^1 L_A(\Omega) \hookrightarrow \exp L^{n'}(\Omega),$$

there exists another, strictly larger Orlicz space, $L_B(\Omega)$, such that

$$W^1 L_B(\Omega) \hookrightarrow \exp L^{n'}(\Omega).$$

A construction of the Young function B which generates such an Orlicz space $L_B(\Omega)$ from a given A, can be found in[18] Theorem 4.5. In a way, this result resembles the unsatisfactory situation with Lebesgue range partners in the limiting embedding (2), where one has an 'open set of range spaces', and illustrates thereby that not even the (apparently rather fine) class of Orlicz spaces is delicate enough to provide satisfactory answers. We can use this as a motivation to look for optimal function spaces in a broader general context.

The last example shows that the investigation of the optimality of domain spaces in well-known embeddings can bring unexpected surprises. Another such situation, although quite different by nature, occurs when we ask about the optimality of the domain $L_n(\Omega)$ in the Trudinger's embedding (3). Indeed, it was shown in[19] that, interestingly, from the scaling property of Lorentz–Zygmund spaces, one can deduce the following embedding:

$$W^1\left(L_{n,1;-\frac{1}{n'}} + L_{n,\infty;\frac{1}{n}}\right)(\Omega) \hookrightarrow \exp L^{n'}(\Omega).$$

Complemented with

$$L_n(\Omega) \hookrightarrow \left(L_{n,1;-\frac{1}{n'}} + L_{n,\infty;\frac{1}{n}}\right)(\Omega),$$

the inclusion being strict, this gives a rather unexpected nontrivial improvement of the domain space in the Trudinger's embedding, quite different from the above-mentioned one, built on Orlicz spaces.

All these examples call for considering some reasonable common environment that would provide a roof for all, or, at least, most, of the function spaces mentioned so far and for considering global optimality within this context. For us, such environment will be that of the so-called rearrangement-invariant spaces.

Furthermore, we should be interested also in higher-order Sobolev embeddings (note that all the illustrative examples mentioned so far were first-order embeddings). Higher-order embeddings are important in applications and, as it turns out, considerably more difficult to handle than the first order ones. This is caused by the fact that, for the first-order embedding, one has the *Pólya–Szegö inequality*, for which there is, regrettably, no equally powerful analogue for higher-order embeddings.

3. Rearrangement-invariant spaces

The context of function spaces in which we shall study optimality of Sobolev embeddings is that of the so-called rearrangement-invariant spaces. Before stating exact definitions, let us just mention that most of the function spaces mentioned above, namely, those of Lebesgue, Orlicz, Zygmund, Lorentz and Lorentz-Zygmund, are, at least for some reasonable parameters, r.i. spaces, with a notable exception of the space $W_p(\Omega)$, which is not even linear. Therefore, r.i. spaces constitute a common roof for many important classes of functions, it is a rich collection of general function spaces, yet they are pleasantly modeled upon the example of Lebesgue spaces, inheriting many of their wonderful properties.

We shall assume throughout, unless stated otherwise, that Ω is an open set in $\mathbb{R}^n$ satisfying $|\Omega| = 1$, where $|\cdot|$ is certain appropriate measure on Ω. In most cases, we shall work either with the usual n-dimensional Lebesgue measure on Ω. We shall also treat special cases when Ω will be the entire $\mathbb{R}^n$ endowed with the Gauss measure or with some more general product probability measure – in such cases we shall clearly specify this fact. We stress that in order to save the reader technicalities, we restrict ourselves to the case when the measure of Ω is equal to 1. If the measure of Ω is finite and different from 1, everything can be easily modified in an obvious way by the change of variables $t \mapsto |\Omega| t$.

A Banach space $X(\Omega)$ of functions defined on Ω, equipped with the norm $\|\cdot\|_{X(\Omega)}$, is said to be *rearrangement-invariant* if the following five axioms hold:

$$0 \leq g \leq f \text{ a.e. implies } \|g\|_{X(\Omega)} \leq \|f\|_{X(\Omega)}; \tag{P1}$$

$$0 \leq f_n \nearrow f \text{ a.e. implies } \|f_n\|_{X(\Omega)} \nearrow \|f\|_{X(\Omega)}; \tag{P2}$$

$$\|\chi_\Omega\|_{X(\Omega)} < \infty, \text{ where } \chi_E \text{ denotes the characteristic function of } E; \tag{P3}$$

$$\text{for every } E \subset \Omega, \text{with } |E| < \infty, \text{ there exists a constant } C_E \tag{P4}$$

$$\text{such that } \int_E f(x)\,dx \leq C_E \|f\|_{X(\Omega)} \quad \text{for every } f \in X(\Omega);$$

$$\|f\|_{X(\Omega)} = \|g\|_{X(\Omega)} \text{ whenever } f^* = g^*. \tag{P5}$$

A basic tool for working with rearrangement-invariant spaces is the *Hardy–Littlewood–Pólya* (HLP) *principle*, treated in[20] Chapter 2, Theorem 4.6). It asserts that $f^{**}(t) \leq g^{**}(t)$ for every $t \in (0,1)$ implies $\|f\|_{X(\Omega)} \leq \|g\|_{X(\Omega)}$ for every r.i. space $X(\Omega)$.

The *inequality of Hardy and Littlewood* states that

$$\int_\Omega |f(x)g(x)|\,dx \leq \int_0^1 f^*(t)g^*(t)\,dt, \qquad f, g \in \mathcal{M}(\Omega). \tag{1}$$

Given an r.i. space $X(\Omega)$, the set

$$X'(\Omega) = \left\{ f \in \mathcal{M}(\Omega); \int_\Omega |f(x)g(x)|\,dx < \infty \text{ for every } g \in X(\Omega) \right\},$$

equipped with the norm

$$\|f\|_{X'(\Omega)} = \sup_{\|g\|_{X(\Omega)} \leq 1} \int_\Omega |fg|,$$

is called the *associate space* of $X(\Omega)$. The space $X'(\Omega)$ is again a rearrangement-invariant space, and one always has $X''(\Omega) = X(\Omega)$. Furthermore, the *Hölder inequality*

$$\int_\Omega |f(x)g(x)|\,dx \le \|f\|_{X(\Omega)}\|g\|_{X'(\Omega)}$$

holds.

For every r.i. space $X(\Omega)$, there exists a unique r.i. space $\overline{X}(0,1)$ on $(0,1)$, satisfying $\|f\|_{X(\Omega)} = \|f^*\|_{\overline{X}(0,1)}$. Such a space, endowed with the norm

$$\|f\|_{\overline{X}(0,1)} = \sup_{\|g\|_{X(\Omega)}\le 1} \int_0^1 f^*(t)g^*(t)\,dt,$$

is called the *representation space* of $X(\Omega)$.

Let $X(\Omega)$ be an r.i. space. Then, the function $\varphi_X : [0,1] \to [0,\infty)$ given by

$$\varphi_X(t) = \begin{cases} \|\chi_{(0,t)}\|_{\overline{X}(0,1)}, & \text{for } t \in (0,1], \\ 0 & \text{for } t = 0, \end{cases}$$

is called the *fundamental function* of $X(\Omega)$. For every r.i. space $X(\Omega)$, its fundamental function φ_X is *quasiconcave* on $[0,1]$, i.e., it is nondecreasing on $[0,1]$, $\varphi_X(0) = 0$, and $\frac{\varphi_X(t)}{t}$ is nonincreasing on $(0,1]$. Moreover, one has

$$\varphi_X(t)\varphi_{X'}(t) = t, \qquad \text{for } t \in [0,1].$$

Given an r.i. space $X(\Omega)$, we can define the *Marcinkiewicz space* $M_X(\Omega)$, corresponding to $X(\Omega)$, as the set of all $f \in \mathcal{M}(\Omega)$ such that

$$\|f\|_{M_X(\Omega)} := \sup_{t\in[0,1]} \varphi_X(t)f^{**}(t) < \infty.$$

Then, $M_X(\Omega)$ is, again, an r.i. space whose fundamental function is φ_X, and it is the largest such r.i. space. In particular, when $Z(\Omega)$ is any other r.i. space whose fundamental function is also φ_X, then, necessarily

$$Z(\Omega) \hookrightarrow M_X(\Omega).$$

For a comprehensive treatment of r.i. spaces we refer the reader to[20].

4. Reduction theorems

In this section we shall treat the first central set of results that will finally lead us to obtaining sharp Sobolev embeddings. The principal goal is to replace the investigation of a (quite complicated in structure) Sobolev-type embedding by a (hopefully easier) task of proving the boundedness of an appropriate integral operator on convenient function spaces. The main point is that the integral operator in question is *one-dimensional* which means that it acts on functions defined on an interval. So rather than having to work with a functions defined on a possibly quite complicated domain Ω in $\mathbb{R}^n$, we then have only to consider functions defined on the interval $(0, |\Omega|)$ (that is, thanks to our convention, $(0,1)$). This step is deep and difficult, but it brings a considerable improvement to our investigation. We shall call such result a *reduction theorem.*

The basic idea is, again, to compare the size of u with that of its *m-th gradient*, $|D^m u|$, in norms of two function spaces, where $D^m u = \left(\frac{\partial^\alpha u}{\partial x^\alpha}\right)_{0\leq|\alpha|\leq m}$ and $|D^m u|$ is its Euclidean length. Thus, more precisely, we will be interested in determining those rearrangement-invariant spaces, $X(\Omega)$ and $Y(\Omega)$ for which one has

$$\|u\|_{Y(\Omega)} \leq C \, \||D^m u|^*\|_{\overline{X}(0,1)}, \qquad u \in W^m X(\Omega),$$

or, written as a Sobolev embedding,

$$W^m X(\Omega) \hookrightarrow Y(\Omega). \tag{1}$$

More specifically, we would like to know that $X(\Omega)$ and $Y(\Omega)$ are *optimal* in the sense that $X(\Omega)$ cannot be replaced by an essentially larger r.i. space and $Y(\Omega)$ cannot be replaced by an essentially smaller one.

For a first-order embedding, the first reduction theorem was obtained in[21].

Theorem 4.1. *Let $X(\Omega)$ and $Y(\Omega)$ be r.i. spaces. Then, in order that the Sobolev embedding*

$$W^1 X(\Omega) \hookrightarrow Y(\Omega)$$

holds, it is necessary and sufficient that there exist $C > 0$ for which

$$\left\| \int_t^1 f(s) s^{\frac{1}{n}-1}\, ds \right\|_{\overline{Y}(0,1)} \leq C \, \|f\|_{\overline{X}(0,1)}, \qquad f \in \mathcal{M}_+(0,1).$$

Although the results in[21] are formulated only for Sobolev spaces of functions vanishing on the boundary of Ω, by the combination of the Stein extension theorem[22] Theorem 5.24 with an interpolation argument based on the De Vore–Scherer theorem[23] or[20] Chapter 5, Theorem 5.12, p. 360, they can be relatively easily extended to bounded domains with Lipschitz boundary. In this approach, the Sobolev space $W^m X(\Omega)$ is extended to $W^m X(\mathbb{R}^n)$ and then restricted again to $W^m X(\Omega_1)$ with $\Omega_1 \supset \Omega$. The details can be found in[24] proof of Theorem 4.1.

A natural important question now is, how to obtain a higher-order version of the reduction theorem. We shall treat this question separately below, however, at this stage let us just mention that, while the 'only if' part is rather straightforward and easily adaptable for all orders, the proof of the 'if' part of Theorem 4.1 involves a version of the Pólya–Szegö inequality due to Talenti[25], whose higher-order version is unavailable without certain restrictions. In 2004, Cianchi[26] obtained the reduction theorem for the case $m = 2$ by overcoming certain considerable technical difficulties and using some special estimates for second-order derivatives. In[24], the following general version of the reduction theorem applicable to all orders up to $n-1$ was obtained by a new method using interpolation techniques and properties of special Hardy-type operators involving suprema.

Theorem 4.2. *Let $X(\Omega)$ and $Y(\Omega)$ be r.i. spaces. Then, the Sobolev embedding* (1) *holds if and only if*

$$\left\| \int_t^1 f(s) s^{\frac{m}{n}-1}\, ds \right\|_{\overline{Y}(0,1)} \le C \|f\|_{\overline{X}(0,1)}, \qquad f \in \mathcal{M}_+(0,1).$$

The proof of Theorem 4.2 is quite involved. We first define the *weighted Hardy operator* $H_{\frac{n}{m}}$, given as

$$(H_{\frac{n}{m}} f)(t) := \int_t^1 f(s) s^{\frac{m}{n}-1}\, ds$$

and its *dual operator* with respect to the L_1 pairing, defined by

$$(H'_{\frac{n}{m}} f)(t) := t^{\frac{m}{n}-1} \int_0^t f(s)\, ds, \qquad t \in (0,1),\ f \in \mathcal{M}_+(0,1).$$

Note that when applied to a nonincreasing function f^*, we get

$$(H'_{\frac{n}{m}} f^*)(t) = t^{\frac{m}{n}} f^{**}(t) \qquad t \in (0,1),\ f \in \mathcal{M}(\Omega).$$

We observe that the functional

$$\left\| t^{\frac{m}{n}} g^{**}(t) \right\|_{\overline{X}'(0,1)}, \quad g \in \mathcal{M}(\Omega),$$

is an r.i. norm on (Ω). This is easy to verify as the only non-trivial part is the triangle inequality, which follows from the well-known subadditivity of the operation $g \to g^{**}$. Therefore, given an r.i. space $X(\Omega)$, we can define the space $X_\omega(\Omega)$, determined by the functional

$$\|f\|_{X_\omega(\Omega)} := \left\|H'_{\frac{n}{m}} f^*\right\|_{\overline{X}(0,1)} = \left\|t^{\frac{m}{n}} f^{**}(t)\right\|_{\overline{X}(0,1)}, \qquad f \in \mathcal{M}(\Omega).$$

Using the same ideas as in[21] Theorem 4.5 it can be shown that $X_\omega(\Omega)$ is also an r.i. space, being, in fact, essentially the largest r.i. space $Y(\Omega)$, satisfying

$$\left\|H'_{\frac{n}{m}} f\right\|_{\overline{X}(0,1)} \leq C\|f\|_{\overline{Y}(0,1)}, \qquad f \in \mathcal{M}_+(0,1).$$

It follows by duality that $(X')'_\omega(\Omega)$, the associate space of $(X')_\omega(\Omega)$, is essentially the smallest r.i. space $Z(\Omega)$ satisfying

$$\left\|H_{\frac{n}{m}} f\right\|_{\overline{Z}(0,1)} \leq C\|f\|_{\overline{X}(0,1)}, \qquad f \in \mathcal{M}_+(0,1).$$

Next, we introduce a special *supremum operator*, $T_{\frac{n}{m}}$, by

$$\left(T_{\frac{n}{m}} f\right)(t) := t^{-\frac{m}{n}} \sup_{t \leq s < 1} s^{\frac{m}{n}} f^*(s), \qquad f \in \mathcal{M}(0,1),\ t \in (0,1).$$

One readily shows that $T_{\frac{n}{m}}$ is bounded on $L_1(0,1)$ and also on the Lorentz space $L_{\frac{n}{m},\infty}(0,1)$. The key result concerning this operator is that it is bounded on $\overline{X_\omega}(0,1)$ for *absolutely arbitrary* r.i. space $X(\Omega)$. As a consequence, we obtain that, for any r.i. space $X(\Omega)$,

$$H_{\frac{n}{m}} := \overline{X}(0,1) \to \overline{(X'_\omega)'}(0,1),$$

and $(X'_\omega)'(\Omega)$ is the optimal (smallest) such r.i. space. The proof of Theorem 4.2 is then finished by combining the obtained estimates with the inequality

$$\int_0^t s^{-\frac{m}{n}} u^*(s)\, ds \leq C \int_0^t s^{-\frac{m}{n}} \int_{\frac{s}{2}}^1 |D^m u|^*(y) y^{\frac{m}{n}-1}\, dy\, ds,$$

$t \in (0,1)$, $u \in W^m X(\Omega)$, which follows from the endpoint Sobolev embeddings, the Holmstedt formulae and the De Vore-Scherer expression for the K-functional between Sobolev spaces.

5. Optimal range and optimal domain r.i. spaces

The principal consequence of any kind of reduction theorem is the characterization of an optimal partner space in the Sobolev embedding. The idea is based on the fact that using duality, it is relatively easy to nail down the optimal partner space corresponding to the integral operator involved. The rest of the burden is then laid on the reduction theorem.

We will demonstrate this process on showing how Theorem 4.2 can be used to characterize the largest r.i. domain space and the smallest r.i. range space in the Sobolev embedding (1). Note that Theorem 4.2 implies the following chain of equivalent statements:

$$\begin{aligned} W^m X(\Omega) \hookrightarrow Y(\Omega) \quad &\Leftrightarrow \quad H_{\frac{n}{m}} : \overline{X}(0,1) \to \overline{Y}(0,1) \\ &\Leftrightarrow \quad H'_{\frac{n}{m}} : \overline{Y}'(0,1) \to \overline{X}'(0,1) \\ &\Leftrightarrow \quad \left\| t^{\frac{m}{n}} g^{**}(t) \right\|_{\overline{X}'(0,1)} \le C \left\| g \right\|_{Y'(\Omega)}, \quad g \in \mathcal{M}(\Omega). \end{aligned}$$

The first equivalence is Theorem 4.2 and the second one is duality (more precisely, we use properties of the associative space). The last equivalence is not entirely obvious; the implication "$\Rightarrow$" is restriction to monotone functions, while the converse one follows from the estimate $\int_0^t g(s)\,ds \le \int_0^t g^*(s)\,ds$, which is just a special case of (1). It is of interest to note that when we replace the operator $H'_{\frac{n}{m}}$ by $H_{\frac{n}{m}}$, then the corresponding equivalence is no longer true. More precisely, the inequality

$$\left\| H_{\frac{n}{m}} g \right\|_{\overline{Y}(0,1)} \le C \|g\|_{\overline{X}(0,1)}, \quad g \in \mathcal{M}(0,1),$$

implies

$$\left\| H_{\frac{n}{m}} g^* \right\|_{\overline{Y}(0,1)} \le C \|g\|_{\overline{X}(0,1)}, \quad g \in \mathcal{M}(0,1),$$

but *not* vice versa. This illustrates that a Sobolev embedding is a rather delicate process that does not permit a direct duality.

All these ideas are summarized in the following theorem.

Theorem 5.1. *Let $X(\Omega)$ be an r.i. space. Let $Y(\Omega)$ be the r.i. space whose associate space, $Y'(\Omega)$, has the norm*

$$\|f\|_{Y'(\Omega)} := \|t^{\frac{m}{n}} f^{**}(t)\|_{\overline{X}'(0,1)}, \qquad f \in \mathcal{M}(\Omega).$$

Then, the Sobolev embedding (1) *holds, and $Y(\Omega)$ is the optimal* (*that is, the smallest possible*) *such r.i. space.*

Theorem 5.1 constitutes an important and rather nice theoretical breakthrough in our search for optimal Sobolev embeddings. On the other hand, it does not easy apply to concrete examples. Generally speaking, in order to determine $Y(\Omega)$, we have to be able to characterize the associate space of the space whose norm is given by a rather complicated functional $g \mapsto \|t^{\frac{m}{n}} g^{**}(t)\|_{\overline{X}'(0,1)}$. That does not have to be easy. Even in the simplest possible instance when $X(\Omega) = L_p(\Omega)$ we can get an explicit formula for $Y(\Omega)$ only by using the duality argument of Sawyer[27], which is highly nontrivial. When $X(\Omega)$ is, for instance, an Orlicz space, the task becomes nearly impossible (however, see Cianchi[28]). In[21], the class of the so-called *Lorentz–Karamata spaces* was introduced, and the explicit formulas for the optimal range space were given in case when the domain space is one of these. The Lorentz–Karamata spaces are a generalization of Lorentz–Zygmund spaces which instead of logarithmic functions involve more general *slowly-varying functions.*

We will now apply Theorem 5.1 to a concrete example, a higher-order version of the Maz'ya–Hansson–Brézis–Wainger embedding (7).

Example 5.1. Let $X(\Omega) = L_{\frac{n}{m}}(\Omega)$. Then, by Theorem 5.1, its optimal range partner $Y(\Omega)$ is the associate space of $Y'(\Omega)$, determined by the norm

$$\|g\|_{Y'(\Omega)} = \|f^{**}(t)t^{\frac{m}{n}}\|_{L_{\frac{n}{n-m}}(0,1)} = \|f\|_{L_{(1,\frac{n}{n-m})}(\Omega)}.$$

Now, by a duality principle of Sawyer[27], we obtain

$$Y(\Omega) = L_{\infty,\frac{n}{m};-1}(\Omega).$$

For $m = 1$, we recover (7). We add a new information that this range space is the best possible among r.i. spaces. As mentioned above already, $W_n(\Omega)$ is still a slightly better range, but it is not an r.i. space for not being linear. An optimality of the range space in a yet broader context was proved by Cwikel and Pustylnik in[14].

Another achievement of the reduction theorem is the following characterization of the optimal domain space in a Sobolev embedding.

Theorem 5.2. *Assume that $Y(\Omega)$ is an r.i. space such that*

$$Y(\Omega) \hookrightarrow L_{\frac{n}{n-m},1}(\Omega).$$

Then, the function space $X(\Omega)$, generated by the norm

$$\|f\|_{X(\Omega)} = \sup_{h^*=f^*} \left\|H_{\frac{n}{m}} h\right\|_{\overline{Y}(0,1)}, \qquad f \in \mathcal{M}(\Omega),\ h \in \mathcal{M}(0,1), \tag{1}$$

is an r.i. space such that

$$H_{\frac{n}{m}} : \overline{X}(0,1) \to \overline{Y}(0,1)$$

(hence $W^m X(\Omega) \hookrightarrow Y(\Omega)$*). Moreover, it is an optimal (largest) such space.*

The requirement of the embedding of $Y(\Omega)$ into $L_{\frac{n}{n-m},1}(\Omega)$ is not restrictive as the space $L_{\frac{n}{n-m},1}(\Omega)$ is the range partner for the space $L_1(\Omega)$, the largest of all r.i. spaces. Therefore, larger spaces than $L_{\frac{n}{n-m},1}(\Omega)$ are not interesting range candidates.

Likewise Theorem 5.1, Theorem 5.2 can hardly be directly applied to a concrete example, since to evaluate $X(\Omega)$ from the quite implicit formula (1) involving supremum over equimeasurable functions is practically impossible. In the search of a simplification, several methods have been applied. Among functions in $\mathcal{M}(0,1)$ that are equimeasurable to a given function f in $\mathcal{M}(\Omega)$, there is one with an exceptional significance, namely f^* itself. So, a natural question is under what conditions one can replace in (1) $\sup_{h^*=f^*} \|H_{\frac{n}{m}} h\|_{\overline{Y}(0,1)}$ by $\|H_{\frac{n}{m}} f^*\|_{\overline{Y}(0,1)}$. If we could do that without loosing, it would mean a great simplification of the formula (1). Of course, only the inequality

$$\sup_{h^*=f^*} \|H_{\frac{n}{m}} h\|_{\overline{Y}(0,1)} \leq C \|H_{\frac{n}{m}} f^*\|_{\overline{Y}(0,1)}$$

is in question, the converse one is trivial. However, this idea contains one hidden danger: the quantity on the right is not necessarily a norm (recall that the operation $f \mapsto f^*$ is not subadditive, so the triangle inequality is not guaranteed), and, indeed, there are r.i. spaces $Y(\Omega)$ for which it isn't. Probably the simplest example of such $Y(\Omega)$ is $L_1(\Omega)$ - it is easy to verify that $\|H_{\frac{n}{m}} f^*\|_{L_1(0,1)}$ is not a norm. In[21], a sufficient condition was established, namely

$$\|H_{\frac{n}{m}} f^{**}\|_{\overline{Y}(0,1)} \leq C \|H_{\frac{n}{m}} f^*\|_{\overline{Y}(0,1)}. \tag{2}$$

Replacing f^* with f^{**} immediately solves the triangle inequality problem, since the operation $f \mapsto f^{**}$ *is* subadditive, but the condition is unsatisfactory (too strong) because it rules out important limiting cases. (It is easy to see that, for example for $Y(\Omega) = L_{\frac{n}{n-m}}(\Omega)$, (2) is not true.) In[29], another approach using special operators was elaborated. Finally, in[30], it was shown that a reasonable sufficient condition is the boundedness of the supremum operator $T_{\frac{n}{m}}$ on an associate space of $\overline{Y}(0,1)$. It was recently communicated to the author by Lenka Slavíková ([31]) that this sufficient condition is also in fact necessary. Hence, we have the following statement.

Theorem 5.3. *Let $Y(\Omega)$ be an r.i. space and let $X(\Omega)$ be the optimal domain r.i. space $X(\Omega)$ corresponding to $Y(\Omega)$ in (1). Then*

$$\|f\|_{X(\Omega)} \approx \left\| \int_t^1 f^*(s) s^{\frac{m}{n}-1}\, ds \right\|_{\overline{Y}'(0,1)}, \quad f \in \mathcal{M}(\Omega),$$

if and only if

$$T_{\frac{n}{m}} : \overline{Y}'(0,1) \to \overline{Y}'(0,1). \tag{3}$$

(Here and below, by $\approx$ we denote the comparability of norms.)

The condition (3) is reasonable and it does not rule out important limiting examples. Moreover, as we shall see, it is quite natural.

Example 5.2. Let us return to the Maz'ya–Hansson–Brézis–Wainger embedding (7), or, more precisely, to its higher-order modification. Starting with $X(\Omega) = L_{\frac{n}{m}}(\Omega)$, then the corresponding optimal range r.i. space $Y(\Omega)$ is $L_{\infty,\frac{n}{m};-1}(\Omega)$, as it was shown in Example 5.1. In order to be able to apply Theorem 5.3, we must show that $T_{\frac{n}{m}}$ is bounded on the associate space of $\overline{Y}(0,1)$, which, as already observed in Example 5.1, happens to be $L_{(1,\frac{n}{n-m})}(0,1)$. In order to prove the boundedness of $T_{\frac{n}{m}}$ on $L_{(1,\frac{n}{n-m})}(0,1)$, we first note that $T_{\frac{n}{m}}$ is bounded on $L_{1,\frac{n}{n-m}}(0,1)$, which is easier and which can be done either by a standard interpolation argument or using conditions for the weighted norm inequalities involving supremum operators from[32] or[33]. Next we show that $(T_{\frac{n}{m}} g)^{**}$ is comparable to $T_{\frac{n}{m}}(g^{**})$. Combining these two facts, we get the desired boundedness of $T_{\frac{n}{m}}$ on $L_{(1,\frac{n}{n-m})}(0,1)$, which is $\overline{Y}'(0,1)$. Hence, according to Theorem 5.3, the optimal r.i. domain partner space, $\widetilde{X}(\Omega)$, has norm

$$\|g\|_{\widetilde{X}(\Omega)} = \|H_{\frac{n}{m}} g^*\|_{L_{\infty,\frac{n}{m};-1}(0,1)}.$$

Now, several interesting facts can be observed about this space. First, it indeed is strictly larger than $X(\Omega) = L_{\frac{n}{m}}(\Omega)$. In fact, it even has an essentially different fundamental function. Moreover, it is a qualitatively new type of function space. In[18], several interesting properties of this space were established, for example its incomparability to several related known function spaces of Orlicz and Lorentz–Zygmund type.

6. Formulas for optimal spaces using the functional $f^{**} - f^*$

In practice, one often wants to solve the following problem: given m and an r.i. space $X(\Omega)$, find its optimal range r.i. partner, let us call it $Y_X(\Omega)$, so that the Sobolev embedding

$$W^m X(\Omega) \hookrightarrow Y_X(\Omega), \tag{1}$$

holds, and $Y_X(\Omega)$ is the smallest possible such r.i. space. A less frequent task, but also of interest, is the converse one; given m and an r.i. space $Y(\Omega)$, find its optimal domain r.i. partner, let us call it $X_Y(\Omega)$, for $Y(\Omega)$, so that

$$W^m X_Y(\Omega) \hookrightarrow Y(\Omega),$$

holds and $X_Y(\Omega)$ is the largest possible such r.i. space.

At this stage, we have formulas for both $Y_X(\Omega)$ and $X_Y(\Omega)$, given by Theorems 5.1 and 5.2, respectively. As we have already noticed, these formulas are too implicit to allow for some practical use. Theorem 5.2 is particularly bad. In this section we will show that significant simplifications of these formulas, such as the one given by Theorem 5.3, are possible if we a-priori know that the given space has been chosen in such a way that it is an optimal partner for some other r.i. space.

We first need to introduce one more supremum operator. Let

$$\left(S_{\frac{n}{m}} f\right)(t) := t^{\frac{m}{n}-1} \sup_{0<s\le t} s^{1-\frac{m}{n}} f^*(s), \quad f \in \mathcal{M}(0,1),\ t \in (0,1).$$

Then, $S_{\frac{n}{m}}$ has the following endpoint mapping properties:

$$S_{\frac{n}{m}} : L_{\frac{n}{n-m},\infty}(0,1) \to L_{\frac{n}{n-m},\infty}(0,1) \quad \text{and} \quad S_{\frac{n}{m}} : L_\infty(0,1) \to L_\infty(0,1).$$

Our point of departure will be the following result from[30].

Theorem 6.1. *Let $X(\Omega)$ be an r.i. space, whose associate space satisfies $X'(\Omega) \hookrightarrow L_{\frac{n}{n-m},\infty}(\Omega)$. Then,*

$$\|f\|_{Y_X(\Omega)} \approx \sup_{\|S_{\frac{n}{m}} g\|_{\overline{X}'(0,1)} \le 1} \int_0^1 t^{-\frac{m}{n}} \left[f^{**}(t) - f^*(t)\right] g^*(t)\, dt + \|f\|_{L_1(\Omega)}, \tag{2}$$

where $f \in \mathcal{M}(\Omega)$, $g \in \mathcal{M}_+(0,1)$.

The most innovative part of Theorem 6.1 is the new formula (2). The L_1-norm has just a cosmetic meaning, it's role is to guarantee that the resulting functional is a norm. The main term is formulated as some kind of duality involving the operator $S_{\frac{n}{m}}$. In cases when $S_{\frac{n}{m}}$ can be peeled off, the whole expression is considerably simpler.

Theorem 6.2. *An r.i. space $X(\Omega)$ is the optimal domain partner in (1) for some other r.i. space $Y(\Omega)$ if and only if*

$$S_{\frac{n}{m}} : \overline{X}'(0,1) \to \overline{X}'(0,1).$$

In that case,

$$\|f\|_{Y_X(\Omega)} \approx \left\|t^{-\frac{m}{n}}\left[f^{**}(t)-f^*(t)\right]\right\|_{\overline{X}(0,1)} + \|f\|_{L_1(\Omega)}, \qquad f \in \mathcal{M}(\Omega).$$

Again, an r.i. space $Y(\Omega)$ is the optimal range partner in (1) *for some other r.i. space $X(\Omega)$ if and only if*

$$T_{\frac{n}{m}} : \overline{Y}'(0,1) \to \overline{Y}'(0,1).$$

In that case,

$$\|f\|_{X_Y(\Omega)} \approx \left\|\int_t^1 f^*(s)s^{\frac{m}{n}-1}\,ds\right\|_{\overline{Y}(0,1)}, \qquad f \in \mathcal{M}(\Omega).$$

This result enables us to apply a new approach. We start with a given r.i. space $X(\Omega)$. We find the corresponding optimal range r.i. partner $Y_X(\Omega)$. Now, the embedding (1) has an optimal range, but it does not necessarily have an optimal domain, as Example 5.2 clearly shows. We thus take one more step in order to get the optimal domain r.i. partner for $Y_X(\Omega)$, let us call it $\widetilde{X}(\Omega)$. At this stage however, instead of the rather unpleasant Theorem 5.2, we can use the far more friendly Theorem 5.3, because $Y_X(\Omega)$ is now already known to be the optimal range partner for $X(\Omega)$, and Theorem 6.2 tells us that this is equivalent to the required boundedness of $T_{\frac{n}{m}}$ on $\overline{Y_X}'(0,1)$. Altogether, we have

$$W^mX(\Omega) \subset W^m\widetilde{X}(\Omega) \hookrightarrow Y(\Omega),$$

and $\widetilde{X}(\Omega)$ now either can be equivalent to $X(\Omega)$, or it can be strictly larger. In any case, after these two steps, the couple $(\widetilde{X}(\Omega), Y(\Omega))$ forms an optimal pair in the Sobolev embedding, and no further iterations of the process can bring anything new.

The functional $f^{**}(t) - f^*(t)$, appearing in (2), should cause some natural concern. It is known ([34]) that function spaces whose norms involve this functional often do not enjoy nice properties such as linearity, lattice property or normability. For example, for $X(\Omega) = L_{\frac{n}{m}}(\Omega)$ (cf.[34] Remark 3.2), all these properties for $Y_X(\Omega)$ are lost. It is instructive to compare this fact with Theorem 5.3, where this case is ruled out by the assumption $S_{\frac{n}{m}} : \overline{X}'(0,1) \to \overline{X}'(0,1)$. This makes the significance of the supremum operator more transparent; $S_{\frac{n}{m}}$ is bounded on $L_{\frac{n}{n-m},\infty}(0,1)$ but *not* on $L_{\frac{n}{n-m}}(0,1)$. This example is typical, and it illustrates the general principle: the boundedness of $S_{\frac{n}{m}}$ on $\overline{X}'(0,1)$ guarantees that $Y_X(\Omega)$ is a norm.

Incidentally, certain care must be exercised always when the norm of a given function space depends on f^*. For illustration of this fact, see[35]. Let us just add that a detailed study of weighted function spaces based on the functional $f^{**} - f^*$ can be found in[34] and[36].

Theorem 6.2 now can be used to obtain a new description of the space $\widetilde{X}(\Omega)$.

Theorem 6.3. *Let $X(\Omega)$ be an r.i. space and let $\widetilde{X}(\Omega)$ be defined as above. Define the space $Z(\Omega)$ by*

$$\|g\|_{Z(\Omega)} := \|S_{\frac{n}{m}} g^{**}\|_{\overline{X}'(0,1)}, \qquad g \in \mathcal{M}(\Omega).$$

Then,

$$\widetilde{X}(\Omega) = Z'(\Omega).$$

The formulas for optimal spaces given by Theorems 6.2 and 6.3 are still not as explicit as one would desire, but, at least, they show the problem in a new light. They also enable us to obtain explicit formulas for some examples such as Orlicz spaces, previously unavailable. We finish this section by an example that can be computed using Theorem 6.2.

Theorem 6.4. *Let A be a Young function for which there exists $r > 1$ with*

$$\tilde{A}(rt) \geq 2r^{\frac{n}{n-m}} \tilde{A}(t), \qquad t \geq 1.$$

Then, the r.i. spaces $X(\Omega) = L_A(\Omega)$ and $Y(\Omega)$, whose norm is given by

$$\|f\|_{Y(\Omega)} := \left\|t^{-\frac{m}{n}} \left[f^{**}(t) - f^*(t)\right]\right\|_{L_A(0,1)} + \|f\|_{L_1(\Omega)}, \qquad f \in \mathcal{M}(\Omega),$$

are optimal in (1).

7. The links between optimality and interpolation

Throughout, we shall write $X \in \mathrm{Int}(Y, Z)$ to denote that X is an interpolation space with respect to the pair (Y, Z).

The proofs of Theorems 6.2 and 6.3 reveal a very interesting link between optimality of r.i. spaces in Sobolev embeddings and their interpolation properties. It is obtained through the following theorem.

Theorem 7.1. *Let $X(\Omega)$ be an r.i. space. Then, the operator $T_{\frac{n}{m}}$ is bounded on $\overline{X}'(0,1)$ if and only if $X(\Omega)$ is an interpolation space with respect to the pair $\left(L_{\frac{n}{n-m},1}(\Omega), L_\infty(\Omega)\right)$, a fact which we will denote as*

$$X(\Omega) \in \mathrm{Int}\left(L_{\frac{n}{n-m},1}(\Omega), L_\infty(\Omega)\right).$$

Similarly, the operator $S_{\frac{n}{m}}$ is bounded on $\overline{X}(0,1)$ if and only if

$$X(\Omega) \in \operatorname{Int}\left(L_{\frac{n}{n-m},\infty}(\Omega), L_\infty(\Omega)\right).$$

In other words, r.i. spaces in a Sobolev embedding can be optimal (domains or range) partners for some other r.i. spaces if and only if they satisfy certain interpolation properties. Of course, for example, a very large space, which does not satisfy the interpolation property, can also be a range in a Sobolev embedding, but not the optimal one.

This result has a very important corollary.

Corollary 7.1. *Assume that the Sobolev embedding*

$$W^m X(\Omega) \hookrightarrow Y(\Omega)$$

holds, and assume that $Y(\Omega)$ is the optimal such range r.i. space. Then,

$$Y(\Omega) \in \operatorname{Int}\left(L_{\frac{n}{n-m},1}, L_\infty\right).$$

To understand this result properly, note that of course one can embed a Sobolev space into some very large space such as L_1, which of course is not an interpolation space between $L_{\frac{n}{n-m},1}$ and L_∞, but then this range space simply *cannot be optimal.* In other words, *interpolation properties have an influence over the optimality ones.*

We shall now sketch the proof of Theorem 7.1 in order to illustrate how the interpolation ideas come to the picture.

Assume T is a linear operator with

$$T : L_{\frac{n}{n-m},1} \to L_{\frac{n}{n-m},1}, \quad \text{and} \quad T : L_\infty \to L_\infty.$$

We must show

$$T : X \to X.$$

A standard argument using the usual decomposition of a function g, yields

$$\int_0^t s^{-\frac{m}{n}} (Tg)^*(s)\,ds \le C \int_0^t s^{-\frac{m}{n}} g^*(s)\,ds, \quad g \in L_{\frac{n}{n-m},1}(I), \quad t \in I.$$

By the HLP Principle, the latter implies

$$\int_0^1 t^{-\frac{m}{n}} (Tg)^*(t) h^*(t)\,dt \le C \int_0^1 t^{-\frac{m}{n}} g^*(t) h^*(t)\,dt,$$

for $g \in L_{\frac{n}{n-m},1}(I)$ and $h \in P(I)$. Taking $f \in X'$ and substituting $h(t) = h^*(t) = \sup_{t\le s<1} s^{\frac{m}{n}} f^*(s)$, we get

$$\begin{aligned}\int_0^1 (Tg)^*(t) f^*(t)\, dt &\le \int_0^1 (Tg)^*(t) T_{\frac{n}{m}} f(t)\, dt \\ &\le C \int_0^1 g^*(t) T_{\frac{n}{m}} f(t)\, dt \\ &\le C\|g\|_X \|T_{\frac{n}{m}} f\|_{X'} \\ &\le C\|g\|_X \|f\|_{X'}, \quad f \in X'.\end{aligned}$$

Taking the supremum over f with $\|f\|_{X'} \le 1$ yields

$$\|Tg\|_X \le C\|g\|_X, \quad g \in L_{\frac{n}{n-m},1},$$

proving Theorem 7.1.

These new facts allow one more interesting return to Example 5.2.

Example 7.1. Let $X = L_{\frac{n}{m}}$. Then, the corresponding optimal range norm Y in the Sobolev inequality satisfies

$$\|f\|_Y \approx \left(\int_0^1 \left(\frac{f^*(t)}{\log(\frac{e}{t})}\right)^{\frac{n}{n-m}} \frac{dt}{t}\right)^{\frac{n-m}{n}}, \quad f \in \mathcal{M}(0,1)$$

(cf. the result of Hansson, Brézis–Wainger and Maz'ya, mentioned above). Now, as we already know, the space X is *not the optimal domain r.i. norm* for Y. Indeed, the optimal one, $\tilde{X}$, has norm

$$\left(\int_0^1 \left(\frac{\int_t^1 s^{\frac{m}{n}-1} g^*(s)\, ds}{\log(\frac{e}{t})}\right)^{\frac{n}{n-m}} \frac{dt}{t}\right)^{\frac{n-m}{n}}, \quad g \in \mathcal{M}(0,1),$$

Observe that, in accordance with our results, the operator $S_{\frac{n}{m}}$ is *not bounded* on $L_{\frac{n}{m}}(0,1)$, while it is bounded on the associate space of $\tilde{X}(0,1)$.

8. Explicit formulas for optimal spaces in Sobolev embeddings

Our aim in this section is to establish explicit formulas for the spaces $Y_X(\Omega)$ and $\widetilde{X}(\Omega)$, given an r.i. space $X(\Omega)$. We recall that the formulas for these spaces which we have so far, are expressed in terms of their associate spaces, namely,

$$\|f\|_{Y_X'(\Omega)} := \|f^{**}(t) t^{\frac{m}{n}}\|_{\overline{X}'(0,1)}, \qquad f \in \mathcal{M}(\Omega), \tag{1}$$

and

$$\|g\|_{\widetilde{X}'(\Omega)} := \|S_{\frac{n}{m}} g^{**}\|_{\overline{X}'(0,1)}, \qquad g \in \mathcal{M}(\Omega). \tag{2}$$

Our focus will be now on the problem how to get these constructions explicit. We first note that the expression for $Y'_X(\Omega)$ turns out to be unsatisfactory in that the function

$$t \to t^{\frac{m}{n}-1} \int_0^t g^*(s)\, ds, \qquad t \in [0,\infty),\ g \in \mathcal{M}(\Omega),$$

need not be nonincreasing. This complicates the construction of explicit formulas for $Y_X(\Omega)$. (However, see[21] Section 4 and[30] Section 4.) Our next theorem from[37] solves this difficulty.

Theorem 8.1. *Suppose $X(\Omega)$ is an r.i. space satisfying*

$$X(\Omega) \supset L_{\frac{n}{m},1}(\Omega).$$

Define the space $Z_X(\Omega)$ by

$$\|g\|_{Z_X(\Omega)} := \left\| t^{\frac{m}{n}-1} \int_0^t g^*(s) s^{-\frac{m}{n}}\, ds \right\|_{\overline{X}'(0,1)}, \qquad g \in \mathcal{M}(\Omega).$$

Then,

$$\|f\|_{Y_X(\Omega)} \approx \|t^{-\frac{m}{n}} f^*(t)\|_{\overline{Z}'_X(0,1)}, \qquad f \in \mathcal{M}(\Omega).$$

We note that this eliminates the above-mentioned problem, since the function

$$t \mapsto t^{\frac{m}{n}-1} \int_0^t g^*(s) s^{-\frac{m}{n}}\, ds, \qquad t \in [0,\infty),$$

is nonincreasing, being a weighted average of a nonincreasing function.

Theorem 8.1 is, again, rather involved. The proof uses delicate estimates and previously obtained optimality results for various integral and supremum operators.

Hence, the remaining task is to compute associate spaces of $\widetilde{X}'(\Omega)$ and $Y'_X(\Omega)$. To this end, we shall use the Brudnyi–Kruglyak duality theory ([38]) and the interpolation methods using the k-functional, elaborated recently in[39].

The main result reads as follows.

Theorem 8.2. *Suppose $X(\Omega)$ is an r.i. space. Define the space $V_X(\Omega)$ by*

$$\|g\|_{V_X(\Omega)} := \|g^{**}(t^{1-\frac{m}{n}})\|_{\overline{X}'(0,1)}, \qquad g \in \mathcal{M}(\Omega).$$

Then,

$$\|g\|_{\widetilde{X}(\Omega)} \approx \left(k(t, g^*; L_1(0,1), L_{\frac{n}{m},1}(0,1))\right\|_{\overline{V}'_X(0,1)}, \qquad g \in \mathcal{M}(\Omega).$$

Moreover,

$$\|f\|_{Y_X(\Omega)} \approx \left\|k(t, s^{-\frac{m}{n}} f^*(s); L_1(0,1), L_{\frac{n}{m},\infty}(0,1))\right\|_{\overline{V}'_X(0,1)}, \qquad f \in \mathcal{M}(\Omega).$$

Theorem 8.2 can be applied to construct the spaces $Y_X(\Omega)$ and $\widetilde{X}(\Omega)$ explicitly.

Let us now briefly indicate how the interpolation K-method comes in.

Let X_1 and X_2 be Banach spaces, compatible in the sense that they are embedded in a common Hausdorff topological vector space H. Suppose $x \in X_1 + X_2$ and $t \in [0, \infty)$. The *Peetre K-functional* is defined by

$$K(t, x; X_1, X_2) := \inf_{x = x_1 + x_2} \left(\|x_1\|_{X_1} + t\|x_2\|_{X_2}\right), \qquad t > 0.$$

It is an increasing concave function of t on $[0, \infty)$, so that

$$k\,(t, x; X_1, X_2) := \frac{d}{dt} K\,(t, x; X_1, X_2)$$

is nonincreasing on $[0, \infty)$.

Given an r.i. space Z on $\mathcal{M}_+([0,\infty))$, for which $\left\|\frac{1}{1+t}\right\|_Z < \infty$, the space X, with $\|x\|_X$ defined at $x \in X_1 + X_2$ by

$$\|x\|_X := \left\|t^{-1} K\,(t, x; X_1, X_2)\right\|_Z,$$

satisfies

$$X_1 \cap X_2 \subset X \subset X_1 + X_2;$$

moreover, for any linear operator T defined on $X_1 + X_2$

$$T : X_i \to X_i, \quad i = 1, 2, \qquad \text{implies} \qquad T : X \to X.$$

We say the space X is generated by the K-method of interpolation.

The asserted connection of the duality theory for the K-method with our task is through certain reformulations of (1) and (2), namely

$$\|f\|_{Y'_X(\Omega)} \approx \left\|t^{\frac{m}{n}-1} K\left(t^{1-\frac{m}{n}}, f; L_{\frac{n}{n-m},\infty}(0,1), L_\infty(0,1)\right)\right\|_{\overline{X}'(0,1)}, \; f \in \mathcal{M}(\Omega),$$

and

$$\|g\|_{\widetilde{X}'(\Omega)} \approx \left\|t^{\frac{m}{n}-1} K\left(t^{1-\frac{m}{n}}, g; L_{\frac{n}{n-m},1}(0,1), L_\infty(0,1)\right)\right\|_{\overline{X}'(0,1)}, \; g \in \mathcal{M}(\Omega).$$

We finish by an example involving Orlicz spaces.

Theorem 8.3. *Let A be a Young function. Assume $A(t) = t^q$ near 0 and suppose*

$$t^{\frac{m}{n}-1} \notin L_{\widetilde{A}}([0,\infty)).$$

Define B through the equation

$$B(\gamma(t)) := \left(\frac{m}{n} - 1\right) \widetilde{A}\left(t^{\frac{m}{n}-1}\right) \frac{\gamma t)}{t\gamma'(t))},$$

in which

$$\gamma(t) := t^{-\frac{m}{n}} \int_t^\infty \widetilde{A}\left(s^{\frac{m}{n}-1}\right) ds, \qquad t \in [0,\infty).$$

Define the space $Z(\Omega)$ by

$$\|g\|_{Z(\Omega)} := \left\| t^{\frac{m}{n}-1} \int_0^{t^{1-\frac{m}{n}}} g^*(s)\, ds \right\|_{L_{\widetilde{A}}(0,1)}, \qquad g \in \mathcal{M}(\Omega).$$

Then, B is a Young function and

$$\|f\|_{Z'(\Omega)} \approx \left\|t^{-\frac{m}{n}} f^*\left(t^{1-\frac{m}{n}}\right)\right\|_{\overline{X}(0,1)}, \qquad f \in \mathcal{M}(\Omega).$$

It is of interest to compare this result with that of Cianchi who obtained, in[28], a description of $Y_X(\Omega)$ different from ours by the use of techniques specific to the Orlicz context.

We note that the results of this section can be applied also to other examples of function spaces such as classical Lorentz spaces of type Gamma and Lambda (details can be found in[37]). These formulas are however rather complicated and therefore omitted here.

9. Compactness of Sobolev embeddings

The most important characteristics of Sobolev spaces is not only whether they embed into other function spaces, but also whether they embed compactly.

Let $X(\Omega)$ and $Y(\Omega)$ be two r.i. spaces. We say that $W^m X(\Omega)$ is *compactly embedded* into $Y(\Omega)$, written,

$$W^m X(\Omega) \hookrightarrow\hookrightarrow Y(\Omega)$$

if, for every sequence $\{f_k\}$ bounded in $W^m X(\Omega)$, there exists a subsequence $\{f_{k_j}\}$ which is convergent in $Y(\Omega)$.

For the case when $X(\Omega)$ and $Y(\Omega)$ are Lebesgue spaces, we have a theorem, which originated in a lemma of Rellich[40] and was proved specifically for Sobolev spaces by Kondrachov[41], and which asserts that

$$W^{m,p}(\Omega) \hookrightarrow\hookrightarrow L_q(\Omega) \tag{1}$$

if $q < p^* = \frac{np}{n-mp}$. Standard examples (see[22]) show it is not compact when $q = \frac{np}{n-mp}$.

As for embeddings into Orlicz spaces, Hempel, Morris and Trudinger[5] showed that the embedding (3) is not compact. By a standard argument using a uniform absolute continuity of a norm it can be proved that

$$W^{1,n}(\Omega) \hookrightarrow\hookrightarrow L_B(\Omega)$$

whenever B is a Young function satisfying, with $A(t) := \exp(t^{n'})$ for large values of t,

$$\lim_{t\to\infty} \frac{A(\lambda t)}{B(t)} = \infty$$

for every $\lambda > 0$.

Considering Lorentz spaces, it is of interest to notice that even the Sobolev embedding

$$W^{m,p}(\Omega) \hookrightarrow\hookrightarrow L_{p^*,\infty}(\Omega)$$

is still not compact. (This is not difficult to verify; in fact, standard examples that demonstrate the noncompactness of (1) with $q = p^*$ (see e.g.[22]), are sufficient.) The space $L_{p^*,\infty}(\Omega)$ is of course considerably larger than $L_{p^*}(\Omega)$, but it simply is not 'larger enough'. This observation is a good point of departure as it raises interesting questions.

For example, we may ask whether the space $L_{p^*,\infty}(\Omega)$ is the *'gateway to compactness'* in the sense that every strictly larger space is already a compact range for $W^{m,p}(\Omega)$. It even makes a good sense to formulate this problem in a broader context of r.i. spaces. (Recall that when the Lebesgue space $L_{p^*}(\Omega)$ is replaced by an arbitrary r.i. space $Y(\Omega)$, then the role of $L_{p^*,\infty}(\Omega)$ is taken over by the endpoint Marcinkiewicz space $M_Y(\Omega)$.)

We can the formulate the following general question (which we have answered for the particular example above).

Let $X(\Omega)$ be an r.i. space and let $Y_X(\Omega)$ be the corresponding optimal range r.i. space. Let $M_{Y_X}(\Omega)$ be the Marcinkiewicz space corresponding to $Y_X(\Omega)$. Then, of course, we have $W^mX(\Omega) \hookrightarrow M_{Y_X}(\Omega)$. Can this embedding ever be compact? If not, is the Marcinkiewicz space the gateway to compactness in the above-mentioned sense?

It is clear that in order to obtain satisfactory answers to these and other questions we need a reasonable characterization of pairs of spaces $X(\Omega), Y(\Omega)$ for which we have the compact Sobolev embedding

$$W^m X(\Omega) \hookrightarrow\hookrightarrow Y(\Omega).$$

From various analogous results in less general situations it can be guessed that one such characterization might be the compactness of $H_{\frac{n}{m}}$ from $\overline{X}(0,1)$ to $\overline{Y}(0,1)$, and another one might be the uniform absolute continuity of the norms of the $H_{\frac{n}{m}}$-image of the unit ball of $\overline{X}(0,1)$ in $\overline{Y}(0,1)$. This guess turns out to be reasonable, but the proof is deep and difficult and contains many unexpected pitfalls. Moreover, the case when the range space is $L_\infty(\Omega)$ must be treated separately.

Theorem 9.1. *Let $X(\Omega)$ and $Y(\Omega)$ be r.i. spaces. Assume $Y(\Omega) \neq L_\infty(\Omega)$. Then, the following three statements are equivalent:*

$$W^m X(\Omega) \hookrightarrow\hookrightarrow Y(\Omega); \tag{2}$$

$$H_{\frac{n}{m}} : \overline{X}(0,1) \to\to \overline{Y}(0,1); \tag{3}$$

$$\lim_{a\to 0+} \sup_{\|f\|_{\overline{X}(0,1)}\leq 1} \left\| \chi_{(0,a)}(t) \int_t^1 f^*(s) s^{\frac{m}{n}-1}\, ds \right\|_{\overline{Y}(0,1)} = 0. \tag{4}$$

The case $Y(\Omega) = L_\infty(\Omega)$ is different and, as such, is treated in

Theorem 9.2. *Let $X(\Omega)$ be an r.i. space. Then, the following three statements are equivalent:*

$$W^m X(\Omega) \hookrightarrow\hookrightarrow L_\infty(\Omega);$$

$$H_{\frac{n}{m}} : \overline{X}(0,1) \to\to L_\infty(0,1);$$

$$\lim_{a\to 0+} \sup_{\|f\|_{\overline{X}(0,1)}\leq 1} \int_0^a f^*(t) t^{\frac{m}{n}-1}\, dt = 0.$$

The most important and involved part is the sufficiency of (4) for (2). When trying to prove this implication, we discovered an unpleasant technical difficulty. All the methods which we tried to apply, and which would naturally solve the problem, seemed to require

$$\overline{Y}(0,1) \in \operatorname{Int}\left(L_{\frac{n}{n-m},1}(0,1), L_\infty(0,1) \right),$$

a restriction that does not offer any obvious circumvention. Such requirement, however, is simply too much to ask. A candidate for a compact range can be as large as it pleases (consider $L_1(\Omega)$, and, in particular, it may by all means lay far outside from the required interpolation sandwich. This obstacle proved to be surprisingly difficult. At the end, it was overcome by the discovery of a useful fact that, given an r.i. space $Y(\Omega)$, we can always construct another one, $Z(\Omega)$, possibly smaller than $Y(\Omega)$, such that the condition (4) is still valid, but which already has the required interpolation properties. We formulate this result as a separate theorem, because it is of independent interest.

Theorem 9.3. *Let $X(\Omega)$ and $Y(\Omega)$ be r.i. spaces satisfying* (4). *Then, there exists another r.i. space, $Z(\Omega)$, with*

$$\overline{Z}(0,1) \in \operatorname{Int}\left(L_{\frac{n}{n-m},1}(0,1), L_\infty(0,1)\right),$$

such that $Z(\Omega) \hookrightarrow Y(\Omega)$ and

$$\lim_{a\to 0+} \sup_{\|f\|_{X(\Omega)}\le 1} \left\| \chi_{(0,a)}(t) \int_t^1 f^*(s) s^{\frac{m}{n}-1}\, ds \right\|_{Z(\Omega)} = 0.$$

The rest of the proof of the main results uses sharp estimates for supremum operators, various optimality results from the preceding sections and the Arzela–Ascoli theorem. The proof of Theorem 9.3 is very involved and delicate and requires extensive preparations. The details can be found in [42].

At one stage of the proof, the necessity of the vanishing *Muckenhoupt condition* is shown.

Theorem 9.4. *Let $X(\Omega)$ and $Y(\Omega)$ be r.i. spaces. Assume $Y(\Omega) \neq L_\infty(\Omega)$. Then, each of* (2), (3) *and* (4) *implies*

$$\lim_{a\to 0+} \|\chi_{(0,a)}\|_{\overline{Y}(0,1)} \|t^{\frac{m}{n}-1}\chi_{(a,1)}(t)\|_{\overline{X}'(0,1)} = 0.$$

This result shows that a candidate $Y(\Omega)$ for a compact range must have an essentially smaller fundamental function than the optimal embedding space $Y_X(\Omega)$, hence also than the Marcinkiewicz space $M_{Y_X}(\Omega)$. In other words, we must have

$$\lim_{t\to 0+} \frac{\varphi_Y(t)}{\varphi_{Y_X}(t)} = 0.$$

This solves the above question: the embedding

$$W^m X(\Omega) \hookrightarrow M_{Y_X}(\Omega)$$

is always true, but *never* (for any choice of $X(\Omega)$) compact.

Likewise, the 'gateway' problem has the negative answer: a counter-example is easily constructed by taking appropriate fundamental functions and using corresponding Marcinkiewicz spaces. It turns out that not even a space which contains $M_{Y_X}(\Omega)$ properly and whose fundamental function is strictly smaller than that of $Y_X(\Omega)$ guarantees compactness.

The connection between a candidate $Y(\Omega)$ for a compact range for a given Sobolev space $W^mX(\Omega)$ and the optimal range $Y_X(\Omega)$ that does imply compactness can be found, but it has to be formulated in terms of a uniform absolute continuity.

Theorem 9.5. *Suppose $X(\Omega)$ and $Y(\Omega)$ are two r.i. spaces. Assume $Y(\Omega) \neq L_\infty(\Omega)$ and let $Y_X(\Omega)$ be the optimal r.i. embedding space for $W^mX(\Omega)$. Then,* (2) *holds if and only if the functions in the unit ball of $Y_X(\Omega)$ have uniformly absolutely continuous norms in $Y(\Omega)$ or, what is the same,*

$$\lim_{a\to 0+} \sup_{\|f\|_{Y_X(\Omega)}\leq 1} \left\|\chi_{(0,a)} f^*\right\|_{\overline{Y}(0,1)} = 0, \qquad f \in \mathcal{M}(\Omega). \tag{5}$$

Theorem 9.5 gives a necessary and sufficient condition for the compactness of a Sobolev embedding. However, an application of the criterion would involve examination of a uniform absolute continuity of many functions, which may be difficult to verify. It is thus worth looking for a more manageable condition, sufficient for the compactness of the embedding and not too far from being also necessary, which could be used in practical examples. Such a condition is provided by our next theorem. In some sense, it substitutes the negative outcome of the gateway problem.

Theorem 9.6. *Let $X(\Omega)$ and $Y(\Omega)$ be r.i. spaces. Set*

$$\phi_R(t) := \frac{dc}{dt},$$

where $c(t)$ is the least concave majorant of

$$t\|s^{\frac{m}{n}-1}\chi_{(t,1)}(s)\|_{\overline{X}'(0,1)}.$$

Then, the condition

$$\lim_{a\to 0+} \left\|\chi_{(0,a)}\phi_R\right\|_{\overline{Y}(0,1)} = 0 \tag{6}$$

suffices for

$$W^mX(\Omega) \hookrightarrow\hookrightarrow Y(\Omega).$$

Observe that the condition (6) can be simply verified in concrete examples, since it requires to consider just one function rather than the whole unit ball as in (5).

Among many examples that can be extracted from these results, we shall present just one, concerning Orlicz spaces.

Theorem 9.7. *Suppose A and $\tilde{A}$ are complementary Young functions and*

$$\int_1^\infty \frac{\tilde{A}(s)}{s^{1+\frac{n}{n-m}}}\, ds = \infty.$$

Define the Young function $A_R(t)$, for t large, by

$$A_R^{-1}(t) := \frac{t^{1-\frac{m}{n}}}{E^{-1}(t)},$$

with

$$E(t) := t^{\frac{n}{n-m}} \int_1^t \frac{\tilde{A}(s)}{s^{1+\frac{n}{n-m}}}\, ds, \qquad t \geq 1.$$

Then,

$$W^m L_A(\Omega) \hookrightarrow\hookrightarrow L_B(\Omega)$$

for a given Young function B if and only if

$$\lim_{t\to\infty} \frac{A_R(\lambda t)}{B(t)} = \infty, \tag{7}$$

for every $\lambda > 0$.

We finally note that in terms of the explicitly known functions B and E, (7) can be expressed by

$$\lim_{t\to\infty} \frac{B\left((\lambda t)^{-1} E(t)^{1-\frac{m}{n}}\right)}{E(t)} = 0, \qquad \text{for every } \lambda > 0.$$

10. Gaussian Sobolev embeddings

In connection with some specific problems in physics such as quantum fields and hypercontractivity semigroups, it turns out that it would be of interest to extend classical Sobolev embeddings in $\mathbb{R}^n$ to an infinite-dimensional space. The motivation for such things stems from the fact that, in certain circumstances, the study of quantum fields can be reduced to operator or semigroup estimates which are in turn equivalent to inequalities of Sobolev type in infinitely many variables (see[43] and the references therein).

However, when we let $n \to \infty$, then $\frac{np}{n-p} \to p+$, and so the gain in integrability will apparently be lost. Even more serious, the Lebesgue measure on an infinite-dimensional space is meaningless.

These problems were overcome in the fundamental paper of L. Gross[44], who replaced the Lebesgue measure by the Gauss one. Note that the Gauss measure γ is defined on $\mathbb{R}^n$ by

$$d\gamma(x) = (2\pi)^{-\frac{n}{2}}\, e^{\frac{-|x|^2}{2}}\, dx.$$

Now, $\gamma(\mathbb{R}^n) = 1$ for every $n \in \mathbb{N}$, hence the extension for $n \to \infty$ is meaningful. The idea was then to seek a version of the Sobolev inequality that would hold on the probability space $(\mathbb{R}^n, \gamma)$ with a constant independent of n. Gross proved in [44] an inequality of this kind, which, in particular, entails that

$$\|u - u_\gamma\|_{L_2\log L(\mathbb{R}^n,\gamma)} \leq C\|\nabla u\|_{L_2(\mathbb{R}^n,\gamma)} \tag{1}$$

for every weakly differentiable function u making the right-hand side finite, where

$$u_\gamma = \int_{\mathbb{R}^n} u(x)d\gamma(x),$$

the mean value of u, and $L_2\log L(\mathbb{R}^n, \gamma)$ is the Orlicz space of those functions u such that $|u|^2|\log|u||$ is integrable in $\mathbb{R}^n$ with respect to γ. Interestingly, (1) still provides some slight gain in integrability from $|\nabla u|$ to u, even though it is no longer a power-gain.

In[45], we studied problems concerning optimality of function spaces in first-order Sobolev embeddings on the Gaussian space, namely

$$\|u - u_\gamma\|_{Y(\mathbb{R}^n,\gamma)} \leq C\|\nabla u\|_{X(\mathbb{R}^n,\gamma)} \tag{2}$$

As usual, we start with a reduction theorem. This time, the role of the operator $H_{\frac{n}{m}}$ is taken by the operator

$$\int_t^1 \frac{f(s)}{s\sqrt{1+\log(1/s)}}\, ds.$$

The reduction theorem then reads as follows.

Theorem 10.1. *Let $X(\mathbb{R}^n, \gamma)$ and $Y(\mathbb{R}^n, \gamma)$ be r.i. spaces. Then,*

$$\|u - u_\gamma\|_{Y(\mathbb{R}^n,\gamma)} \leq C\,\|\nabla u\|_{X(\mathbb{R}^n,\gamma)}$$

for every $u \in W^1X(\mathbb{R}^n,\gamma)$ *if and only if*

$$\left\| \int_t^1 \frac{f(s)}{s\sqrt{1+\log(1/s)}}\, ds \right\|_{\overline{Y}(0,1)} \leq C\, \|f\|_{\overline{X}(0,1)}$$

for every $f \in \overline{X}(0,1)$.

Then, the characterization of the optimal range r.i. space for the Gaussian Sobolev embedding when the domain space is obtained via the usual scheme.

Theorem 10.2. *Let* $X(\mathbb{R}^n,\gamma)$ *be an r.i. space, and let* $Z(\mathbb{R}^n,\gamma)$ *be the r.i. space equipped with the norm*

$$\|g\|_{Z(\mathbb{R}^n,\gamma)} := \left\| \frac{g^{**}(s)}{\sqrt{1+\log\frac{1}{s}}} \right\|_{\overline{X}'(0,1)}$$

for any measurable function u *on* $\mathbb{R}^n$. *Let* $Y(\mathbb{R}^n,\gamma) = Z'(\mathbb{R}^n,\gamma)$. *Then* $Y(\mathbb{R}^n,\gamma)$ *is the optimal range space in the Gaussian Sobolev embedding* (2).

The role of the operator $T_{\frac{n}{m}}$ is in the Gaussian setting taken over by the operator

$$(Tf)(t) = \sqrt{1+\log\frac{1}{t}}\, \sup_{t\leq s\leq 1} \frac{f^*(s)}{\sqrt{1+\log\frac{1}{s}}}, \qquad \text{for } t \in (0,1).$$

With the help of the operator T, we can characterize the optimal domain space.

Theorem 10.3. *Let* $Y(\mathbb{R}^n,\gamma)$ *be an r.i. space such that:*

$$\exp L^2(\mathbb{R}^n,\gamma) \hookrightarrow Y(\mathbb{R}^n,\gamma) \hookrightarrow L\,(\log L)^{\frac{1}{2}}\,(\mathbb{R}^n,\gamma),$$

and

$$T \text{ is bounded on } \overline{Y}'(0,1).$$

Let $X(\mathbb{R}^n,\gamma)$ *be the r.i. space equipped with the norm*

$$\|u\|_{X(\mathbb{R}^n,\gamma)} = \left\| \int_t^1 \frac{u^*(s)}{s\sqrt{1+\log\frac{1}{s}}}\, ds \right\|_{\overline{Y}(0,1)}.$$

Then $X(\mathbb{R}^n,\gamma)$ *is the optimal domain space for* $Y(\mathbb{R}^n,\gamma)$ *in the Gaussian Sobolev embedding* (2).

We shall now collect the basic examples.

Example 10.1. (i) Let $1 \le p < \infty$. Then the spaces

$$X(\mathbb{R}^n,\gamma) = L_p(\mathbb{R}^n,\gamma),\ Y(\mathbb{R}^n,\gamma) = L_p(\log L)^{\frac{p}{2}}(\mathbb{R}^n,\gamma)$$

form an optimal pair in the Gaussian Sobolev embedding (2).
(ii) The spaces $X(\mathbb{R}^n,\gamma) = L_\infty(\mathbb{R}^n,\gamma)$, $Y(\mathbb{R}^n,\gamma) = \exp L^2(\mathbb{R}^n,\gamma)$ form an optimal pair in the Gaussian Sobolev embedding (2).
(iii) Let $\beta > 0$. Then the spaces $(\exp L^\beta(\mathbb{R}^n,\gamma), \exp L^{\frac{2\beta}{2+\beta}}(\mathbb{R}^n,\gamma))$ form an optimal pair in the Gaussian Sobolev embedding (2).

These examples demonstrate a surprising phenomenon: while there is a *gain* in integrability when the domain space is a Lebesgue space, there is actually a *loss* near L_∞. This fact is caused by the nature of the Gaussian measure, which rapidly decreases at infinity.

11. Sobolev embeddings via isoperimetric inequalities

In this section we address the following two questions:

1) how to obtain sharp results for higher-order Sobolev embeddings when first-order ones are known,

2) what can we say when Ω is not a Lipschitz domain but some information about its regularity is given in terms of its isoperimetric function.

The intimate connection of Sobolev spaces to isoperimetric inequalities was first observed by Maz'ya[46,47], who proved that quite general Sobolev inequalities are equivalent to either isoperimetric or isocapacitary inequalities. Before the two problems had been investigated separately, by Sobolev ([48,49]), Gagliardo ([50]) and Nirenberg ([51]) on the one hand, and by De Giorgi ([52]) on the other. Independently, Federer and Fleming[53] also exploited De Giorgi's isoperimetric inequality to exhibit the best constant in the special case of the Sobolev inequality for functions whose gradient is integrable with power 1 in $\mathbb{R}^n$ via De Giorgi's isoperimetric inequality. These results led to an extensive research of the interplay between isoperimetric and Sobolev inequalities and also to many applications.

The approach to Sobolev embeddings via isoperimetric inequalities has some considerable advantages. First it covers a fairly broad range of situations, including Lipschitz domains, John domains, Maz'ya domains, probability measure spaces, and more. Second, it typically leads to sharp results. The drawback is that the existing literature is often restricted to the first-order derivatives, except perhaps for a very few quite specific instances while many important problems require higher order of the derivatives involved. This is mainly caused by the techniques used such as truncation, symmetrization, Pólya–Szegö principles etc., which do not allow a direct generalization to the higher-order case. Alternative methods which can be employed to handle higher-order Sobolev inequalities such as representation formulas, potential estimates, Fourier transforms or atomic decomposition are not flexible enough to provide us with optimal results in sufficient generality.

Recently, in[54], we adopted a new approach which shows that in fact isoperimetric inequalities do imply optimal higher-order Sobolev embeddings in a very general framework. Moreover, we employ Sobolev-type spaces built upon any rearrangement-invariant Banach function norms. The central ingredient of our approach is the combination of the first-order reduction principle with an iteration method, which is sharp enough to produce optimal higher-order reduction principles from the first-order ones. As a consequence, optimal results within the class of rearrangement-invariant spaces are obtained for Sobolev embeddings of any order. More precisely, we can characterize the best possible target for arbitrary-order Sobolev embeddings, in the class of all rearrangement-invariant Banach function spaces. A key step of the method is the development of a sharp iteration method involving subsequent applications of optimal Sobolev embeddings. We consider this method of independent interest for its possible use in different problems.

Let Ω be a domain in $\mathbb{R}^n$, $n \geq 1$, equipped with a finite measure ν which is absolutely continuous with respect to the Lebesgue measure, with density ω. Namely, $d\nu(x) = \omega(x)\, dx$, where ω is a Borel function such that $\omega(x) > 0$ a.e. in Ω. We assume, for simplicity, that $\nu(\Omega) = 1$. The basic case when ν is the Lebesgue measure will be referred to as Euclidean. Sobolev embeddings of arbitrary order for functions defined in Ω, with unconstrained values on $\partial\Omega$, will be considered.

The isoperimetric function $I_{\Omega,\nu}$ is precisely known only in some rather rare situations. This is so e.g. when Ω is an Euclidean ball[55] or when it is the entire $\mathbb{R}^n$ equipped with the Gauss measure[56]. For our purpose, however, only the asymptotic behavior of $I_{\Omega,\nu}$ near 0 is needed, and that can be evaluated for various classes of domains, for example for Lipschitz domains, John domains or for the space $\mathbb{R}^n$ equipped with product probability measures.

Let $m \in \mathbb{N}$ and let $X(\Omega, \nu)$ be a rearrangement-invariant space. We define the m-th order Sobolev space $V^m X(\Omega, \nu)$ as

$$V^m X(\Omega, \nu) = \{u : u \text{ is } m\text{-times weakly differentiable in } \Omega, \\ \text{and } |\nabla^m u| \in X(\Omega, \nu)\}. \quad (1)$$

We also define the subspace $V_\perp^m X(\Omega, \nu)$ of $V^m X(\Omega, \nu)$ as

$$V_\perp^m X(\Omega, \nu) = \left\{u \in V^m X(\Omega, \nu) : \int_\Omega \nabla^k u \, d\nu(x) = 0, \right. \\ \left. \text{for } k = 0, \dots, m-1 \right\}. \quad (2)$$

Here, $\nabla^m u$ denotes the vector of all m-th order weak derivatives of u. Let us notice that in the definition of $V^m X(\Omega, \nu)$ it is only required that the derivatives of the highest order m of u belong to $X(\Omega, \nu)$. This assumption does not entail, in general, that also u and its derivatives up to the order $m-1$ belong to $X(\Omega, \nu)$, and even to $L_1(\Omega, \nu)$. Thus, it may happen that $V^m X(\Omega, \nu) \nsubseteq V^k X(\Omega, \nu)$ for $m > k$.

As already mentioned, our approach to reduction principles for higher-order Sobolev embeddings relies on the iteration of first-order results. A natural worry here is of course the question whether optimality is preserved under iteration. Examples show that when optimality is considered within standard families of function spaces, this is not necessarily so, even in the basic setting of Euclidean domains with Lipschitz boundaries. For instance, for a regular domain Ω in $\mathbb{R}^2$, one has

$$V^2 L_1(\Omega) \to L_\infty(\Omega). \quad (3)$$

On the other hand, iteration of two consequent first-order Sobolev embeddings, optimal within Lebesgue spaces, only gives

$$V^2 L_1(\Omega) \to V^1 L_2(\Omega) \to L_q(\Omega) \quad (4)$$

for every $q < \infty$. The space $V^1 L_2(\Omega)$ contains unbounded functions, hence there must be a loss of information in the iteration process.

One might relate the loss of optimality in the chain of embeddings (4) to the lack of an optimal Lebesgue target space for the first-order Sobolev embedding of $V^1 L_2(\Omega)$ when $n = 2$. However, similar loss can happen in situations where the optimal first-order target spaces do exist. Consider, for example, Euclidean Sobolev embeddings involving Orlicz spaces. In this setting the optimal target space always exists, and it can be explicitly determined[6,57]. Indeed, if Ω is a regular domain in $\mathbb{R}^n$ and $1 \leq m < n$, then

$$V^m L_{\frac{n}{m}}(\Omega) \to \exp L_{\frac{n}{n-m}}(\Omega) \tag{5}$$

[2,4,58]; see also[1] for $m = 1$. Here, $\exp L_\alpha(\Omega)$, with $\alpha > 0$, denotes the Orlicz space associated with the Young function given by $e^{t^\alpha} - 1$ for $t \geq 0$. The target space in (5) is known to be optimal in the class of all Orlicz spaces[6,28]. Assume, for example, that $n \geq 3$ and $m = 2$. Then (5) reduces to

$$V^2 L_{\frac{n}{2}}(\Omega) \to \exp L_{\frac{n}{n-2}}(\Omega). \tag{6}$$

However, iterating first-order embeddings, optimal in Orlicz spaces, one gets only

$$V^2 L_{\frac{n}{2}}(\Omega) \to V^1 L_n(\Omega) \to \exp L_{\frac{n}{n-1}}(\Omega) \supsetneqq \exp L_{\frac{n}{n-2}}(\Omega). \tag{7}$$

Thus, subsequent applications of first-order optimal Sobolev embeddings even in the class of Orlicz spaces, where optimal target space always exists, need not result in optimal higher-order Sobolev embeddings.

The underlying idea behind the method that we shall develop is that such a loss of optimality of the target space under iteration does not occur, provided that first-order (in fact, any-order) Sobolev embeddings whose targets are *optimal among all rearrangement-invariant spaces are iterated.* We thus proceed via a two-step argument, which can be outlined as follows. Firstly, given any r.i. space $X(\Omega)$ and the isoperimetric function $I_{\Omega,\nu}$ of (Ω, ν), the optimal target among all rearrangement-invariant function norms for the first-order Sobolev space $V^1X(\Omega,\nu)$ is characterized; secondly, first-order Sobolev embeddings with an optimal target are iterated to derive optimal targets in arbitrary-order Sobolev embeddings.

Returning back to our examples, we may recall that the optimal r.i. range partner for $V^1L_1(\Omega)$ is not the Lebesgue space $L^2(\Omega)$ but the (essentially smaller) Lorentz space $L_{2,1}(\Omega)$, and then we get

$$V^2 L_1(\Omega) \to V^1 L_{2,1}(\Omega) \to L_\infty(\Omega), \tag{8}$$

with no loss of information or optimality. Similarly, iterating the (optimal) embeddings

$$V^1 L_{\frac{n}{2}}(\Omega) \to V^1 L_{n,\frac{n}{2}}(\Omega) \quad \text{and} \quad V^1 L_{n,\frac{n}{2}}(\Omega) \to L^{\infty,\frac{n}{2};-1}(\Omega),$$

where $L^{\infty,\frac{n}{2};-1}(\Omega)$ is the so-called *Lorentz–Zygmund space*, we get

$$V^2 L_1(\Omega) \to L^{\infty,\frac{n}{2};-1}(\Omega),$$

which, thanks to the (strict) embedding $L^{\infty,\frac{n}{2};-1}(\Omega) \hookrightarrow \exp L_{\frac{n}{n-2}}(\Omega)$ is even better result than (6), we see that, again, no information has been lost.

It turns out that the Lipschitz property of the underlying domain Ω is not necessary for obtaining sharp Sobolev embeddings. Both the reduction principle and the optimal range construction work just fine with no changes when Ω is only required that its isoperimetric function $I_\Omega(s)$ satisfies

$$I_\Omega(s) \approx s^{\frac{1}{n'}} \tag{9}$$

near 0, where $n' = \frac{n}{n-1}$. This condition, in fact, defines the class of the so-called *John domains*, which are more general than the Lipschitz ones.

The isoperimetric inequality relative to (Ω, ν) tells us that

$$P_\nu(E,\Omega) \geq I_{\Omega,\nu}(\nu(E)), \tag{10}$$

where E is any measurable set $E \subset \Omega$, and $P_\nu(E,\Omega)$ stands for its perimeter in Ω with respect to ν. Moreover, $I_{\Omega,\nu}$ denotes the largest non-decreasing function in $[0, \frac{1}{2}]$ for which (10) holds, called the isoperimetric function (or isoperimetric profile) of (Ω, ν), which was introduced in [46].

The results concerning optimality of function spaces in Sobolev embeddings depend only on a lower bound for the isoperimetric function $I_{\Omega,\nu}$ of (Ω, ν) in terms of some other non-decreasing function

$$I : [0,1] \to [0,\infty).$$

Precisely, on the existence of a positive constant c such that

$$I_{\Omega,\nu}(s) \geq cI(cs) \quad \text{for } s \in [0, \tfrac{1}{2}]. \tag{11}$$

First, it can be observed that if $I_{\Omega,\nu}(s)$ does not decay at 0 faster than linearly, namely if there exists a positive constant C such that

$$I_{\Omega,\nu}(s) \geq Cs \quad \text{for } s \in [0, \tfrac{1}{2}], \tag{12}$$

then any function $u \in V^m X(\Omega, \nu)$ does at least belong to $L_1(\Omega, \nu)$, together with all its derivatives up to the order $m-1$. In the light of (12), we can safely assume that

$$\inf_{t \in (0,1)} \frac{I(t)}{t} > 0. \tag{13}$$

We shall now state the main general reduction principle.

Theorem 11.1. *Assume that* (Ω, ν) *fulfils* (11) *for some non-decreasing function* I *satisfying* (13). *Let* $m \in \mathbb{N}$, *and let* $\|\cdot\|_{X(0,1)}$ *and* $\|\cdot\|_{Y(0,1)}$ *be* *rearrangement-invariant function norms. If there exists a constant* C_1 *such that*

$$\left\| \int_t^1 \frac{f(s)}{I(s)} \left(\int_t^s \frac{dr}{I(r)} \right)^{m-1} ds \right\|_{Y(0,1)} \leq C_1 \|f\|_{X(0,1)} \tag{14}$$

for every nonnegative $f \in X(0,1)$, *then*

$$V^m X(\Omega,\nu) \to Y(\Omega,\nu), \tag{15}$$

and there exists a constant C_2 *such that*

$$\|u\|_{Y(\Omega,\nu)} \leq C_2 \|\nabla^m u\|_{X(\Omega,\nu)} \tag{16}$$

for every $u \in V_\perp^m X(\Omega,\nu)$.

It turns out that inequality (14) holds for every nonnegative $f \in X(0,1)$ if and only if it just holds for every nonnegative and non-increasing $f \in X(0,1)$.

A major feature of Theorem 11.1 is the difference occurring in (14) between the first-order case ($m = 1$) and the higher-order case ($m > 1$). Indeed, the integral operator appearing in (14) when $m = 1$ is just a weighted Hardy-type operator, namely a primitive of f times a weight, whereas, in the higher-order case, a genuine kernel, with a more complicated structure, comes into play. In fact, this seems to be the first known instance where such a kernel operator is needed in a reduction principle for Sobolev-type embeddings.

As we shall see, the Sobolev embedding (15) (or the Poincaré inequality (16)) and inequality (14) are actually equivalent in customary families of measure spaces (Ω,ν). This fact allows us to determine the optimal rearrangement-invariant target spaces in Sobolev embeddings for these measure spaces. Incidentally, let us mention that when $m = 1$, this is the case whenever the geometry of (Ω,ν) allows the construction of a family of trial functions u in (15) or (16) characterized by the following properties: the level sets of u are isoperimetric (or almost isoperimetric) in (Ω,ν); $|\nabla u|$ is constant (or almost constant) on the boundary of the level sets of u. If $m > 1$, then the latter requirement has to be complemented by requiring that the derivatives of u up to the order m restricted to the boundary of the level sets satisfy certain conditions depending on I.

Such conditions have, however, a technical nature, and it is not worth to state them explicitly. In fact, heuristically speaking, properties (14), (16) and (15) turn out to be equivalent for every $m \geq 1$ on the same measure spaces (Ω,ν) as for $m = 1$. Such an equivalence certainly holds in any customary, non-pathological situation, including the three frameworks to which our results will be applied, namely John domains, Euclidean domains from Maz'ya classes and product probability spaces in $\mathbb{R}^n$ extending the Gauss space. In all these cases, we can characterize optimal arbitrary-order rearrangement-invariant target spaces.

Now we are in a position to characterize the space which is the optimal rearrangement-invariant target space in the Sobolev embedding (15). Such an optimal space is the one associated with the rearrangement-invariant function norm $\|\cdot\|_{X_{m,I}(0,1)}$, whose associate norm is defined as

$$\|f\|_{X'_{m,I}(0,1)} = \left\| \frac{1}{I(s)} \int_0^s \left(\int_t^s \frac{dr}{I(r)} \right)^{m-1} f^*(t)dt \right\|_{X'(0,1)} \tag{17}$$

for $f \in \mathcal{M}_+(0,1)$.

Theorem 11.2. *Assume that* (Ω, ν), m, I *and* $\|\cdot\|_{X(0,1)}$ *are as in Theorem 11.1. Then the functional* $\|\cdot\|_{X'_{m,I}(0,1)}$, *given by* (17), *is a rearrangement-invariant function norm, whose associate norm* $\|\cdot\|_{X_{m,I}(0,1)}$ *satisfies*

$$V^m X(\Omega, \nu) \to X_{m,I}(\Omega, \nu), \tag{18}$$

and there exists a constant C *such that*

$$\|u\|_{X_{m,I}(\Omega,\nu)} \le C \, \|\nabla^m u\|_{X(\Omega,\nu)} \tag{19}$$

for every $u \in V_\perp^m X(\Omega, \nu)$.

Moreover, if (Ω, ν) *is such that* (15), *or equivalently* (16), *implies* (14), *and hence* (14), (15) *and* (16) *are equivalent, then the function norm* $\|\cdot\|_{X_{m,I}(0,1)}$ *is optimal in* (18) *and* (19) *among all rearrangement-invariant norms.*

An important special case of Theorems 11.1 and 11.2 is enucleated in the following corollary.

Corollary 11.1. *Assume that* (Ω, ν), m, I *and* $\|\cdot\|_{X(0,1)}$ *are as in Theorem 11.1. If*

$$\left\| \frac{1}{I(s)} \left(\int_0^s \frac{dr}{I(r)} \right)^{m-1} \right\|_{X'(0,1)} < \infty, \tag{20}$$

then

$$V^m X(\Omega, \nu) \to L_\infty(\Omega, \nu), \tag{21}$$

and there exists a constant C such that

$$\|u\|_{L_\infty(\Omega,\nu)} \le C \,\|\nabla^m u\|_{X(\Omega,\nu)} \tag{22}$$

for every $u \in V_\perp^m X(\Omega,\nu)$.

Moreover, if (Ω,ν) is such that (15)*, or equivalently* (16)*, implies* (14)*, and hence* (14)*,* (15) *and* (16) *are equivalent, then* (20) *is necessary for* (21) *or* (22) *to hold.*

If (Ω,ν) is such that (15), or equivalently (16), implies (14), and hence (14), (15) and (16) are equivalent, then (21) cannot hold, whatever $\|\cdot\|_{X(0,1)}$ is, if I decays so fast at 0 that

$$\int_0 \frac{dr}{I(r)} = \infty.$$

We shall now point out the preservation of optimality in targets among all rearrangement-invariant spaces under iteration of Sobolev embeddings of arbitrary order.

Theorem 11.3. *Assume that (Ω,ν), I and $\|\cdot\|_{X(0,1)}$ are as in Theorem 11.1. Let $k,h \in \mathbb{N}$. Then*

$$(X_{k,I})_{h,I}(\Omega,\nu) = X_{k+h,I}(\Omega,\nu), \tag{23}$$

up to equivalent norms.

In many instances in practice, the function I satisfies the estimate

$$\int_0^s \frac{dr}{I(r)} \approx \frac{s}{I(s)} \quad \text{for } s \in (0,1). \tag{24}$$

We note that (24) is not true for every relevant case. It holds for instance for John domains and for domains from Maz'ya classes $\mathcal{J}_\alpha$ with $\alpha < 1$, but it does not hold for domains in $\mathcal{J}_1$ or for the Gaussian space.

It is useful to treat the cases for which (24) holds separately because then the results of Theorems 11.1, 11.2 and 11.3 can be considerably simplified. For example, the reduction theorem then reads as follows.

Corollary 11.2. *Let (Ω,ν), m, I, $\|\cdot\|_{X(0,1)}$ and $\|\cdot\|_{Y(0,1)}$ be as in Theorem 11.1. Assume, in addition, that I fulfills* (24)*. If there exists a constant C_1 such that*

$$\left\| \int_t^1 f(s) \frac{s^{m-1}}{I(s)^m}\, ds \right\|_{Y(0,1)} \le C_1 \,\|f\|_{X(0,1)} \tag{25}$$

for every nonnegative $f \in X(0,1)$*, then*

$$V^m X(\Omega,\nu) \to Y(\Omega,\nu), \tag{26}$$

and there exists a constant C_2 *such that*

$$\|u\|_{Y(\Omega,\nu)} \leq C_2 \|\nabla^m u\|_{X(\Omega,\nu)} \tag{27}$$

for every $u \in V_\perp^m X(\Omega,\nu)$*.*

Given an r.i. space $X(\Omega)$, the corresponding optimal range partner in the Sobolev embedding on a John domain is then the space $X^\sharp_{m,I}(\Omega)$, whose associate space has norm

$$\|f\|_{(X^\sharp_{m,I})'(\Omega)} = \left\| \frac{t^{m-1}}{I(t)^m} \int_0^t f^*(s)\,ds \right\|_{X'(0,1)} \tag{28}$$

for every $f \in \mathcal{M}_+(0,1)$.

Let us now deal with specific situations separately.

We say that a bounded open set Ω in $\mathbb{R}^n$ is called a *John domain* if there exist a constant $c \in (0,1)$ and a point $x_0 \in \Omega$ such that for every $x \in \Omega$ there exists a rectifiable curve $\varpi : [0,l] \to \Omega$, parameterized by the arc length and such that $\varpi(0) = x$, $\varpi(l) = x_0$, and

$$\operatorname{dist}(\varpi(r), \partial\Omega) \geq cr \qquad \text{for } r \in [0,l].$$

The class of John domains includes other more classical families of domains, such as Lipschitz domains, and domains with the cone property. The John domains arise in connection with the study of holomorphic dynamical systems and quasiconformal mappings. John domains are known to support a first-order Sobolev inequality with the same exponents as in the standard Sobolev inequality[59–61]. In fact, being a John domain is a necessary condition for such a Sobolev inequality to hold in the class of two-dimensional simply connected open sets, and in quite general classes of higher dimensional domains[62]. We note that, as a consequence of (9), the corresponding function I now satisfies (24). The reduction principle thus takes the following form.

Theorem 11.4. *Let* $n \in \mathbb{N}$*,* $n \geq 2$*, and let* $m \in \mathbb{N}$*. Assume that* Ω *is a John domain in* $\mathbb{R}^n$*. Let* $\|\cdot\|_{X(0,1)}$ *and* $\|\cdot\|_{Y(0,1)}$ *be rearrangement-invariant function norms. Then the following assertions are equivalent.*

(i) *The Hardy type inequality*

$$\left\| \int_t^1 f(s) s^{-1+\frac{m}{n}}\,ds \right\|_{Y(0,1)} \leq C_1 \|f\|_{X(0,1)} \tag{29}$$

holds for some constant C_1, *and for every nonnegative* $f \in X(0,1)$.

(ii) *The Sobolev embedding*

$$V^m X(\Omega) \to Y(\Omega) \tag{30}$$

holds.

(iii) *The Poincaré inequality*

$$\|u\|_{Y(\Omega)} \le C_2 \|\nabla^m u\|_{X(\Omega)} \tag{31}$$

holds for some constant C_2 *and every* $u \in V_\perp^m X(\Omega)$.

Given an r.i. space $X(\Omega)$ where Ω is a John domain, the corresponding optimal range partner in the Sobolev embedding on a John domain is then the space $X_{m,John}(\Omega)$, whose associate space has norm

$$\|f\|_{X'_{m,John}(\Omega)} = \left\| s^{-1+\frac{m}{n}} \int_0^s f^*(r)dr \right\|_{X'(0,1)} \tag{32}$$

Our next set of instances will be *Maz'ya classes* of Euclidean domains. Given $\alpha \in [\frac{1}{n'}, 1]$, we denote by $\mathcal{J}_\alpha$ the *Maz'ya class* of all Euclidean domains Ω satisfying (11), with $I(s) = s^\alpha$ for $s \in [0, \frac{1}{2}]$, namely domains Ω in $\mathbb{R}^n$ such that

$$I_\Omega(s) \ge C s^\alpha \quad \text{for } s \in [0, \tfrac{1}{2}], \tag{33}$$

for some positive constant C. Thanks to (9), any John domain belongs to the class $\mathcal{J}_{\frac{1}{n'}}$.

The reduction theorem in the class $\mathcal{J}_\alpha$ takes the following form.

Theorem 11.5. *Let* $n \in \mathbb{N}$, $n \ge 2$, $m \in \mathbb{N}$, *and* $\alpha \in [\frac{1}{n'}, 1]$. *Let* $\|\cdot\|_{X(0,1)}$ *and* $\|\cdot\|_{Y(0,1)}$ *be rearrangement-invariant function norms. Assume that either* $\alpha \in [\frac{1}{n'}, 1)$ *and there exists a constant* C_1 *such that*

$$\left\| \int_t^1 f(s) s^{-1+m(1-\alpha)}\, ds \right\|_{Y(0,1)} \le C_1 \|f\|_{X(0,1)} \tag{34}$$

for every nonnegative $f \in X(0,1)$, *or* $\alpha = 1$ *and there exists a constant* C_1 *such that*

$$\left\| \int_t^1 f(s) \frac{1}{s} \left(\log \frac{s}{t} \right)^{m-1} ds \right\|_{Y(0,1)} \le C_1 \|f\|_{X(0,1)} \tag{35}$$

for every nonnegative $f \in X(0,1)$. *Then the Sobolev embedding*

$$V^m X(\Omega) \to Y(\Omega) \tag{36}$$

holds for every $\Omega \in \mathcal{J}_\alpha$ and, equivalently, the Poincaré inequality

$$\|u\|_{Y(\Omega)} \le C_2 \|\nabla^m u\|_{X(\Omega)} \tag{37}$$

holds for every $\Omega \in \mathcal{J}_\alpha$, for some constant C_2 and every $u \in V_\perp^m X(\Omega)$.

Conversely, if the Sobolev embedding (36), or, equivalently, the Poincaré inequality (37), holds for every $\Omega \in \mathcal{J}_\alpha$, then either inequality (34), or (35) holds, according to whether $\alpha \in [\frac{1}{n'}, 1)$ or $\alpha = 1$.

Given an r.i. space $X(\Omega)$ where $\Omega \in \mathcal{J}_\alpha$, the corresponding optimal range partner in the Sobolev embedding on a John domain is then the space $X_{m,\alpha}(\Omega)$, whose associate space has norm

$$\|f\|_{X'_{m,\alpha}(\Omega)} = \begin{cases} \left\| s^{-1+m(1-\alpha)} \int_0^s f^*(r)dr \right\|_{X'(0,1)} & \text{if } \alpha \in [\frac{1}{n'}, 1), \\ \left\| \frac{1}{s} \int_0^s \left(\log \frac{s}{r}\right)^{m-1} f^*(r)dr \right\|_{X'(0,1)} & \text{if } \alpha = 1. \end{cases} \tag{38}$$

We shall now apply the general results to some concrete function spaces. We shall mainly focus on Lebesgue, Lorentz and Orlicz spaces.

Theorem 11.6. *Let $n \in \mathbb{N}$, $n \ge 2$, and let $\Omega \in \mathcal{J}_\alpha$ for some $\alpha \in [\frac{1}{n'}, 1)$. Let $m \in \mathbb{N}$ and $p \in [1, \infty]$. Then*

$$V^m L_p(\Omega) \to \begin{cases} L_{\frac{p}{1-mp(1-\alpha)}}(\Omega) & \textit{if } m(1-\alpha) < 1 \\ & \textit{and } 1 \le p < \frac{1}{m(1-\alpha)}, \\ L_r(\Omega) & \textit{for any } r \in [1, \infty), \\ & \textit{if } m(1-\alpha) < 1 \textit{ and } p = \frac{1}{m(1-\alpha)}, \\ L_\infty(\Omega) & \textit{otherwise.} \end{cases} \tag{39}$$

Moreover, in the first and the third cases, the target spaces in (39) are optimal among all Lebesgue spaces, as Ω ranges in $\mathcal{J}_\alpha$.

Although the target spaces in (39) cannot be improved in the class of Lebesgue spaces, the conclusions of (39) can be strengthened if more general rearrangement-invariant spaces are employed. Such a strengthening can be obtained as a special case of a Sobolev embedding for Lorentz spaces which reads as follows.

Theorem 11.7. *Let $n \in \mathbb{N}$, $n \ge 2$, and let $\Omega \in \mathcal{J}_\alpha$ for some $\alpha \in [\frac{1}{n'}, 1)$. Let $m \in \mathbb{N}$ and $p, q \in [1, \infty]$. Assume that one of the conditions in (6)* *holds. Then*

$$V^m L_{p,q}(\Omega) \to \begin{cases} L_{\frac{p}{1-mp(1-\alpha)},q}(\Omega) & \text{if } m(1-\alpha)<1 \\ & \text{and } 1\le p<\frac{1}{m(1-\alpha)}, \\ L_{\infty,q;-1}(\Omega) & \text{if } m(1-\alpha)<1,\ p=\frac{1}{m(1-\alpha)} \\ & \text{and } q>1, \\ L_\infty(\Omega) & \text{otherwise.} \end{cases} \tag{40}$$

Moreover, the target spaces in (40) *are optimal among all rearrangement-invariant spaces, as* Ω *ranges in* $\mathcal{J}_\alpha$.

The particular choice of parameters $p = q$, $1 \le p < \frac{1}{m(1-\alpha)}$ in Theorem 11.7 shows that

$$V^m L^p(\Omega) \hookrightarrow L_{\frac{p}{1-mp(1-\alpha)},p}(\Omega).$$

This is a non-trivial strengthening of the first embedding in (39), since $L_{\frac{p}{1-mp(1-\alpha)},p}(\Omega) \subsetneq L_{\frac{p}{1-mp(1-\alpha)}}$.

Likewise, the choice $m(1-\alpha) < 1$ and $p = q = \frac{1}{m(1-\alpha)}$ shows that also the second embedding in (39) can be in fact essentially improved to

$$V^m L^p(\Omega) \hookrightarrow L^{\infty,p;-1}(\Omega).$$

Assume now that $\alpha = 1$. The embedding theorem in Lebesgue spaces takes the following form.

Theorem 11.8. *Let* $n \in \mathbb{N}$, $n \ge 2$, *and let* $\Omega \in \mathcal{J}_1$. *Let* $m \in \mathbb{N}$ *and* $p \in [1,\infty]$. *Then*

$$V^m L_p(\Omega) \to \begin{cases} L_p(\Omega) & \text{if } 1 \le p < \infty, \\ L_r(\Omega) & \text{for any } r \in [1,\infty), \text{ if } p=\infty. \end{cases} \tag{41}$$

In the former case, the target space is optimal in (41) *among all Lebesgue spaces, as* Ω *ranges in* $\mathcal{J}_1$.

Optimal embeddings for Lorentz-Sobolev spaces are provided in the next theorem.

Theorem 11.9. *Let* $n \in \mathbb{N}$, $n \ge 2$, *and let* $\Omega \in \mathcal{J}_1$. *Let* $m \in \mathbb{N}$ *and* $p, q \in [1,\infty]$. *Assume that one of the conditions in* (6) *holds. Then*

$$V^m L_{p,q}(\Omega) \to \begin{cases} L_{p,q}(\Omega) & \text{if } 1 \le p < \infty, \\ \exp L^{\frac{1}{m}}(\Omega) & \text{if } p=q=\infty. \end{cases} \tag{42}$$

The target spaces are optimal in (42) *among all rearrangement-invariant spaces, as* Ω *ranges in* $\mathcal{J}_1$.

Our last application in this section concerns Orlicz-Sobolev spaces. Let $n \in \mathbb{N}$, $n \geq 2$, $m \in \mathbb{N}$, $\alpha \in [\frac{1}{n'}, 1)$, and let A be a Young function. We may assume, without loss of generality, that

$$\int_0 \left(\frac{t}{A(t)}\right)^{\frac{m(1-\alpha)}{1-m(1-\alpha)}} dt < \infty. \tag{43}$$

Indeed, the function A can be modified near 0, if necessary, in such a way that (43) is fulfilled, on leaving the space $V^m L_A(\Omega)$ unchanged (up to equivalent norms).

If $m < \frac{1}{1-\alpha}$ and

$$\int^\infty \left(\frac{t}{A(t)}\right)^{\frac{m(1-\alpha)}{1-m(1-\alpha)}} dt = \infty, \tag{44}$$

we define the function $H_{m,\alpha} : [0, \infty) \to [0, \infty)$ as

$$H_{m,\alpha}(s) = \left(\int_0^s \left(\frac{t}{A(t)}\right)^{\frac{m(1-\alpha)}{1-m(1-\alpha)}} dt\right)^{1-m(1-\alpha)} \qquad \text{for } s \geq 0, \tag{45}$$

and the Young function $A_{m,\alpha}$ as

$$A_{m,\alpha}(t) = A(H_{m,\alpha}^{-1}(t)) \qquad \text{for } t \geq 0. \tag{46}$$

Theorem 11.10. *Assume that* $n \in \mathbb{N}$, $n \geq 2$, $m \in \mathbb{N}$, $\alpha \in [\frac{1}{n'}, 1)$ *and* $\Omega \in \mathcal{J}_\alpha$. *Let* A *be a Young function fulfilling* (43). *Then*

$$V^m L_A(\Omega) \to \begin{cases} L_{A_{m,\alpha}}(\Omega) & \text{if } m < \frac{1}{1-\alpha} \text{ and (44) holds,} \\ L_\infty(\Omega) & \text{if either } m \geq \frac{1}{1-\alpha} \\ & \text{or } m < \frac{1}{1-\alpha} \text{ and (44) fails.} \end{cases} \tag{47}$$

Moreover, the target spaces in (47) *are optimal among all Orlicz spaces, as* Ω *ranges in* $\mathcal{J}_\alpha$.

Example 11.1. Consider the case when $A(t) \approx t^p(\log t)^\beta$ near infinity, where either $p > 1$ and $\beta \in R$, or $p = 1$ and $\beta \geq 0$. Hence,

$L_A(\Omega) = L_p \log^\beta(\Omega)$. An application of Theorem 11.10 tells us that

$$V^m L_p \log^\beta(\Omega) \to \begin{cases} L_{\frac{p}{1-pm(1-\alpha)}} \log^{\frac{\beta}{1-pm(1-\alpha)}}(\Omega) & \text{if } mp(1-\alpha) < 1, \\ \exp L_{\frac{1}{1-(1+\beta)m(1-\alpha)}}(\Omega) & \text{if } mp(1-\alpha) = 1 \\ & \text{and } \beta < \frac{1-m(1-\alpha)}{m(1-\alpha)}, \\ \exp \exp L_{\frac{1}{1-m(1-\alpha)}}(\Omega) & \text{if } mp(1-\alpha) = 1 \\ & \text{and } \beta = \frac{1-m(1-\alpha)}{m(1-\alpha)}, \\ L_\infty(\Omega) & \text{if either } mp(1-\alpha) > 1, \\ & \text{or } mp(1-\alpha) = 1 \\ & \text{and } \beta > \frac{1-m(1-\alpha)}{m(1-\alpha)}. \end{cases} \tag{48}$$

Moreover, the target spaces in (48) are optimal among all Orlicz spaces, as Ω ranges in $\mathcal{J}_\alpha$.

The first three embeddings in (48) can be improved on allowing more general rearrangement-invariant target spaces. Indeed, we have that

$$V^m L_p \log^\beta(\Omega) \to \begin{cases} L_{\frac{p}{1-pm(1-\alpha)}, p; \frac{\beta}{p}}(\Omega) & \text{if } mp(1-\alpha) < 1, \\ L_{\infty, \frac{1}{m(1-\alpha)}; m(1-\alpha)\beta - 1}(\Omega) & \text{if } mp(1-\alpha) = 1 \\ & \text{and } \beta < \frac{1-m(1-\alpha)}{m(1-\alpha)}, \\ L_{\infty, \frac{1}{m(1-\alpha)}; -m(1-\alpha), -1}(\Omega) & \text{if } mp(1-\alpha) = 1 \\ & \text{and } \beta = \frac{1-m(1-\alpha)}{m(1-\alpha)}, \end{cases} \tag{49}$$

the targets being optimal among all rearrangement-invariant spaces in (49) as Ω ranges among all domains in $\mathcal{J}_\alpha$.

Our final set of examples will concern the *product probability spaces.*

The class of product probability measures in $\mathbb{R}^n$, $n \geq 1$, arises in connection with the study of generalized hypercontractivity theory and integrability properties of the associated heat semigroups. The isoperimetric problem in the corresponding probability spaces was studied in [63], see also [64–67].

Assume that $\Phi : [0,\infty) \to [0,\infty)$ is a strictly increasing, twice continuously differentiable convex function in $(0,\infty)$ such that $\sqrt{\Phi}$ is concave, and $\Phi(0) = 0$. Let μ_Φ be the probability measure on $\mathbb{R}$ given by

$$d\mu_\Phi(x) = c_\Phi e^{-\Phi(|x|)}\, dx, \tag{50}$$

where c_Φ is a constant chosen in such a way that $\mu_\Phi(\mathbb{R}) = 1$. The product measure $\mu_{\Phi,n}$ on $\mathbb{R}^n$, $n \geq 1$, generated by μ_Φ, is then defined as

$$\mu_{\Phi,n} = \underbrace{\mu_\Phi \times \cdots \times \mu_\Phi}_{n-\text{times}}. \tag{51}$$

Clearly, $\mu_{\Phi,1} = \mu_\Phi$, and $(\mathbb{R}^n, \mu_{\Phi,n})$ is a probability space for every $n \in \mathbb{N}$.

The main example of a measure μ_Φ is obtained by taking

$$\Phi(t) = \tfrac{1}{2}t^2.$$

This choice yields $\mu_{\Phi,n} = \gamma_n$, the Gauss measure which obeys

$$d\gamma_n(x) = (2\pi)^{-\frac{n}{2}} e^{-\frac{|x|^2}{2}} dx. \tag{52}$$

More generally, the *Boltzmann measures* $\gamma_{n,\beta}$ are associated with

$$\Phi(t) = \tfrac{1}{\beta}t^\beta$$

for some $\beta \in [1,2]$. Let $H : \mathbb{R} \to (0,1)$ be defined as

$$H(t) = \int_t^\infty c_\Phi e^{-\Phi(|r|)}\, dr \quad \text{for } t \in \mathbb{R}, \tag{53}$$

and let $F_\Phi : [0,1] \to [0,\infty)$ be given by

$$F_\Phi(s) = c_\Phi e^{-\Phi(|H^{-1}(s)|)} \quad \text{for } s \in (0,1), \quad \text{and} \quad F_\Phi(0) = F_\Phi(1) = 0. \tag{54}$$

Since μ_Φ is a probability measure and $\mu_{\Phi,n}$ is defined by (51), it is easily seen that, for each $i = 1, \dots, n$,

$$\mu_{\Phi,n}(\{(x_1, \dots, x_n) : x_i > t\}) = H(t) \quad \text{for } t \in \mathbb{R}, \tag{55}$$

and

$$P_{\mu_{\Phi,n}}(\{(x_1, \dots, x_n) : x_i > t\}, \mathbb{R}^n) = c_\Phi e^{-\Phi(|t|)} = -H'(t) \quad \text{for } t \in \mathbb{R}. \tag{56}$$

Hence, $F_\Phi(s)$ agrees with the perimeter of any half-space of the form $\{x_i > t\}$, whose measure is s.

Next, define $L_\Phi : [0,1] \to [0,\infty)$ as

$$L_\Phi(s) = s\Phi'\big(\Phi^{-1}(\log\big(\tfrac{2}{s}\big))\big) \qquad \text{for } s \in (0,1], \quad \text{and} \quad L_\Phi(0) = 0. \tag{57}$$

Then the isoperimetric function of $(\mathbb{R}^n, \mu_{\Phi,n})$ satisfies

$$I_{(\mathbb{R}^n, \mu_{\Phi,n})}(s) \approx F_\Phi(s) \approx L_\Phi(s) \quad \text{for } s \in [0, \tfrac{1}{2}] \tag{58}$$

(see[63] Proposition 13 and Theorem 15). Furthermore, half-spaces, whose boundary is orthogonal to a coordinate axis, are “approximate solutions” to the isoperimetric problem in $(\mathbb{R}^n, \mu_{\Phi,n})$ in the sense that there exist constants C_1 and C_2, depending on n, such that, for every $s \in (0,1)$, any such half-space V with measure s satisfies

$$C_1 P_{\mu_{\Phi,n}}(V,\mathbb{R}^n) \le I_{(\mathbb{R}^n,\mu_{\Phi,n})}(s) \le C_2 P_{\mu_{\Phi,n}}(V,\mathbb{R}^n).$$

In the special case when $\mu_{\Phi,n} = \gamma_n$, the Gauss measure, equation (58) yields

$$I_{(\mathbb{R}^n,\gamma_n)}(s) \approx s\big(\log \tfrac{2}{s}\big)^{\frac{1}{2}} \quad \text{for } s \in (0,\tfrac{1}{2}]. \tag{59}$$

Note that now (24) is not satisfied. Moreover, any half-space is, in fact, an exact minimizer in the isoperimetric inequality[59,68].

The reduction theorem for Sobolev embeddings in product probability spaces reads as follows.

Theorem 11.11. *Let $n \in \mathbb{N}$, $m \in \mathbb{N}$, let $\mu_{\Phi,n}$ be the probability measure defined by* (51)*, and let $\|\cdot\|_{X(0,1)}$ and $\|\cdot\|_{Y(0,1)}$ be rearrangement-invariant function norms. Then the following facts are equivalent.*

(i) *The inequality*

$$\left\| \int_s^1 f(r) \frac{\left(\Phi^{-1}\left(\log \frac{2}{s}\right) - \Phi^{-1}\left(\log \frac{2}{r}\right)\right)^{m-1}}{r\Phi'\left(\Phi^{-1}\left(\log \frac{2}{r}\right)\right)}\, dr \right\|_{Y(0,1)} \le C_1 \|f\|_{X(0,1)} \tag{60}$$

holds for some constant C_1, and for every nonnegative $f \in X(0,1)$.

(ii) *The embedding*

$$V^m X(\mathbb{R}^n,\mu_{\Phi,n}) \to Y(\mathbb{R}^n,\mu_{\Phi,n}) \tag{61}$$

holds.

(iii) *The Poincaré inequality*

$$\|u\|_{Y(\mathbb{R}^n,\mu_{\Phi,n})} \le C_2 \|\nabla^m u\|_{X(\mathbb{R}^n,\mu_{\Phi,n})} \tag{62}$$

holds for some constant C_2, and for every $u \in V^m_\perp X(\mathbb{R}^n,\mu_{\Phi,n})$.

When $m = 1$ and the measure $\mu_{\Phi,n}$ agrees with the Gauss measure γ_n, the result of Theorem 11.11 is by now standard (see e.g.[45]).

The rearrangement-invariant function norm $\|\cdot\|_{X_{m,\Phi}(0,1)}$ which yields the optimal rearrangement-invariant target space $Y(\mathbb{R}^n,\mu_{\Phi,n})$ in embedding (61) is defined as follows. Let $\|\cdot\|_{X(0,1)}$ be a rearrangement-invariant function norm, and let $n, m \in \mathbb{N}$. Then $\|\cdot\|_{X_{m,\Phi}(0,1)}$ is the rearrangement-invariant function norm whose associate function norm is given by

$$\|f\|_{X'_{m,\Phi}(0,1)} = \left\| \int_0^r f^*(s) \frac{\left(\Phi^{-1}\left(\log \frac{2}{s}\right) - \Phi^{-1}\left(\log \frac{2}{r}\right)\right)^{m-1}}{r\Phi'\left(\Phi^{-1}\left(\log \frac{2}{r}\right)\right)}\, ds \right\|_{X'(0,1)} \tag{63}$$

for $f \in \mathcal{M}_+(0,1)$.

The reduction theorem takes a simpler form in the case of Gaussian measure.

Theorem 11.12. *Let $X(\mathbb{R}^n,\gamma_n)$ and $Y(\mathbb{R}^n,\gamma_n)$ be r.i. spaces, and let $m \geq 1$. There exists a constant C_1 such that*

$$\|u\|_{Y(\mathbb{R}^n,\gamma_n)} \leq C_1 \|\nabla^m u\|_{X(\mathbb{R}^n,\gamma_n)}$$

for every $u \in V_\perp^m X(\mathbb{R}^n,\gamma_n)$ if and only if there exists a constant C_2 such that

$$\left\| \frac{1}{\left(1+\log\frac{1}{s}\right)^{\frac{m-1}{2}}} \int_s^1 f(r) \frac{\left(\log\frac{r}{s}\right)^{m-1}}{r\left(1+\log\frac{1}{r}\right)^{1/2}}\, dr \right\|_{Y(0,1)} \leq C_2 \|f\|_{X(0,1)}$$

for every $f \in X(0,1)$.

Given $n,m \in \mathbb{N}$, and a rearrangement-invariant function norm $\|\cdot\|_{X(0,1)}$, define $\|\cdot\|_{X_{m,G}(0,1)}$ as the rearrangement-invariant function norm whose associate function norm is given by

$$\|f\|_{X'_{m,G}(0,1)} = \left\| \frac{1}{r\left(\log\frac{2}{r}\right)^{1/2}} \int_0^r f^*(s) \frac{\left(\log\frac{r}{s}\right)^{m-1}}{\left(\log\frac{2}{s}\right)^{\frac{m-1}{2}}}\, ds \right\|_{X'(0,1)} \tag{64}$$

for $f \in \mathcal{M}_+(0,1)$.

Theorem 11.13. *Let $n \in \mathbb{N}$, $m \in \mathbb{N}$, and let $\|\cdot\|_{X(0,1)}$ be a rearrangement-invariant function norm. Then the functional $\|\cdot\|_{X'_{m,G}(0,1)}$, given by (64), is a rearrangement-invariant function norm, whose associate norm $\|\cdot\|_{X_{m,G}(0,1)}$ satisfies*

$$V^m X(\mathbb{R}^n,\gamma_n) \to X_{m,G}(\mathbb{R}^n,\gamma_n) \tag{65}$$

with norm independent of n, and

$$\|u\|_{X_{m,G}(\mathbb{R}^n,\gamma_n)} \leq C \|\nabla^m u\|_{X(\mathbb{R}^n,\gamma_n)} \tag{66}$$

for some constant C independent of n, for every $u \in V_\perp^m X(\mathbb{R}^n,\gamma_n)$. Moreover, the function norm $\|\cdot\|_{X_{m,G}(0,1)}$ is optimal in (65) and (66) among all rearrangement-invariant norms.

We finish this section with an application of the results of this section to the particular case when $\mu_{\Phi,n}$ is a Boltzmann measure, and the norms are of Lorentz–Zygmund type.

Theorem 11.14. *Let $n, m \in \mathbb{N}$, let $\beta \in [1,2]$ and let $p, q \in [1,\infty]$ and $\alpha \in \mathbb{R}$ be such that one of the conditions in (6) is satisfied. Then*

$$V^m L^{p,q;\alpha}(\mathbb{R}^n, \gamma_{n,\beta}) \to \begin{cases} L^{p,q;\alpha+\frac{m(\beta-1)}{\beta}}(\mathbb{R}^n, \gamma_{n,\beta}) & \text{if} \quad p < \infty; \\ L^{\infty,q;\alpha-\frac{m}{\beta}}(\mathbb{R}^n, \gamma_{n,\beta}) & \text{if} \quad p = \infty. \end{cases} \tag{67}$$

Moreover, in both cases, the target space is optimal among all rearrangement-invariant spaces.

When $\beta = 2$, Theorem 11.14 yields the following sharp Sobolev type embeddings in Gauss space.

Theorem 11.15. *Let $n, m \in \mathbb{N}$, and let $p, q \in [1,\infty]$ and $\alpha \in \mathbb{R}$ be such that one of the conditions in (6) is satisfied. Then*

$$V^m L^{p,q;\alpha}(\mathbb{R}^n, \gamma_n) \to \begin{cases} L^{p,q;\alpha+\frac{m}{2}}(\mathbb{R}^n, \gamma_n) & \text{if} \quad p < \infty; \\ L^{\infty,q;\alpha-\frac{m}{2}}(\mathbb{R}^n, \gamma_n) & \text{if} \quad p = \infty. \end{cases}$$

Moreover, in both cases, the target space is optimal among all rearrangement-invariant spaces.

A further specialization of the indices p, q, α appearing in Theorem 11.15 leads to the following basic embeddings.

Corollary 11.3. *Let $n, m \in \mathbb{N}$.*

(i) *Assume that $p \in [1,\infty)$. Then*

$$V^m L^p(\mathbb{R}^n, \gamma_n) \to L^p(\log L)^{\frac{mp}{2}}(\mathbb{R}^n, \gamma_n),$$

and the target space is optimal among all rearrangement-invariant spaces.

(ii) *Assume that $\gamma > 0$. Then*

$$V^m \exp L_\gamma(\mathbb{R}^n, \gamma_n) \to \exp L_{\frac{2\gamma}{2+m\gamma}}(\mathbb{R}^n, \gamma_n),$$

and the target space is optimal among all rearrangement-invariant spaces.

(iii)

$$V^m L_\infty(\mathbb{R}^n, \gamma_n) \to \exp L_{\frac{2}{m}}(\mathbb{R}^n, \gamma_n),$$

and the target space is optimal among all rearrangement-invariant spaces.

12. Optimal trace embeddings

One of the main applications of Sobolev space techniques is in the field of traces of functions defined on domains.

One important property enjoyed by functions from the Sobolev space $W^{m,p}(\mathbb{R}^n)$, $m \in \mathbb{N}$, $p \in [1,\infty]$, is that their restrictions, called traces, to lower dimensional spaces can be properly defined, provided that the dimension d of the relevant subspaces is not too small, depending on n, m and p. The trace of a function $u \in W^{m,p}(\mathbb{R}^n)$ turns out to be measurable with respect to the d-dimensional measure on the relevant subspaces, and also integrable to some power q, depending on n, m, p and d. Loosely speaking, increasing the values of m and p causes u to be more regular, and hence allows smaller values of d and larger values of q.

To be more specific, let $n, d \in \mathbb{N}$, and let $n \geq 2$ and $1 \leq d < n$. Since $\mathbb{R}^n = \mathbb{R}^d \times \mathbb{R}^{n-d}$, any point $x \in \mathbb{R}^n$ can be represented as $x = (y, z)$, with $y \in \mathbb{R}^d$ and $z \in \mathbb{R}^{n-d}$. Moreover, $\mathbb{R}^d$ can be identified with the subspace of those points in $\mathbb{R}^n$ having the form $(y, 0)$ for some $y \in \mathbb{R}^d$.

Various approaches to traces of functions are available in the literature. We shall adopt the following definition, which extends more customary notions of traces of functions in Sobolev spaces – see[69] Chapt. 5. A function $u \in L_{1,\mathrm{loc}}(\mathbb{R}^n)$ is said to have a trace $\operatorname{Tr} u \in L_{1,\mathrm{loc}}(\mathbb{R}^d)$ on $\mathbb{R}^d$ if there exists a function $\overline{u}$, equivalent to u on $\mathbb{R}^n$, such that

$$\lim_{z\to 0} \overline{u}(\cdot, z) = \operatorname{Tr} u(\cdot) \qquad \text{in } L_{1,\mathrm{loc}}(\mathbb{R}^d) .$$

A standard trace embedding theorem (see e.g.[70] Theorem 5.4 or[10] Corollary 1.4.1), combined with[69] Cor. 1, Chapt. 5, tells us that if $1 \leq m < n$ and either

$$d \geq n - m \quad \text{and} \quad p \geq 1, \tag{1}$$

or

$$d < n - m \quad \text{and} \quad p > \frac{n-d}{m}, \tag{2}$$

then every function $u \in W^{m,p}_{\mathrm{loc}}(\mathbb{R}^n)$ has a trace on $\mathbb{R}^d$. Moreover, the operator Tr, which associates $\operatorname{Tr} u$ with u, is linear, and, if $p < \frac{n}{m}$, then

$$\operatorname{Tr} : W^{m,p}(\mathbb{R}^n) \to L_{\frac{dp}{n-mp}}(\mathbb{R}^d) , \tag{3}$$

where the arrow "$\to$" stands for bounded operator. In particular,

$$\| \operatorname{Tr} u\|_{L_{\frac{dp}{n-mp}}(\mathbb{R}^d)} \leq C\|\nabla^m u\|_{L_p(\mathbb{R}^n)} \tag{4}$$

for every $u \in W^{m,p}(\mathbb{R}^n)$. Note that the case when $m \geq n$ is uninteresting, since any function $u \in W^{m,p}_{\mathrm{loc}}(\mathbb{R}^n)$, $p \geq 1$, is continuous, and hence $\operatorname{Tr} u$ trivially exists on $\mathbb{R}^d$ for every $d \in [1, n-1]$.

Unlike (1), in the limiting case when $p = \frac{n-d}{m} > 1$, functions from the Sobolev space $W^{m,\frac{n-d}{m}}_{\mathrm{loc}}(\mathbb{R}^n)$ need not admit a trace on $\mathbb{R}^d$.

In[71] we filled in this gap and exhibited an optimal (largest possible) Sobolev-type space when $d < n - m$ such that all of its functions admit a trace on $\mathbb{R}^d$. This was accomplished on calling into play the scale of Lorentz-Sobolev spaces. Indeed, it turns out that existence of traces can be restored when $p = \frac{n-d}{m}$, provided that $W^{m,\frac{n-d}{m}}_{\mathrm{loc}}(\mathbb{R}^n)$ is replaced by the Sobolev type space $W^m_{\mathrm{loc}} L_{\frac{n-d}{m},1}(\mathbb{R}^n)$ built upon the Lorentz space $L_{\frac{n-d}{m},1}(\mathbb{R}^n)$. Such a space is slightly smaller than $W^{m,\frac{n-d}{m}}_{\mathrm{loc}}(\mathbb{R}^n)$ if $\frac{n-d}{m} > 1$, but agrees with the standard space $W^{m,1}_{\mathrm{loc}}(\mathbb{R}^n)$ when $\frac{n-d}{m} = 1$. Moreover, $L^{\mathrm{loc}}_{\frac{n-d}{m},1}(\mathbb{R}^n)$ is optimal among all rearrangement-invariant spaces.

Observe that, since $L_{p,q}(\mathbb{R}^n) \subsetneq L_{p,r}(\mathbb{R}^n)$ if $1 \leq q < r \leq \infty$, and $L_{p,p}(\mathbb{R}^n) = L_p(\mathbb{R}^n)$, actually

$$W^m L_{\frac{n-d}{m},1}(\mathbb{R}^n) \subsetneq W^{m,\frac{n-d}{m}}(\mathbb{R}^n) \qquad \text{if } \tfrac{n-d}{m} > 1\,, \tag{5}$$

whereas

$$W^m L_{1,1}(\mathbb{R}^n) = W^{m,1}(\mathbb{R}^n)\,. \tag{6}$$

Relations (5) and (6) continue to hold if all the Sobolev spaces are replaced by their local versions.

Theorem 12.1. *Assume that $n \geq 2$, $1 \leq m < n$ and $1 \leq d \leq n - m$. Then any function from $W^m_{\mathrm{loc}} L_{\frac{n-d}{m},1}(\mathbb{R}^n)$ admits a trace on $\mathbb{R}^d$. Moreover, $L_{\frac{n-d}{m},1;\mathrm{loc}}(\mathbb{R}^n)$ is the optimal (largest) rearrangement-invariant space enjoying this property, in the sense that if $X(\mathbb{R}^n)$ is another r.i. space such that any function from $W^m_{\mathrm{loc}} X(\mathbb{R}^n)$ admits a trace on $\mathbb{R}^d$, then, necessarily, $X_{\mathrm{loc}}(\mathbb{R}^n) \subset L_{\frac{n-d}{m},1;\mathrm{loc}}(\mathbb{R}^n)$.*

In an analogy with the classical situation described in (3)–(4), we can establish a trace embedding for $W^m L_{\frac{n-d}{m},1}(\mathbb{R}^n)$. In fact, we can find the optimal (smallest) range space in the class of Lorentz spaces for trace embeddings of $W^m L_{\frac{n-d}{m},1}(\mathbb{R}^n)$. Interestingly enough, the optimal range space in this endpoint trace embedding turns out not to be the genuine Lorentz space $L^{\frac{n-d}{m},1}(\mathbb{R}^d)$ as one would expect in the light of other known optimal Sobolev embeddings such as those treated in[3,7,12,72] (see[14,21] for the optimality), and trace embeddings ([73]), but merely the (strictly larger) Lebesgue space $L_{\frac{n-d}{m}}(\mathbb{R}^d)$.

Theorem 12.2. *Assume that $n \geq 2$, $1 \leq m < n$ and $1 \leq d \leq n-m$. Then*

$$\operatorname{Tr}: W^m L_{\frac{n-d}{m},1}(\mathbb{R}^n) \to L_{\frac{n-d}{m}}(\mathbb{R}^d). \tag{7}$$

In particular, a constant $C = C(n,m,d)$ exists such that

$$\|\operatorname{Tr} u\|_{L_{\frac{n-d}{m}}(\mathbb{R}^d)} \leq C \|\nabla^m u\|_{L_{\frac{n-d}{m},1}(\mathbb{R}^n)} \tag{8}$$

for every $u \in W^m L_{\frac{n-d}{m},1}(\mathbb{R}^n)$.

Moreover, $L_{\frac{n-d}{m}}(\mathbb{R}^d)$ is optimal on the left-hand side of (8) among all Lorentz spaces.

A result on a related topic has been also established in[74], where a characterization of Sobolev inequalities involving general measures and Lorentz norms is given in terms of capacitary inequalities.

Remark 12.1. Embedding (7) continues to hold provided that the whole of $\mathbb{R}^n$ is replaced by any extension domain Ω (see e.g.[69,70,75,76] for a definition). Of course, $\mathbb{R}^d$ has to be replaced by $\Omega \cap \mathbb{R}^d$ in this case. Another generalization of embedding (7) concerns the case when traces on d-dimensional subspaces are replaced by (suitably defined – see e.g.[69]) traces on smooth d-dimensional Riemannian submanifolds of $\mathbb{R}^n$.

13. Boundary traces

The situation is much easier when the *boundary* traces on regular domains are studied. In such case, one has $d = n-1$, and many of above-mentioned technical problems disappear. Since these results are of particular interest in applications, we shall single them out here, too.

The theory of boundary traces in Sobolev spaces has a number of applications, especially to boundary-value problems for partial differential equations, in particular when the Neumann problem is studied. The trace operator can be defined just by

$$\operatorname{Tr}\ u = u_{|\partial\Omega}$$

for a continuous function u on $\overline{\Omega}$, and then extended to a bounded linear operator

$$\operatorname{Tr}: W^{1,1}(\Omega) \to L_1(\partial\Omega),$$

where, $L_1(\partial\Omega)$ denotes the Lebesgue space of summable functions on $\partial\Omega$ with respect to the $(n-1)$–dimensional Hausdorff measure $\mathcal{H}^{n-1}$. There exist many powerful methods for proving trace embedding theorems for the trace operator Tr, usually however quite dependent on a concrete norms involved. For specific limiting situations, other (e.g. potential) methods have been used, but there does not seem to exist a unified flexible approach that would cover the whole range of situations of interest in applications.

In[73], we developed a new method for obtaining sharp trace inequalities in a general context based on the ideas elaborated in the preceding sections.

Assume that $1 \leq m < n$. The standard trace embedding theorem (cf. e.g.[70]) tells us that if $1 \leq p < \frac{n}{m}$, then a constant $C = C(\Omega, p, m)$ exists such that

$$\|\operatorname{Tr} u\|_{L^{\frac{p(n-1)}{n-mp}}(\partial\Omega)} \leq C\|u\|_{W^{m,p}(\Omega)} \tag{1}$$

for every $u \in W^{m,p}(\Omega)$, whereas, if $p > \frac{n}{m}$, then a constant $C = C(\Omega, p, m)$ exists such that

$$\|\operatorname{Tr} u\|_{L^\infty(\partial\Omega)} \leq C\|u\|_{W^{m,p}(\Omega)} \tag{2}$$

for every $u \in W^{m,p}(\Omega)$.

A classical approach to trace inequalities such as (1) and (2) makes use of local coordinates on $\partial\Omega$ and of Sobolev inequalities in Ω, and heavily relies upon the fact that Lebesgue norms of $\operatorname{Tr} u$ and of $D^m u$ are involved. On the other hand, such a method does not seem to allow extensions to trace inequalities when more general norms are in play. Alternative techniques can be used to deal with specific instances. For example, a representation formula for Sobolev functions in terms of Riesz potentials enables one to effectively treat the limiting case when $p = \frac{n}{m}$, and to prove that a constant $C = C(\Omega, m)$ exists such that

$$\|\operatorname{Tr} u\|_{\exp L^{\frac{n}{n-m}}(\partial\Omega)} \leq C\|u\|_{W^{m,\frac{n}{m}}(\Omega)} \tag{3}$$

for every $u \in W^{m,\frac{n}{m}}(\Omega)$ (see e.g.[77] 7.6.4). Here, $\exp L^{\frac{n}{n-m}}(\partial\Omega)$ is the Orlicz space associated with the Young function $\exp(t^{\frac{n}{n-m}}) - 1$. The shortcoming of such an approach is that function spaces defined in terms of potentials are not always equivalent to corresponding spaces defined by derivatives. In fact, none of the available methods seems to cover the whole range of situations of interest in applications.

Thus, the problem can be raised of a unified flexible approach to trace inequalities fitting a fairly general class of function spaces. A powerful tool in the study of (first-order) Sobolev type inequalities in the whole domain Ω for functions vanishing on $\partial\Omega$ is symmetrization. The strength of this elegant technique, which relies upon the Pólya–Szegö principle on the decrease of gradient norms under radially non-increasing rearrangement ([25,78,79]) and has led to such results as those of Aubin[80] and Talenti[25], and of Moser[81], is in that the relevant Sobolev inequalities are reduced to one-dimensional inequalities.

Again, the key result is a reduction theorem.

Theorem 13.1. *Let $X(\Omega)$ and $Y(\partial\Omega)$ be r.i. spaces. Then*

$$\left\| \int_{t^{n'}}^{1} f(s) s^{\frac{m}{n}-1} \, ds \right\|_{\overline{Y}(0,1)} \leq C \|f\|_{\overline{X}(0,1)}, \qquad f \in \mathcal{M}_{+}(0,1),$$

if and only if

$$\| \operatorname{Tr} u \|_{Y(\partial\Omega)} \leq C \|u\|_{W^m X(\Omega)} \tag{4}$$

for every $u \in W^m X(\Omega)$.

Thus, when dealing with boundary traces, the role of the operator $H_{\frac{n}{m}}$ is taken over by the operator $\int_{t^{n'}}^{1} f(s) s^{\frac{m}{n}-1} \, ds$. Using appropriate interpolation methods, we can characterize the optimal trace range on $\partial\Omega$.

Theorem 13.2. *Let $X(\Omega)$ be an r.i. space. Then, the r.i. space $Y(\partial\Omega)$, whose associate norm is given by*

$$\|g\|_{Y'(\partial\Omega)} = \left\| t^{\frac{m-1}{n}} g^{**}(t^{\frac{1}{n'}}) \right\|_{\overline{X}'(0,1)} \tag{5}$$

for every $\mathcal{H}^{n-1}$–measurable function g on $\partial\Omega$, is optimal in (4).

Our trace results recover many known examples, prove their optimality that had not been known before, and bring new ones. For details, see[73].

Theorem 13.2 can be used to derive optimal trace embeddings whenever the associate norms appearing in (5) can be explicitly computed. For instance, in the next theorem, we present an application of Theorem 13.2 to the proof of trace inequalities in Lorentz spaces, which, in particular, improve (1) and (2). This can be regarded as a counterpart of results dealing with Sobolev embeddings in Ω which go back to[3,7] in the standard case, and to[10,12,72] in the limiting case. In what follows, Lorentz spaces on a measure space (R,ν) are denoted by $L^{p,q}(R,\nu)$; the notation $L^{p,q}(\log L)^{\alpha}(R,\nu)$ is adopted for the more general Lorentz–Zygmund spaces (see Section 2 for definitions).

Theorem 13.3. *Let Ω, n and m be as in* Theorem 13.1.

(i) *If* $1 \le p < \frac{n}{m}$ *and* $1 \le q \le \infty$, *then a constant* $C = C(\Omega, p, q, m)$ *exists such that*

$$\|\operatorname{Tr} u\|_{L^{\frac{p(n-1)}{n-mp},q}(\partial\Omega)} \le C\|u\|_{W^m L^{p,q}(\Omega)} \tag{6}$$

for every $u \in W^m L^{p,q}(\Omega)$.

(ii) *If* $p = \frac{n}{m}$ *and* $1 < q \le \infty$, *then a constant* $C = C(\Omega, q, m)$ *exists such that*

$$\|\operatorname{Tr} u\|_{L^{\infty,q}(\log L)^{-1}(\partial\Omega)} \le C\|u\|_{W^m L^{\frac{n}{m},q}(\Omega)} \tag{7}$$

for every $u \in W^m L^{\frac{n}{m}}(\Omega)$.

(iii) *If either* $p = \frac{n}{m}$ *and* $q = 1$ *or* $p > \frac{n}{m}$ *and* $1 \le q \le \infty$, *then a constant* $C = C(\Omega, p, q, m)$ *exists such that*

$$\|\operatorname{Tr} u\|_{L^\infty(\partial\Omega)} \le C\|u\|_{W^m L^{p,q}(\Omega)} \tag{8}$$

for every $u \in W^{m,p}(\Omega)$.

The spaces $L^{\frac{p(n-1)}{n-mp},q}(\partial\Omega)$, $L^{\infty,q}(\log L)^{-1}(\partial\Omega)$ *and* $L^\infty(\partial\Omega)$ *are optimal among all r.i. spaces on* $\partial\Omega$ *in* (6), (7) *and* (8), *respectively.*

Notice that inequalities (1) and (2) are not only generalized, but also improved by (6) and (7), respectively. Indeed, $L^{p,p}(\Omega) = L^p(\Omega)$, whence $W^m L^{p,p}(\Omega) = W^{m,p}(\Omega)$, and, on the other hand,

$$L^{\frac{p(n-1)}{n-mp},p}(\partial\Omega) \subsetneq L^{\frac{p(n-1)}{n-mp}}(\partial\Omega) \text{ and } L^{\infty,\frac{n}{m}}(\log L)^{-1}(\partial\Omega) \subsetneq \exp L^{\frac{n}{n-m}}(\partial\Omega).$$

It is of interest that to prove Theorem 13.1, one has first to prove its particular case corresponding to first-order inequalities, namely to the case when $m = 1$. Another approach to higher-order trace embeddings will be taken in the following section.

Theorem 13.4. *Let* Ω, $X(\Omega)$ *and* $Y(\partial\Omega)$ *be as in* Theorem 13.1. *If there exists a constant* C_1 *such that*

$$\left\|\int_{t^{n'}}^1 f(s)s^{-\frac{1}{n'}}\,ds\right\|_{\overline{Y}(0,1)} \le C_1\|f\|_{\overline{X}(0,1)} \tag{9}$$

for every nonnegative non-increasing function $f \in \overline{X}(0,1)$, *then a constant* $C_2 = C_2(C_1, \Omega)$ *exists such that*

$$\|\operatorname{Tr} u\|_{Y(\partial\Omega)} \le C_2\|u\|_{W^1 X(\Omega)} \tag{10}$$

for every $u \in W^1 X(\Omega)$.

Conversely, if (10) *holds for every* $u \in W^1 X(\Omega)$, *then there exists a constant* $C_1 = C_1(C_2, \Omega)$ *such that* (9) *holds for every* $f \in \overline{X}(0,1)$.

The key step towards the proofs of the main results is the following auxiliary result in the spirit of[21] Theorem 4.5 and[24] Theorem 3.2.

Proposition 13.1. *Let $X(0,1)$ be an r.i. space. Then the space $X_\sigma(0,1)$ of all measurable functions f on $(0,1)$ such that the quantity*

$$\|f\|_{X_\sigma(0,1)} = \left\| t^{\frac{m}{n}-1} \int_0^{t^{\frac{1}{n'}}} f^*(s)\,ds \right\|_{X'(0,1)} \tag{11}$$

is finite is an r.i. space on $(0,1)$, and $\|\cdot\|_{X_\sigma(0,1)}$ is a norm. Its associate space $X'_\sigma(0,1)$ satisfies

$$\left\| \int_{t^{n'}}^1 f(s) s^{\frac{m}{n}-1}\,ds \right\|_{X'_\sigma(0,1)} \leq \|f\|_{X(0,1)} \tag{12}$$

for every $f \in X(0,1)$. Moreover, $X'_\sigma(0,1)$ is the optimal r.i. space in (12), in the sense that if $Z(0,1)$ is an r.i. space such that

$$\left\| \int_{t^{n'}}^1 f(s) s^{\frac{m}{n}-1}\,ds \right\|_{Z(0,1)} \leq C\|f\|_{X(0,1)} \tag{13}$$

for some constant C and for every $f \in X(0,1)$, then

$$\|g\|_{Z(0,1)} \leq C\|g\|_{X'_\sigma(0,1)} \tag{14}$$

for every $g \in X'_\sigma(0,1)$.

14. Higher-order trace embeddings

We shall finally concentrate on trace embeddings for Sobolev functions of higher-order. We shall combine the techniques of previous sections, although a direct use of the iteration method from Section 11 is not convenient here. For the results in this section we need certain regularity of the domain in question. In particular, we shall throughout this section assume that, with $m \in \mathbb{N}$ fixed, Ω is a bounded *extension domain* for both $W^{m,1}(\Omega)$ and $W^{m,\infty}(\Omega)$ (meaning that there exists a bounded linear extension operator $E : W^{m,1}(\Omega) \to W^{m,1}(\mathbb{R}^n)$ and $E : W^{m,\infty}(\Omega) \to W^{m,\infty}(\mathbb{R}^n)$). In particular, Ω can be, for example, a Lipschitz domain.

The main result is, as usual, a reduction theorem.

Theorem 14.1. *Let Ω be an extension domain in $\mathbb{R}^n$, $n \geq 2$. Let $m \in \mathbb{N}$ and $d \in \mathbb{N}$ be such that $1 \leq d \leq n$ and $n - d \leq m$. Let Ω_d be the (non empty) intersection of Ω with any d-dimensional affine subspace of $\mathbb{R}^n$. Let*

$\|\cdot\|_{X(0,1)}$ and $\|\cdot\|_{Y(0,1)}$ be rearrangement-invariant function norms. Then the following facts are equivalent.

(i) *The Sobolev trace embedding*

$$\mathrm{Tr}: W^m X(\Omega) \to Y(\Omega_d) \tag{1}$$

holds. Here, $X(\Omega)$ denotes the rearrangement-invariant space on Ω associated with $\|\cdot\|_{X(0,1)}$, and $Y(\Omega_d)$ the rearrangement-invariant space on Ω_d, with respect to the d-dimensional Hausdorff measure $\mathcal{H}^d$, associated with $\|\cdot\|_{Y(0,1)}$.

(ii) *The Poincaré trace inequality*

$$\|\mathrm{Tr}\, u\|_{Y(\Omega_d)} \le C_1 \|\nabla^m u\|_{X(\Omega)} \tag{2}$$

holds for some constant C_1, and for every $u \in W^m_\perp X(\Omega)$.

(iii) *The inequality*

$$\left\| \int_{t^{\frac{n}{d}}}^1 f(s)\, s^{-1+\frac{m}{n}}\, ds \right\|_{Y(0,1)} \le C_2 \|f\|_{X(0,1)} \tag{3}$$

holds for some constant C and for every $f \in X(0,1)$.

A characterization of the optimal target space $Y(\Omega_d)$ in the trace embedding (1), for any given rearrangement-invariant space $X(\Omega)$ is provided by the next result. Its statement requires the following definition. Given a rearrangement-invariant function norm $\|\cdot\|_{X(0,1)}$, and $n, m, d \in \mathbb{N}$ be such that $1 \le d \le n$ and $n - d \le m$, we call $\|\cdot\|_{X^m_{d,n}(0,1)}$ the rearrangement-invariant function norm whose associate function norm is given by

$$\|f\|_{(X^m_{d,n})'(0,1)} = \left\| s^{-1+\frac{m}{n}} \int_0^{s^{\frac{d}{n}}} f^*(r)dr \right\|_{X'(0,1)} \tag{4}$$

for $f \in \mathcal{M}(0,1)$.

Theorem 14.2. *Let Ω be an extension domain in $\mathbb{R}^n$, $n \ge 2$. Let $m, d \in \mathbb{N}$ be such that $1 \le d \le n$ and $n - d \le m$. Let Ω_d be the (non empty) intersection of Ω with any d-dimensional affine subspace of $\mathbb{R}^n$. Let $\|\cdot\|_{X(0,1)}$ be a rearrangement-invariant function norm. Then*

$$\mathrm{Tr}: W^m X(\Omega) \to X^m_{d,n}(\Omega_d). \tag{5}$$

Here, $X(\Omega)$ denotes the rearrangement-invariant space on Ω associated with $\|\cdot\|_{X(0,1)}$, and $X^m_{d,n}(\Omega_d)$ the rearrangement-invariant space on Ω_d associated with $\|\cdot\|_{(X^m_{d,n})'(0,1)}$.

Moreover, the space $X^m_{d,n}(\Omega_d)$ is optimal in (5) *among all rearrangement-invariant spaces.*

It is of interest to note that in case when $m \geq n$ one gets, for arbitrary space X, automatically the embedding into L_∞. In fact, a more general result is available.

Corollary 14.1. *Let n, d, m, Ω, Ω_d and $\|\cdot\|_{X(0,1)}$ be as in Theorem 14.2. Then the following facts are equivalent:*

$$\mathrm{Tr}: W^m X(\Omega) \to L_\infty(\Omega_d); \tag{6}$$

$$X^m_{d,n}(\Omega_d) = L_\infty(\Omega_d); \tag{7}$$

$$\|s^{-1+\frac{m}{n}}\|_{X'(0,1)} < \infty. \tag{8}$$

In particular, (6) *and* (7) *hold for any rearrangement invariant function norm $\|\cdot\|_{X(0,1)}$, provided that $m \geq n$.*

As above, we present a theorem on sharpness of the iteration of optimal targets. Let us emphasize that, in this framework, iteration not only involves the order of differentiation, but also the dimension of subspaces. This is a very important observation.

Theorem 14.3. *Let Ω be an extension domain domain in $\mathbb{R}^n$, $n \geq 2$. Let $k, h, d, \ell \in \mathbb{N}$, be such that $1 \leq d \leq \ell \leq n$, $n - \ell \leq k$ and $\ell - d \leq h$. Let Ω_ℓ be the (non empty) intersection of Ω with any ℓ-dimensional affine subspace S_ℓ of $\mathbb{R}^n$, and let Ω_d be the (non empty) intersection of Ω with any d-dimensional affine subspace of S_ℓ. Let $\|\cdot\|_{X(0,1)}$ be a rearrangement-invariant function norm. Then*

$$(X^k_{\ell,n})^h_{d,\ell}(\Omega_d) = X^{k+h}_{d,n}(\Omega_d). \tag{9}$$

We shall now present some examples. The next theorem concerns the situation when the source space is a Lorentz space.

Theorem 14.4. *Let Ω be an extension domain in $\mathbb{R}^n$, $n \geq 2$. Let $m \in \mathbb{N}$ and $d \in \mathbb{N}$, be such that $1 \leq d \leq n$ and $n - d \leq m$. Let Ω_d be the (non empty) intersection of Ω with any d-dimensional affine subspace of $\mathbb{R}^n$. Assume that either $p = q = 1$ or $p = q = \infty$ or $p \in (1, \infty)$ and $q \in [1, \infty]$. Then*

$$\mathrm{Tr}: W^m L_{p,q}(\Omega) \to \begin{cases} L_{\frac{pd}{n-mp},q}(\Omega_d) & \text{if } m < n \text{ and } p \in [1, \frac{n}{m}), \\ L_{\infty,q;-1}(\Omega_d) & \text{if } m < n,\ p = \frac{n}{m} \text{ and } q > 1, \\ L_\infty(\Omega_d) & \text{otherwise.} \end{cases} \tag{10}$$

Moreover, the target spaces in (10) *are optimal among all rearrangement-invariant spaces on* Ω_d.

Our next set of examples involves Orlicz spaces. Let $n, m, d \in \mathbb{N}$ be as in the statement of Theorem 14.1. Let A be a Young function such that

$$\int_0 \left(\frac{t}{A(t)}\right)^{\frac{m}{n-m}} dt < \infty. \tag{11}$$

If $m < n$, assume that the integral

$$\int^{\infty} \left(\frac{t}{A(t)}\right)^{\frac{m}{n-m}} dt \tag{12}$$

diverges, let $H_m : [0, \infty) \to [0, \infty)$ be the function defined as

$$H_m(s) = \left(\int_0^s \left(\frac{t}{A(t)}\right)^{\frac{m}{n-m}} dt\right)^{\frac{n}{n-m}} \qquad \text{for } s \geq 0, \tag{13}$$

and let $A_{m,d}$ be the Young function given by

$$A_{m,d}(t) = \int_0^{H_m^{-1}(t)} \left(\frac{A(s)}{s}\right)^{\frac{d-m}{n-m}} H_m(s)^{\frac{d-n}{n-m}} ds \qquad \text{for } t \geq 0. \tag{14}$$

Theorem 14.5. *Let* Ω *be an extension domain in* $\mathbb{R}^n$, $n \geq 2$. *Let* $m \in \mathbb{N}$ *and* $d \in \mathbb{N}$ *be such that* $1 \leq d \leq n$ *and* $n - d \leq m$. *Let* Ω_d *be the (non empty) intersection of* Ω *with any* d*-dimensional affine subspace of* $\mathbb{R}^n$. *Let* A *be a Young function fulfilling* (11). *Then*

$$\mathrm{Tr} : W^m L_A(\Omega) \to \begin{cases} L_{A_{m,d}}(\Omega_d) & \text{if } m < n, \text{ and the integral} \\ & \text{(12) diverges,} \\ L_\infty(\Omega_d) & \text{if either } m \geq n, \text{ or } m < n \\ & \text{and the integral (12) converges.} \end{cases} \tag{15}$$

Moreover, the target spaces in (15) *are optimal among all Orlicz spaces.*

Theorem 14.5 follows from Theorem 14.1 and[82] Theorem 3.5.

It turns out that the optimal rearrangement-invariant target space in the first case of (15) is an Orlicz-Lorentz space. Under the assumption that $m < n$ and that the integral in (12) is divergent, let $a : [0, \infty) \to [0, \infty]$ be a non-decreasing, left-continuous function which is neither identically equal to 0 nor to ∞ and which satisfies

$$A(t) = \int_0^t a(\tau) d\tau \qquad \text{for } t \geq 0, \tag{16}$$

let E_m be the Young function given by

$$E_m(t) = \int_0^t e_m(s)ds \qquad \text{for } t \geq 0, \tag{17}$$

where e_m is the non-decreasing, left-continuous function in $[0, \infty)$ obeying

$$e_m^{-1}(s) = \left(\int_{a^{-1}(s)}^{\infty} \left(\int_0^{\tau} \left(\frac{1}{a(t)} \right)^{\frac{m}{n-m}} dt \right)^{-\frac{n}{m}} \frac{d\tau}{a(\tau)^{\frac{n}{n-m}}} \right)^{\frac{m}{m-n}} \tag{18}$$

for $s \geq 0$. $L(\frac{n}{m}, \frac{n}{d}, E_m)(\Omega_d)$ is the Orlicz-Lorentz space equipped with the norm given by

$$\|g\|_{L(\frac{n}{m}, \frac{n}{d}, E_m)(\Omega_d)} = \|s^{-\frac{m}{n}} g^*(\mathcal{H}^d(\Omega_d) s^{\frac{d}{n}})\|_{L_{E_m}(0,1)} \tag{19}$$

for any function $g \in \mathcal{M}(\Omega \cap S_d)$.

Theorem 14.6. *Let Ω be an extension domain in $\mathbb{R}^n$, $n \geq 2$. Let $m \in \mathbb{N}$ and $d \in \mathbb{N}$ be such that $1 \leq d \leq n$ and $n - d \leq m$. Let Ω_d be the (non empty) intersection of Ω with any d-dimensional affine subspace of $\mathbb{R}^n$. Let A be a Young function fulfilling* (11). *Assume that $m < n$, and the integral in* (12) *diverges. Then*

$$\mathrm{Tr} : W^m L_A(\Omega) \to L(\frac{n}{m}, \frac{n}{d}, E_m)(\Omega_d), \tag{20}$$

and the target space in (20) *is optimal among all rearrangement-invariant spaces.*

Embedding (20) follows from Theorem 14.1, via an analogous argument as in the proof of[82] Theorem 4.1.

Acknowledgment. It was a great honor and pleasure for me to be invited to deliver a minicourse at the Fifth CIDAMA.

I would like first to express my deep gratitude to the editors of this volume, in particular to Juan Carlos Navarro Pascual, for his collaboration and patience.

The lecture notes are based on the results of a long-time research partly carried out in collaboration with some of my great friends and colleagues. Most of the results presented were obtained jointly with either Andrea Cianchi or Ron Kerman. Apart from them, I have also used ideas of María Carro, Michael Cwikel, David Edmunds, Des Evans, Amiran Gogatishvili, Petr Gurka, Desmond Harris, Jan Malý, Mirek Opic, Colin Phipps, Evgenniy Pustylnik, Lenka Slavíková, Javier Soria, Vlad Stepanov and more.

Over the years, the research has been also partly supported by the grants no. MSM 113200007 and MSM 0021620839 of the Czech Ministry of Education, research grants no. 201/94/1066, 201/97/0744, 201/01/0333, 201/03/0935, 201/05/2033, 201/07/0388, 201/08/0383 and 13-14743S of the Grant Agency of the Czech Republic and NATO grants no. CRG 930 358 and OUTR.CRG 9700071, and by several institutions, of which I name at least Mathematical Institute of the Czech Academy of Sciences, University of Wales College of Cardiff, Faculty of Mathematics and Physics of the Charles University in Prague, Brock University, University of Sussex, the CRM in Barcelona, the EPSRC, and the Royal Society.

References

1. N. S. Trudinger, On imbeddings into Orlicz spaces and some applications, *J. Math. Mech.* **17**, 473 (1967).
2. V. I. Yudovič, Some estimates connected with integral operators and with solutions of elliptic equations, *Soviet Math. Doklady* **2**, 746 (1961).
3. J. Peetre, Espaces d'interpolation et théorème de Soboleff, *Ann. Inst. Fourier* **16**, 279 (1966).
4. S. I. Pokhozhaev, On eigenfunctions of the equation $\delta u + \lambda f(u) = 0$, *Dokl. Akad. Nauk SSSR* **165**, 36 (1965).
5. J. A. Hempel, G. R. Morris and N. S. Trudinger, On the sharpness of a limiting case of the Sobolev imbedding theorem., *Bull. Austral. Math. Soc.* **3**, 369 (1970).
6. A. Cianchi, A sharp embedding theorem for Orlicz-Sobolev spaces, *Indiana Univ. Math. J.* **45**, 39 (1996).
7. R. O'Neil, Convolution operators and $L(p, q)$ spaces, *Duke Math. J.* **30**, 129 (1963).
8. R. A. Hunt, On $L(p, q)$ spaces, *Enseignement Math.* **12**, 249 (1966).
9. C. Bennett and K. Rudnick, On Lorentz-Zygmund spaces, *Dissertationes Math. (Rozprawy Mat.)* **175**, p. 67 pp. (1980).
10. V. G. Maz'ya, *Sobolev spaces* (Springer, Berlin, 1985).
11. T. Holmstedt, Interpolation of quasi-normed spaces, *Math. Scand.* **26**, 177 (1970).
12. H. Brézis and S. Wainger, A note on limiting cases of Sobolev embeddings and convolution inequalities, *Comm. Partial Differential Equations* **5**, 773 (1980).
13. J. A. Brudnyĭ, Rational approximation and imbedding theorems, *Sov. Math Dokl.* **247**, 681 (1979).

14. M. Ćwikel and E. Pustylnik, Weak type interpolation near "endpoint" spaces, *J. Funct. Anal.* **171**, 235 (2000).
15. J. Malý and L. Pick, An elementary proof of sharp Sobolev embeddings, *Proc. Amer. Math. Soc.* **130**, 555 (2002).
16. J. Bastero, M. Milman and F. J. Ruiz Blasco, A note on $L(\infty, q)$ spaces and Sobolev embeddings, *Indiana Univ. Math. J.* **52**, 1215 (2003).
17. C. Bennett, R. A. DeVore and R. Sharpley, Weak-L^∞ and BMO, *Ann. of Math. (2)* **113**, 601 (1981).
18. L. Pick, *Optimal Sobolev Embeddings* (Rudolph - Lipschitz - Vorlesungsreihe no. 43, Rheinische Friedrich - Wilhelms - Universität Bonn, x+144 p, 2002).
19. W. D. Evans, B. Opic and L. Pick, Interpolation of operators on scales of generalized Lorentz-Zygmund spaces, *Math. Nachr.* **182**, 127 (1996).
20. C. Bennett and R. Sharpley, *Interpolation of operators*, Pure and Applied Mathematics, Vol. 129 (Academic Press, Inc., Boston, MA, 1988).
21. D. E. Edmunds, R. Kerman and L. Pick, Optimal Sobolev imbeddings involving rearrangement-invariant quasinorms, *J. Funct. Anal.* **170**, 307 (2000).
22. R. A. Adams and J. J. F. Fournier, *Sobolev spaces*, Pure and Applied Mathematics (Amsterdam), Vol. 140, second edn. (Elsevier/Academic Press, Amsterdam, 2003).
23. R. DeVore and K. Scherer, Interpolation of linear operators on Sobolev spaces, *Ann. of Math. (2)* **109**, 583 (1979).
24. R. Kerman and L. Pick, Optimal Sobolev imbeddings, *Forum Math.* **18**, 535 (2006).
25. G. Talenti, Best constant in Sobolev inequality, *Ann. Mat. Pura Appl.* **110**, 353 (1976).
26. A. Cianchi, Symmetrization and second-order Sobolev inequalities, *Ann. Mat. Pura Appl.* **183**, 45 (2004).
27. E. Sawyer, Boundedness of classical operators on classical Lorentz spaces, *Studia Math.* **96**, 145 (1990).
28. A. Cianchi, Optimal Orlicz-Sobolev embeddings, *Rev. Mat. Iberoamericana* **20**, 427 (2004).
29. L. Pick, Supremum operators and optimal Sobolev inequalities, in *Function spaces, differential operators and nonlinear analysis (Pudasjärvi, 1999)*, (Acad. Sci. Czech Repub., Prague, 2000).
30. R. Kerman and L. Pick, Optimal Sobolev imbedding spaces, *Studia Math.* **192**, 195 (2009).

31. L. Slavíková, unpublished and untitled manuscript, Prague, (2013).
32. A. Cianchi, R. Kerman, B. Opic and L. Pick, A sharp rearrangement inequality for the fractional maximal operator, *Studia Math.* **138**, 277 (2000).
33. A. Gogatishvili, B. Opic and L. Pick, Weighted inequalities for Hardy-type operators involving suprema, *Collect. Math.* **57**, 227 (2006).
34. M. J. Carro, A. Gogatishvili, J. Martín and L. Pick, Functional properties of rearrangement invariant spaces defined in terms of oscillations, *J. Funct. Anal.* **229**, 375 (2005).
35. M. Ćwikel, A. Kamińska, L. Maligranda and L. Pick, Are generalized Lorentz "spaces" really spaces?, *Proc. Amer. Math. Soc.* **132**, 3615 (2004).
36. M. Carro, A. Gogatishvili, J. Martín and L. Pick, Weighted inequalities involving two Hardy operators with applications to embeddings of function spaces, *J. Operator Theory* **59**, 309 (2008).
37. R. Kerman and L. Pick, Explicit formulas for optimal rearrangement-invariant norms in Sobolev imbedding inequalities, *Studia Math.* **206**, 97 (2011).
38. Y. A. Brudnyĭ and N. Y. Krugljak, *Interpolation functors and interpolation spaces*, North-Holland Mathematical Library, Vol. 1 (North-Holland Publishing Co., Amsterdam, 1991).
39. R. Kerman, M. Milman and G. Sinnamon, On the Brudnyĭ-Krugljak duality theory of spaces formed by the K-method of interpolation, *Rev. Mat. Complut.* **20**, 367 (2007).
40. F. Rellich, Ein satz über mittlere konvergenz, *Göttingen Nachr.* , 30 (1930).
41. W. Kondrachov, Certain properties of functions in the space l^p, *Dokl. Akad. Nauk SSSR* **48**, 535 (1945).
42. R. Kerman and L. Pick, Compactness of Sobolev imbeddings involving rearrangement-invariant norms, *Studia Math.* **186**, 127 (2008).
43. E. Nelson, The free Markoff field, *J. Functional Analysis* **12**, 211 (1973).
44. L. Gross, Logarithmic Sobolev inequalities, *Amer. J. Math.* **97**, 1061 (1975).
45. A. Cianchi and L. Pick, Optimal Gaussian Sobolev embeddings, *J. Funct. Anal.* **256**, 3588 (2009).
46. V. G. Maz'ya, Classes of domains and imbedding theorems for function spaces, *Soviet Math. Dokl.* **1**, 882 (1960).
47. V. G. Maz'ya, p-conductivity and theorems on imbedding certain functional spaces into a C-space, *Dokl. Akad. Nauk SSSR* **140**, 299

(1961).

48. S. Sobolev, On some estimates relating to families of functions having derivatives that are square integrable, *Dokl. Akad. Nauk. SSSR* **1**, 267 (1936), (Russian).
49. S. Sobolev, On a theorem in functional analysis, *Sb. Math.* **4**, 471 (1938), (Russian); English translation: Am. Math. Soc. Trans. **34** (1963), 39–68.
50. E. Gagliardo, Proprietà di alcune classi di funzioni in più variabili, *Ricerche Mat.* **7**, 102 (1958).
51. L. Nirenberg, On elliptic partial differential equations, *Ann. Scuola Norm. Sup. Pisa* **13**, 115 (1959).
52. E. De Giorgi, Sulla proprietà isoperimetrica dell'ipersfera, nella classe degli insiemi aventi frontiera orientata di misura finita, *Atti Accad. Naz. Lincei. Mem. Cl. Sci. Fis. Mat. Nat. Sez. I (8)* **5**, 33 (1958).
53. H. Federer and W. H. Fleming, Normal and integral currents, *Ann. of Math. (2)* **72**, 458 (1960).
54. A. Cianchi, L. Pick and L. Slavíková, Higher-order Sobolev embeddings and isoperimetric inequalities, *Adv. Math.* **273**, 568 (2015).
55. V. G. Maz'ya, *Sobolev spaces with applications to elliptic partial differential equations* (Springer, Berlin, 2011).
56. C. Borell, The Brunn-Minkowski inequality in Gauss space, *Invent. Math.* **30**, 207 (1975).
57. A. Cianchi, Higher-order Sobolev and Poincaré inequalities in Orlicz spaces, *Forum Math.* **18**, 745 (2006).
58. R. S. Strichartz, A note on Trudinger's extension of Sobolev's inequalities, *Indiana Univ. Math. J.* **21**, 841 (1971/72).
59. B. Bojarski, Remarks on sobolev imbedding inequalities, in *Complex analysis, Joensuu 1987*, , Lecture Notes in Math. Vol. 1351 (Springer, Berlin, 1988).
60. P. Hajłasz and P. Koskela, Isoperimetric inequalities and imbedding theorems in irregular domains, *J. London Math. Soc.* **58**, 425 (1998).
61. T. Kilpeläinen and J. Malý, Sobolev inequalities on sets with irregular boundaries, *Z. Anal. Anwendungen* **19**, 369 (2000).
62. S. Buckley and P. Koskela, Sobolev-Poincaré implies John, *Math. Res. Lett.* **2**, 577 (1995).
63. F. Barthe, P. Cattiaux and C. Roberto, Isoperimetry between exponential and Gaussian, *Electron. J. Probab.* **12**, no. 44, 1212 (2007).
64. F. Barthe, P. Cattiaux and C. Roberto, Interpolated inequalities between exponential and Gaussian, Orlicz hypercontractivity and

isoperimetry, *Rev. Mat. Iberoam.* **22**, 993 (2006).
65. S. G. Bobkov and C. Houdré, Some connections between isoperimetric and Sobolev-type inequalities, *Mem. Amer. Math. Soc.* **129**, viii+111 (1997).
66. M. Ledoux, *The concentration of measure phenomenon*, Mathematical Surveys and Monographs, Vol. 89 (American Mathematical Society, Providence, RI, 2001).
67. M. Ledoux, Spectral gap, logarithmic sobolev constant and geometric bounds, in *Eigenvalues of Laplacians and other geometric operators*, eds. A. Grigor'yan and S.-T. Yau, Surveys in Differential Geometry, Vol. IX (Int. Press, Somerville, MA, 2004) pp. 219–240.
68. V. N. Sudakov and B. S. Tsirel'son, Extremal properties of half-spaces for spherically invariant measures, *Zap. Naučn. Sem. Leningrad. Otdel. Mat. Inst. Steklov. (LOMI)* **41**, 14 (1974), English translation: J. Soviet Math. 9 (1978), 9–18.
69. V. I. Burenkov, *Sobolev spaces on domains* (B. G. Teubner Verlagsgesellschaft mbH, Stuttgart, 1998).
70. R. A. Adams, *Sobolev spaces* (Academic Press, New York-London, 1975), Pure and Applied Mathematics, Vol. 65.
71. A. Cianchi and L. Pick, An optimal endpoint trace embedding, *Ann. Inst. Fourier* **60**, 939 (2010).
72. K. Hansson, Imbedding theorems of Sobolev type in potential theory, *Math. Scand.* **45**, 77 (1979).
73. A. Cianchi, R. Kerman and L. Pick, Boundary trace inequalities and rearrangements, *J. Anal. Math.* **105**, 241 (2008).
74. S. Costea and V. Maz'ya, *Conductor inequalities and criteria for Sobolev-Lorentz two-weight inequalities*, Int. Math. Ser. (N. Y.), Vol. 9 (Springer, New York, 2009).
75. E. M. Stein, *Singular integrals and differentiability properties of functions* (Princeton University Press, Princeton, 1970).
76. W. P. Ziemer, *Weakly differentiable functions* (Springer-Verlag, New York, 1989).
77. D. R. Adams and L. I. Hedberg, *Function spaces and potential theory* (Springer-Verlag, Berlin, 1996).
78. J. E. Brothers and W. P. Ziemer, Minimal rearrangements of Sobolev functions, *J. Reine Angew. Math.* **384**, 153 (1988).
79. B. Kawohl, *Rearrangements and convexity of level sets in PDE*, Lecture Notes in Mathematics, Vol. 1150 (Springer-Verlag, Berlin, 1985).
80. T. Aubin, Problèmes isopérimétriques et espaces de Sobolev, *C. R.*

Acad. Sci. Paris Sér. A-B **280**, Aii, A279 (1975).

81. J. Moser, A sharp form of an inequality by N. Trudinger, *Indiana Univ. Math. J.* **20**, 1077 (1970/71).
82. A. Cianchi, Orlicz-Sobolev boundary trace embeddings, *Math. Z.* **266**, 431 (2010).

Derivations and projections on Jordan triples: An introduction to nonassociative algebra, continuous cohomology, and quantum functional analysis

Bernard Russo

Department of Mathematics, University of California, Irvine,
525 Rowland Hall, Irvine, CA 92697-3875, USA
E-mail: brusso@math.uci.edu

This paper is an elaborated version of the material presented by the author in a three-hour minicourse at *V International Course of Mathematical Analysis in Andalusia*, at Almeria, Spain September 12-16, 2011. The author wishes to thank the scientific committee for the opportunity to present the course and to the organizing committee for their hospitality. The author also personally thanks Antonio Peralta for his collegiality and encouragement.

The minicourse on which this paper is based had its genesis in a series of talks the author had given to undergraduates at Fullerton College in California. I thank my former student Dana Clahane for his initiative in running the remarkable undergraduate research program at Fullerton College of which the seminar series is a part. With their knowledge only of the product rule for differentiation as a starting point, these enthusiastic students were introduced to some aspects of the esoteric subject of non associative algebra, including triple systems as well as algebras. Slides of these talks and of the minicourse lectures, as well as other related material, can be found at the author's website (www.math.uci.edu/~brusso). Conversely, these undergraduate talks were motivated by the author's past and recent joint works on derivations of Jordan triples ([1],[2],[3]), which are among the many results discussed here.

Part I (*Derivations*) is devoted to an exposition of the properties of derivations on various algebras and triple systems in finite and infinite dimensions, the primary questions addressed being whether the derivation is automatically continuous and to what extent it is an inner derivation.

Part II (*Cohomology*) discusses cohomology theory of algebras and triple systems, in both finite and infinite dimensions. Although the cohomology of associative and Lie algebras is substantially developed, in both finite and infinite dimensions ([4],[5],[6]; see[7] for a review of[4]), the same could not be said for Jordan algebras. Moreover, the cohomology of triple systems has a rather sparse literature which is essentially non-existent in infinite dimensions. Thus, one of the goals of this paper is to encourage the study of continuous cohomology of some Banach triple systems. Occasionally, an idea for a research project is mentioned (especially in subsection 6.2). Readers are invited to contact the author, at brusso@uci.edu, if they share this interest.

Part III (*Quantum Functional Analysis*) begins with the subject of contractive projections, which plays an important role in the structure theory of Jordan triples. The remainder of Part III discusses three topics, two very recent, which involve the interplay between Jordan theory and operator space theory (quantum functional analysis). The first one, a joint work of the author[8], discusses the structure theory of contractively complemented Hilbertian operator spaces, and is instrumental to the third topic, which is concerned with some recent work on enveloping TROs and K-theory for JB*-triples[9],[10],[11]. The second topic presents some very recent joint work by the author concerning quantum operator algebras which leans very heavily on contractive projection theory[12].

Parts I and II, and the first section of Part III of this paper were covered in the three lectures making up the minicourse. The rest of Part III of this paper, although not covered in the lectures, is included here as significant application of contractive projections and because there is some implicit reference to it in part I.

A few proofs have been included but for the most part, theorems and other results are just stated and a reference for the proof and in some cases for a definition, is provided. Brief biographical information on some mathematicians, obtained primarily from Wikipedia, is provided in footnotes.

As part of the undergraduate lectures mentioned above, a set of exercises was developed, proceeding from elementary verifications to nontrivial theorems in the literature, with appropriate references. These occur at the end of sections 1 and 2.

Since this paper is meant to introduce a large number of topics, many of which are purely algebraical in nature, and others which the author has not fully studied yet, he has taken the liberty occasionally to rely on *Mathematical Reviews*, as well as authors' introductions for some of the summaries.

Contents

Part I
Derivations

1. Algebras

Let $\mathcal{C}$ denote the algebra of continuous complex valued functions on a locally compact Hausdorff space.

Definition 1.1. A derivation on $\mathcal{C}$ is a linear mapping $\delta : \mathcal{C} \to \mathcal{C}$ satisfying the "product" rule:

$$\delta(fg) = \delta(f)g + f\delta(g).$$

Theorem 1.1. *There are no (non-zero) derivations on $\mathcal{C}$. In other words, every derivation of $\mathcal{C}$ is identically zero.*

The proof of this theorem can be broken up into two parts of independent interest. The first theorem, due to Singer[1] and Wermer[2], is valid for commutative semisimple Banach algebras. The second one is due to Sakai[3] and is valid for all C^*-algebras.

Theorem 1.2 (Singer and Wermer 1955[13]). *Every continuous derivation on $\mathcal{C}$ is zero.*

Theorem 1.3 (Sakai 1960[14]). *Every derivation on $\mathcal{C}$ is continuous.*

Singer and Wermer actually proved that every continuous derivation on a commutative Banach algebra maps the algebra into the radical and they conjectured that the continuity assumption was unnecessary and could be removed. Johnson showed this for commutative semisimple Banach algebras[15] and Thomas gave an affirmative answer to the Singer-Wermer conjecture for all commutative Banach algebras[16]. There are various noncommutative versions of the Singer-Wermer theorem for Banach algebras, for example[17–19]. This is the first ripe area for exploration in the non associative contexts being considered in this paper.

[1]**Isadore Singer** (b. 1924). Isadore Manuel Singer is an Institute Professor in the Department of Mathematics at the Massachusetts Institute of Technology. He is noted for his work with Michael Atiyah in 1962, which paved the way for new interactions between pure mathematics and theoretical physics
[2]**John Wermer** (b. 1925), retired from Brown University
[3]**Soichiro Sakai** (b. 1928), retired from University of Pennsylvania and returned to Japan

1.1. *Derivations on finite dimensional algebras*

1.1.1. *Matrix multiplication—Associative algebras*

Let $M_n(\mathbf{C})$ denote the algebra of all n by n complex matrices, or more generally any finite dimensional semisimple associative algebra.

Definition 1.2. A derivation on $M_n(\mathbf{C})$ with respect to matrix multiplication is a linear mapping δ which satisfies the product rule

$$\delta(AB) = \delta(A)B + A\delta(B).$$

Proposition 1.1. *Fix a matrix A in $M_n(\mathbf{C})$ and define*

$$\delta_A(X) = AX - XA.$$

Then δ_A is a derivation with respect to matrix multiplication.

A more general form of the following theorem is due to Hochschild[4] [20]. For an interesting summary of Hochschild's career see[21] and[22]. Theorem 1.4 is probably due to either Wedderburn[5] or Noether[6]. The proof we present here is due to Jacobson[23].

Theorem 1.4. *Every derivation on $M_n(\mathbf{C})$ with respect to matrix multiplication is of the form δ_A for some A in $M_n(\mathbf{C})$.*

Proof. If δ is a derivation, consider the two representations of $M_n(\mathbf{C})$

$$z \mapsto \begin{bmatrix} z & 0 \\ 0 & z \end{bmatrix} \text{ and } z \mapsto \begin{bmatrix} z & 0 \\ \delta(z) & z \end{bmatrix}.$$

---

[4]**Gerhard Hochschild** (1915–2010) Gerhard Paul Hochschild was an American mathematician who worked on Lie groups, algebraic groups, homological algebra and algebraic number theory

[5]**Joseph Henry Maclagan Wedderburn** (1882–1948). Scottish mathematician, who taught at Princeton University for most of his career. A significant algebraist, he proved that a finite division algebra is a field, and part of the Artin–Wedderburn theorem on simple algebras. He also worked on group theory and matrix algebra

[6]**Amalie Emmy Noether** (1882–1935) Amalie Emmy Noether was an influential German mathematician known for her groundbreaking contributions to abstract algebra and theoretical physics. Described as the most important woman in the history of mathematics, she revolutionized the theories of rings, fields, and algebras. In physics, Noether's theorem explains the fundamental connection between symmetry and conservation laws

The first is a direct sum of two copies of the identity representation; but so is the second, since

$$\begin{bmatrix} 0 & 0 \\ \delta(z) & z \end{bmatrix} \text{ is equivalent to } \begin{bmatrix} 0 & 0 \\ 0 & z \end{bmatrix}.$$

So $\begin{bmatrix} z & 0 \\ \delta(z) & 0 \end{bmatrix} \begin{bmatrix} a & b \\ c & d \end{bmatrix} = \begin{bmatrix} a & b \\ c & d \end{bmatrix} \begin{bmatrix} z & 0 \\ 0 & z \end{bmatrix}$. Thus $az = za,\ bz = zb$ and

$$\delta(z)a = cz - zc \text{ and } \delta(z)b = dz - zd.$$

Since a and b are multiples of I they can't both be zero.

1.1.2. *Bracket multiplication—Lie algebras*

We next consider a multiplication on matrices (or associative algebras) which leads to the theory of Lie algebras[7].

Definition 1.3. A derivation on $M_n(\mathbf{C})$ with respect to bracket multiplication

$$[X, Y] = XY - YX$$

is a linear mapping δ which satisfies the product rule

$$\delta([A, B]) = [\delta(A), B] + [A, \delta(B)].$$

Proposition 1.2. *Fix a matrix A in $M_n(\mathbf{C})$ and define*

$$\delta_A(X) = [A, X] = AX - XA.$$

Then δ_A is a derivation with respect to bracket multiplication.

An algebra $\mathcal{L}$ with multiplication $(x, y) \mapsto [x, y]$ is a Lie algebra if

$$[x, x] = 0 \text{ and } [[x, y], z] + [[y, z], x] + [[z, x], y] = 0.$$

Left multiplication in a Lie algebra is denoted by $\operatorname{ad} x$: $\operatorname{ad} x(y) = [x, y]$. An associative algebra A becomes a Lie algebra A^- under the product, $[x, y] = xy - yx$.

The first axiom implies that $[x, y] = -[y, x]$ and the second (called the *Jacobi identity*) implies that $\operatorname{ad} x$ is a derivation, called an *inner derivation*, and $x \mapsto \operatorname{ad} x$ is a homomorphism of $\mathcal{L}$ into the Lie algebra $(\operatorname{End} \mathcal{L})^-$,

[7]**Sophus Lie** (1842–1899) Marius Sophus Lie was a Norwegian mathematician. He largely created the theory of continuous symmetry, and applied it to the study of geometry and differential equations

that is, $\mathrm{ad}\,[x,y] = [\mathrm{ad}\,x, \mathrm{ad}\,y]$. Assuming that $\mathcal{L}$ is finite dimensional, the Killing form is defined by $\lambda(x,y) = \mathrm{tr}\,\mathrm{ad}(x)\mathrm{ad}(y)$. The proof of the following theorem is not given in Meyberg's notes[24], but can be found in many places, for example[25].

Theorem 1.5 (CARTAN criterion[24]). *A finite dimensional Lie algebra $\mathcal{L}$ over a field of characteristic 0 is semisimple if and only if the Killing form is nondegenerate.*

A more general form of the following theorem is due to Hochschild[20]. Theorem 1.6 has been attributed to Zassenhaus[8] in[24], and to E. Cartan[9] in[26]. We have taken the proof from[24] p.42.

Theorem 1.6. *If the finite dimensional Lie algebra $\mathcal{L}$ over a field of characteristic 0 is semisimple (that is, its Killing form is nondegenerate), then every derivation is inner.*

Proof. Let D be a derivation of $\mathcal{L}$. Then $D \cdot \mathrm{ad}(x) = \mathrm{ad}(Dx) + \mathrm{ad}(x) \cdot D$. Since $x \mapsto \mathrm{tr}\,D \cdot \mathrm{ad}\,(x)$ is a linear form, there exists $d \in \mathcal{L}$ such that

$$\mathrm{tr}\,D \cdot \mathrm{ad}\,(x) = \lambda(d,x) = \mathrm{tr}\,\mathrm{ad}\,(d) \cdot \mathrm{ad}\,(x).$$

Let E be the derivation $E = D - \mathrm{ad}\,(d)$ so that

$$\mathrm{tr}\,E \cdot \mathrm{ad}\,(x) = 0. \tag{1}$$

Note next that

$$\begin{aligned} E \cdot [\mathrm{ad}\,(x), \mathrm{ad}\,(y)] &= E \cdot \mathrm{ad}\,(x) \cdot \mathrm{ad}\,(y) - E \cdot \mathrm{ad}\,(y) \cdot \mathrm{ad}\,(x) \\ &= (\mathrm{ad}\,(x) \cdot E + [E, \mathrm{ad}\,(x)]) \cdot \mathrm{ad}\,(y) - E \cdot \mathrm{ad}\,(y) \cdot \mathrm{ad}\,(x) \end{aligned}$$

so that

$$\begin{aligned} &[E, \mathrm{ad}\,(x)] \cdot \mathrm{ad}\,(y) \\ &= E \cdot [\mathrm{ad}\,(x), \mathrm{ad}\,(y)] - \mathrm{ad}\,(x) \cdot E \cdot \mathrm{ad}\,(y) + E \cdot \mathrm{ad}\,(y) \cdot \mathrm{ad}\,(x) \\ &= E \cdot [\mathrm{ad}\,(x), \mathrm{ad}\,(y)] + [E \cdot \mathrm{ad}\,(y), \mathrm{ad}\,(x)] \end{aligned}$$

[8] **Hans Zassenhaus** (1912–1991) Hans Julius Zassenhaus was a German mathematician, known for work in many parts of abstract algebra, and as a pioneer of computer algebra

[9] **Elie Cartan** (1869–1951) Elie Joseph Cartan was an influential French mathematician, who did fundamental work in the theory of Lie groups and their geometric applications. He also made significant contributions to mathematical physics, differential geometry, and group theory. He was the father of another influential mathematician, Henri Cartan.

and $\operatorname{tr}[E, \operatorname{ad}(x)] \cdot \operatorname{ad}(y) = \operatorname{tr} E \cdot [\operatorname{ad}(x), \operatorname{ad}(y)]$. However, since E is a derivation

$$\begin{aligned}[E, \operatorname{ad}(x)] \cdot \operatorname{ad}(y) &= E \cdot \operatorname{ad}(x) \cdot \operatorname{ad}(y) - \operatorname{ad}(x) \cdot E \cdot \operatorname{ad}(y) \\ &= (\operatorname{ad}(Ex) + \operatorname{ad}(x) \cdot E) \cdot \operatorname{ad}(y) - \operatorname{ad}(x) \cdot E \cdot \operatorname{ad}(y) \\ &= \operatorname{ad}(Ex) \cdot \operatorname{ad}(y).\end{aligned}$$

Thus

$$\begin{aligned}\lambda(Ex, y) &= \operatorname{tr} \operatorname{ad}(Ex) \cdot \operatorname{ad}(y) \\ &= \operatorname{tr}[E, \operatorname{ad}(x)] \cdot \operatorname{ad}(y) \\ &= \operatorname{tr} E \cdot [\operatorname{ad}(x), \operatorname{ad}(y)] = 0 \text{ by } (1).\end{aligned}$$

Since x and y are arbitrary, $E = 0$ and so $D - \operatorname{ad}(d) = 0$.

1.1.3. *Circle multiplication—Jordan algebras*

We next consider a multiplication on matrices (or associative algebras) which leads to the theory of Jordan algebras[10]

Definition 1.4. A derivation on $M_n(\mathbf{C})$ with respect to circle multiplication

$$X \circ Y = (XY + YX)/2$$

is a linear mapping δ which satisfies the product rule

$$\delta(A \circ B) = \delta(A) \circ B + A \circ \delta(B)$$

Proposition 1.3. *Fix a matrix A in $M_n(\mathbf{C})$ and define*

$$\delta_A(X) = AX - XA.$$

Then δ_A is a derivation with respect to circle multiplication.

The first part of the following theorem of Sinclair is valid for all semisimple Banach algebras (and continuous Jordan derivations).

Theorem 1.7 (Sinclair 1970[27]). *Every derivation on $M_n(\mathbf{C})$ with respect to circle multiplication is also a derivation with respect to matrix multiplication, and is therefore of the form δ_A for some A in $M_n(\mathbf{C})$.*

---

[10]**Pascual Jordan** (1902–1980) Pascual Jordan was a German theoretical and mathematical physicist who made significant contributions to quantum mechanics and quantum field theory

According to the pioneering work of Jacobson[11] the above proposition and theorem need to be modified for the real subalgebra (with respect to circle multiplication) of Hermitian matrices.

Proposition 1.4. *Fix two hermitian matrices A, B in $M_n(\mathbf{C})$ and define*

$$\delta_{A,B}(X) = A \circ (B \circ X) - B \circ (A \circ X).$$

Then $\delta_{A,B}$ is a derivation of the real Jordan algebra of hermitian matrices with respect to circle multiplication.

In preparation for the next theorem, we first note that for any algebra, D is a derivation if and only if $[R_a, D] = R_{Da}$, where R_a denotes the operator of multiplication (on the right) by a. If you polarize the Jordan axiom $(a^2b)a = a^2(ba)$, you get $[R_a, [R_b, R_c]] = R_{A(b,a,c)}$ where

$$A(b, a, c) = (ba)c - b(ac)$$

is the "associator". From the commutative law $ab = ba$, you get

$$A(b, a, c) = [R_b, R_c]a$$

and so $[R_b, R_c]$ is a derivation, sums of which are called **inner**, forming an ideal in the Lie algebra of all derivations. The **Lie multiplication algebra** L of the Jordan algebra A is the Lie algebra generated by the multiplication operators R_a. It is given by

$$L = \{R_a + \sum_i [R_{b_i}, R_{c_i}] : a, b_i, c_i \in A\}$$

so that L is the sum of a Lie triple system (see subsection 2.2.2) and the ideal of inner derivations.

Theorem 1.8 (1949 Jacobson[26]). *Every derivation of a finite dimensional semisimple Jordan algebra J (in particular $M_n(\mathbf{C})$ with circle multiplication) is a sum of derivations of the form $\delta_{A,B}$ for some A's and B's elements of J.*

Proof. Let D be a derivation of a semisimple finite dimensional unital Jordan algebra A. Then $\tilde{D} : X \mapsto [X, D]$ is a derivation of L. It is well known to algebraists that $L = L' + C$ where L' (the derived algebra $[L, L]$)

[11]**Nathan Jacobson** (1910–1999) Nathan Jacobson was an American mathematician who was recognized as one of the leading algebraists of his generation, and he was also famous for writing more than a dozen standard monographs

is semisimple and C is the center of L. Also $\tilde{D}$ maps L' into itself and C to zero.

By Theorem 1.6, $\tilde{D}$ is an inner derivation of L' and hence also of L, so there exists $U \in L$ such that $[X, D] = [X, U]$ for all $X \in L$ and in particular $[R_a, D] = [R_a, U]$. Then

$$Da = R_{Da}1 = [R_a, D]1 = [R_a, U]1 = (R_a U - U R_a)1 = a \cdot U1 - Ua$$

so that $D = R_{U1} - U \in L$. Thus, $D = R_a + \sum [R_{b_i}, R_{c_i}]$ and so

$$0 = D1 = a + 0 = a.$$

We summarize the previous three theorems, insofar as they concern $M_n(\mathbf{C})$, in the following table. Since $M_n(\mathbf{C})$ is not a semisimple Lie algebra, its derivations must be taken modulo the center (See 3.5).

Table 1

$M_n(\mathbf{C})$ (SEMISIMPLE ALGEBRAS)

matrix	bracket	circle
$ab = a \times b$	$[a, b] = ab - ba$	$a \circ b = ab + ba$
Th. 1.4	Th. 1.6	Th. 1.7
$\delta_a(x)$ $=$ $ax - xa$	$\delta_a(x)$ $=$ $ax - xa$	$\delta_a(x)$ $=$ $ax - xa$

The following table shows some properties of the three kinds of multiplication considered and thus provides the axioms for various kinds of algebras which we will consider. These two tables also provide a model for the discussion of various kinds of triple systems which will be considered later.

Table 2

ALGEBRAS

commutative algebras

$ab = ba$

associative algebras

$a(bc) = (ab)c$

Lie algebras

$$a^2 = 0$$
$$(ab)c + (bc)a + (ca)b = 0$$

Jordan algebras

$$ab = ba$$
$$a(a^2b) = a^2(ab)$$

The algebra $M_n(\mathbf{C})$, with matrix multiplication, as well as the algebra $\mathcal{C}$, with ordinary multiplication, are examples of C^*-algebras (finite dimensional and commutative respectively). The following theorem, which is due to Kadison[12] and Sakai, and was preceded by some preliminary results of Kaplansky[13] thus explains Theorems 1.1 and 1.4. We will have much more to say about C*-algebras in subsection 3.1.

Theorem 1.9 (Sakai, Kadison 1966[28],[29]). *Every derivation of a C^*-algebra is of the form $x \mapsto ax - xa$ for some a in the weak operator closure of the C^*-algebra.*

Why are derivations important? One answer is given by cohomology, which is introduced here and revisited in later sections as the principal purpose of this paper.

Let M be a Banach algebra and X a Banach M-module. For every $n \geq 1$, let $L^n(M, X)$ = all continuous n-linear maps ($L^0(M, X) = X$). The coboundary operator is $\partial : L^n \to L^{n+1}$ (for $n \geq 1$), defined by

$$\begin{aligned}\partial\phi(a_1, \cdots, a_{n+1}) &= a_1\phi(a_2, \cdots, a_{n+1}) \\ &\quad + \sum(-1)^j\phi(a_1, \cdots, a_{j-1}, a_ja_{j+1}, \cdots, a_{n+1}) \\ &\quad + (-1)^{n+1}\phi(a_1, \cdots, a_n)a_{n+1}.\end{aligned}$$

For $n = 0$, $\partial : X \to L(M, X)$, $\partial x(a) = ax - xa$ and so $\mathrm{Im}\partial$ is the space of inner derivations. Since $\partial \circ \partial = 0$,

$$\mathrm{Im}(\partial : L^{n-1} \to L^n) \subset \ker(\partial : L^n \to L^{n+1}),$$

and $H^n(M, X) = \ker\partial/\mathrm{Im}\partial$ is a vector space. For $n = 1$,

$$\ker\partial = \{\phi : M \to X : a_1\phi(a_2) - \phi(a_1a_2) + \phi(a_1)a_2 = 0\}$$

---

[12]**Richard Kadison** (b. 1925) Richard V. Kadison is an American mathematician known for his contributions to the study of operator algebras

[13]**Irving Kaplansky** (1917–2006) Kaplansky made major contributions to group theory, ring theory, the theory of operator algebras and field theory

is the space of continuous derivations from M to X. Thus,

$$H^1(M,X) = \frac{\text{derivations from } M \text{ to } X}{\text{inner derivations from } M \text{ to } X}$$

measures how close continuous derivations are to inner derivations.

Later we shall discuss what the spaces $H^2(M,X)$, $H^3(M,X),\ldots$ measure. In anticipation of this, we present some miscellaneous remarks.

(1) One of the earliest papers on continuous cohomology is the pioneering[30] in which it is shown that $H^1(C(\Omega),E) = H^2(C(\Omega),E) = 0$.
(2) "The major open question in the theory of derivations on C*-algebras" is whether $H^1(A,B(H)) = 0$ $(A \subset B(H))$.
(3) A derivation from A into $B(H)$ is inner if and only if it is completely bounded. ([31] Theorem 3.1)
(4) The origin of continuous cohomology is the memoir of Barry Johnson[32] [14].

1.2. *Exercises (Gradus ad Parnassum)—Algebras*

(1) Prove the analog of Proposition 1.1 for associative algebras: Fix an element a in an associative algebra A and define

$$\delta_a(x) = [a,x] = ax - xa.$$

Then δ_a is a derivation with respect to the product of A.

(2) Prove the analog of Proposition 1.2 for associative algebras: Fix an element a in an associative algebra A and define

$$\delta_a(x) = [a,x] = ax - xa.$$

Then δ_a is a derivation with respect to bracket multiplication

$$[x,y] = xy - yx.$$

(3) Prove the analog of Proposition 1.3 for associative algebras: Fix an element a in an associative algebra A and define

$$\delta_a(x) = [a,x] = ax - xa.$$

Then δ_a is a derivation with respect to circle multiplication

$$x \circ y = xy + yx.$$

---

[14] **Barry Johnson** (1937–2002) see[33] for his fundamental contributions to cohomology of Banach algebras, among other things

(4) Let a and b be two fixed elements in an associative algebra A. Show that the linear mapping $\delta_{a,b}(x) = a \circ (b \circ x) - b \circ (a \circ x)$ is a derivation of A with respect to circle multiplication $x \circ y = xy + yx$. (cf. Proposition 1.4)

(5) Show that an associative algebra is a Lie algebra with respect to bracket multiplication.

(6) Show that an associative algebra is a Jordan algebra with respect to circle multiplication.

(7) Let us write $\delta_{a,b}$ for the linear mapping $\delta_{a,b}(x) = a(bx) - b(ax)$ in a Jordan algebra. Show that $\delta_{a,b}$ is a derivation of the Jordan algebra by following the outline below. (cf. problem 4 above.)

(a) In the Jordan algebra axiom

$$u(u^2v) = u^2(uv),$$

replace u by $u + w$ to obtain the equation

$$2u((uw)v) + w(u^2v) = 2(uw)(uv) + u^2(wv) \tag{2}$$

(b) In (1), interchange v and w and subtract the resulting equation from (1) to obtain the equation

$$2u(\delta_{v,w}(u)) = \delta_{v,w}(u^2). \tag{3}$$

(c) In (2), replace u by $x + y$ to obtain the equation

$$\delta_{v,w}(xy) = y\delta_{v,w}(x) + x\delta_{v,w}(y),$$

which is the desired result.

2. Triple systems

In this section, all algebras and triple systems will generally be finite dimensional. The first subsection parallels subsection 1.1. Infinite dimensional cases and references for the proofs will be given in section 3.

2.1. *Derivations on finite dimensional triple systems*

2.1.1. *Triple matrix multiplication*

Let $M_{m,n}(\mathbf{C})$ denote the linear space of rectangular m by n complex matrices and note that it is closed under the operation $(A, B, C) \mapsto AB^*C$, which we will call triple matrix multiplication.

Definition 2.1. A <u>derivation</u> on $M_{m,n}(\mathbf{C})$ with respect to triple matrix multiplication is a linear mapping δ which satisfies the triple product rule $\delta(AB^*C) = \delta(A)B^*C + A\delta(B)^*C + AB^*\delta(C)$.

Proposition 2.1. *For two matrices, $A \in M_m(\mathbf{C}), B \in M_n(\mathbf{C})$, such that $A^* = -A, B^* = -B$, define $\delta_{A,B}(X) = AX + XB$. Then $\delta_{A,B}$ is a derivation with respect to triple matrix multiplication.*

Theorem 2.1. *Every derivation on $M_{m,n}(\mathbf{C})$ with respect to triple matrix multiplication is of the form $\delta_{A,B}$.*

The proof of Theorem 2.1 can be obtained by applying the result of Theorem 2.3 below to the symmetrized product $(AB^*C + CB^*A)/2$.

Remark 2.1. These results hold true and are of interest for $m = n$.

2.1.2. *Triple bracket multiplication*

Let's go back for a moment to square matrices and the bracket multiplication. Motivated by the last remark, we define the triple bracket multiplication to be $(X, Y, Z) \mapsto [[X, Y], Z]$.

Definition 2.2. A <u>derivation</u> on $M_n(\mathbf{C})$ with respect to triple bracket multiplication is a linear mapping δ which satisfies the triple product rule $\delta([[A, B], C]) = [[\delta(A), B], C] + [[A, \delta(B)], C] + [[A, B], \delta(C)]$.

Proposition 2.2. *Fix two matrices A, B in $M_n(\mathbf{C})$ and define*

$$\delta_{A,B}(X) = [[A, B], X].$$

Then $\delta_{A,B}$ is a derivation with respect to triple bracket multiplication.

$M_n(\mathbf{C})$ is not a semisimple Lie algebra or semisimple Lie triple system, so a modification needs to be made when considering Lie derivations or Lie triple derivations on it, as in Theorems 3.17 and 3.18. The proof of the following theorem is taken from[24] Chapter 6. We do not define Lie triple system or semisimple here.

Theorem 2.2. *Every derivation of a finite dimensional semisimple Lie triple system F is a sum of derivations of the form $\delta_{A,B}$, for some A's and B's in the triple system. These derivations are called inner derivations and their set is denoted Inder$F$.*

Proof. Let F be a finite dimensional semisimple Lie triple system (over a field of characteristic 0) and suppose that D is a derivation of F. Let L be the Lie algebra $(\operatorname{Inder} F) \oplus F$ with product

$$[(H_1, x_1), (H_2, x_2)] = ([H_1, H_2] + L(x_1, x_2), H_1 x_2 - H_2 x_1).$$

A derivation of L is defined by $\delta(H \oplus a) = [D, H] \oplus Da$. Together with the definition of semisimple Lie triple system, it is proved in[24] that F semisimple implies L semisimple. Thus there exists $U = H_1 \oplus a_1 \in L$ such that $\delta(X) = [U, X]$ for all $X \in L$. Then

$$0 \oplus Da = \delta(0 \oplus a) = [H_1 + a_1, 0 \oplus a] = L(a_1, a) \oplus H_1 a$$

so $L(a_1, a) = 0$ and $D = H_1 \in \operatorname{Inder} F$.

2.1.3. *Triple circle multiplication*

Let's now return to rectangular matrices and form the triple circle multiplication $(AB^*C + CB^*A)/2$. For sanity's sake, let us write this as

$$\{A, B, C\} = (AB^*C + CB^*A)/2.$$

Definition 2.3. A <u>derivation</u> on $M_{m,n}(\mathbf{C})$ with respect to triple circle multiplication is a linear mapping δ which satisfies the triple product rule $\delta(\{A, B, C\}) = \{\delta(A), B, C\} + \{A, \delta(B), C\} + \{B, A, \delta(C)\}$.

Proposition 2.3. *Fix two matrices A, B in $M_{m,n}(\mathbf{C})$ and define*

$$\delta_{A,B}(X) = \{A, B, X\} - \{B, A, X\}.$$

Then $\delta_{A,B}$ is a derivation with respect to triple circle multiplication.

The following theorem is a special case of Theorem 2.8 below.

Theorem 2.3. *Every derivation of $M_{m,n}(\mathbf{C})$ with respect to triple circle multiplication is a sum of derivation of the form $\delta_{A,B}$.*

We summarize Theorems 2.1, 2.2 and 2.3, insofar as they concern $M_n(\mathbf{C})$, in the following table. Since $M_n(\mathbf{C})$ is not a semisimple Lie triple system, its derivations must be taken modulo the center (cf. 3.5).

Table 3

$M_{m,n}(\mathbf{C})$ (SS TRIPLE SYSTEMS)

triple matrix	triple bracket	triple circle
ab^*c	$[[a,b],c]$	ab^*c+cb^*a
Th. 2.1	Th. 2.2	Th. 2.3
$\delta_{a,b}(x)$ $=$ ab^*x $+xb^*a$ $-ba^*x$ $-xa^*b$	$\delta_{a,b}(x)$ $=$ abx $+xba$ $-bax$ $-xab$	$\delta_{a,b}(x)$ $=$ ab^*x $+xb^*a$ $-ba^*x$ $-xa^*b$
(sums)	(sums) $(m=n)$	(sums)

2.2. *Axiomatic approach for triple systems*

Definition 2.4. A triple system is defined to be a vector space with one ternary operation called triple multiplication. Addition is denoted by $a+b$ and is commutative and associative:

$$a+b=b+a, \quad (a+b)+c=a+(b+c).$$

Triple multiplication is denoted (simply and temporarily) by abc and is required to be linear (conjugate-linear in some cases) in each of its three variables:

$$(a+b)cd=acd+bcd, \quad a(b+c)d=abd+acd, \quad ab(c+d)=abc+abd.$$

Simple but important examples of triple systems can be formed from any algebra: if ab denotes the algebra product, just define a triple multiplication to be $(ab)c$

Let's see how this works in the algebras we introduced in subsection 1.1.

$\mathcal{C}$; $fgh=(fg)h$, OR $fgh=(f\overline{g})h$

$(M_n(\mathbf{C}),\times)$; $abc=abc$ OR $abc=ab^*c$

$(M_n(\mathbf{C}),[,])$; $abc=[[a,b],c]$

$(M_n(\mathbf{C}),\circ)$; $abc=(a\circ b)\circ c$

All of these except the last one are useful. It turns out that the appropriate form of the triple product for Jordan algebras is

$$abc=(a\circ b)\circ c+(c\circ b)\circ a-(a\circ c)\circ b,$$

since then you obtain a Jordan triple system (see subsection 2.2.3). Also for the second example, by taking the symmetrized triple product, in each case you also obtain a Jordan triple system.

Definition 2.5. A triple system is said to be associative if the triple multiplication is associative in the sense that $ab(cde) = (abc)de = a(bcd)e$ (**first kind**) or $ab(cde) = (abc)de = a(dcb)e$ (**second kind**); and commutative if $abc = cba$.

2.2.1. *Associative triple systems*

The axiom which characterizes triple matrix multiplication is

$$(abc)de = ab(cde) = a(dcb)e.$$

The triple systems so defined are called **associative triple systems** (of the second kind) or **Hestenes**[15] **algebras.**

Theorem 2.4 (Lister 1971[34]**).** *Every derivation of a finite dimensional semisimple associative triple system (of the first kind) is inner.*

Theorem 2.5 (Carlsson 1976[35]**).** *Every derivation of a finite dimensional semisimple associative triple system (first or second kind) into a module, is inner.*

The following theorem really belongs in section 3 since it concerns Banach associative triple systems. TROs play an important role in Part III, the definition is in subsection 8.4.1.

Theorem 2.6 (Zalar 1995[36]**).** *Let $W \subset B(H,K)$ be a TRO which contains all the compact operators. If D is a derivation of W with respect to the associative triple product ab^*c then there exist $a = -a^* \in B(K)$ and $b = -b^* \in B(H)$ such that $Dx = ax + xb$.*

This result has been extended to $B(X,Y)$ (X, Y Banach spaces) in[37].

[15]**Magnus Hestenes** (1906–1991) Magnus Rudolph Hestenes was an American mathematician. Together with Cornelius Lanczos and Eduard Stiefel, he invented the conjugate gradient method

2.2.2. *Lie triple systems*

The axioms which characterize triple bracket multiplication are

$$aab = 0, \quad abc + bca + cab = 0 \quad \text{and} \quad de(abc) = (dea)bc + a(deb)c + ab(dec).$$

The triple systems so defined are called **Lie triple systems** and were developed by Jacobson and Koecher[16].

Theorem 2.7 (Lister 1952[38]). *Every derivation of a finite dimensional semisimple Lie triple system is inner.*

2.2.3. *Jordan triple systems*

The axioms which characterize triple circle multiplication are

$$abc = cba \quad \text{and} \quad de(abc) = (dea)bc - a(edb)c + ab(dec).$$

The triple systems so defined are called **Jordan triple systems** and were developed principally by **Kurt Meyberg, Ottmar Loos and Erhard Neher**, among many others.

Theorem 2.8. *Every derivation of a finite dimensional semisimple Jordan triple system is inner.*

We outline a proof following a construction in[39] (cf.[24] Chapter 11). For simplicity, we assume non-degeneracy of a Jordan triple system[39] p. 25. Let V be a Jordan triple and let $\mathcal{L}(V)$ be its TKK Lie algebra (**Tits-Kantor-Koecher**). $\mathcal{L}(V) = V \oplus V_0 \oplus V$ and the Lie product is given by

$$[(x,h,y),(u,k,v)] = (hu - kx,\ [h,k] + x\square v - u\square y,\ k^\natural y - h^\natural v).$$

Here, $a\square b$ is the left multiplication operator $x \mapsto \{abx\}$ (also called the box operator), $V_0 = \text{span}\{V\square V\}$ is a Lie subalgebra of $\mathcal{L}(V)$ and for $h = \sum_i a_i \square b_i \in V_0$, the map $h^\natural : V \to V$ is defined by

$$h^\natural = \sum_i b_i \square a_i.$$

[16]**Max Koecher** (1924–1990) Max Koecher was a German mathematician. His main research area was the theory of Jordan algebras, where he introduced the Kantor-Koecher-Tits construction

Let $\theta : \mathcal{L}(V) \to \mathcal{L}(V)$ be the main involution $\theta(x \oplus h \oplus y) = y \oplus -h^\natural \oplus x$. Let $\delta : V \to V$ be a derivation of a Jordan triple V, with TKK Lie algebra $(\mathcal{L}(V), \theta)$. Given $a, b \in V$, we define

$$\begin{aligned} D(a,0,0) &= (\delta a, 0, 0) \\ D(0,0,b) &= (0,0,\delta b) \\ D(0, a\square\, b, 0) &= (0,\ \delta a \square\, b + a \square\, \delta b,\ 0) \end{aligned}$$

and extend D linearly on $\mathcal{L}(V)$. For the details of the proof of the following theorem and other results on Jordan triple cohomology, see the forthcoming paper[40].

Theorem 2.9. *Let V be a Jordan triple with TKK Lie algebra $(\mathcal{L}(V), \theta)$. There is a one-one correspondence between the triple derivations of V and the Lie derivations $D : \mathcal{L}(V) \to \mathcal{L}(V)$ satisfying $D(V) \subset V$ and $D\theta = \theta D$. Under this correspondence, D is an inner Lie derivation if and only if $D|_V$ is an inner triple derivation.*

Since the TKK Lie algebra $\mathcal{L}(V)$ of a semisimple Jordan triple system is semisimple, Theorem 2.8 follows from Theorem 1.6 and Theorem 2.9.

We summarize the previous three definitions in the following table

Table 4

TRIPLE SYSTEMS

associative triple systems

$$(abc)de = ab(cde) = a(dcb)e$$

Lie triple systems

$$aab = 0$$
$$abc + bca + cab = 0$$
$$de(abc) = (dea)bc + a(deb)c + ab(dec)$$

Jordan triple systems

$$abc = cba$$
$$de(abc) = (dea)bc - a(edb)c + ab(dec)$$

2.3. *Exercises (Gradus ad Parnassum)—Triple systems*

(1) Prove the analog of Proposition 2.1 for associative algebras A with involution: For two elements $a, b \in A$ with $a^* = -a, b^* = -b$, define $\delta_{a,b}(x) = ax + xb$. Then $\delta_{a,b}$ is a derivation with respect to the triple multiplication ab^*c. (Use the notation $\langle abc\rangle$ for ab^*c)

(2) Prove the analog of Proposition 2.2 for associative algebras A: Fix two elements $a, b \in A$ and define $\delta_{a,b}(x) = [[a,b],x]$. Then $\delta_{a,b}$ is a derivation with respect to the triple multiplication $[[a,b],c]$. (Use the notation $[abc]$ for $[[a,b],c]$)

(3) Prove Proposition 2.3: Fix two matrices a, b in $M_{m,n}(\mathbf{C})$ and define $\delta_{a,b}(x) = \{a,b,x\} - \{b,a,x\}$. Then $\delta_{a,b}$ is a derivation with respect to triple circle multiplication. (Use the notation $\{abc\}$ for $(ab^*c+cb^*a)/2$)

(4) Show that $M_n(\mathbf{C})$ is a Lie triple system with respect to triple bracket multiplication. In other words, show that the three axioms for Lie triple systems in Table 4 are satisfied if abc denotes

$$[[a,b],c] = (ab-ba)c - c(ab-ba) \quad (a, b \text{ and } c \text{ denote matrices}).$$

(Use the notation $[abc]$ for $[[a,b],c]$)

(5) Show that $M_{m,n}(\mathbf{C})$ is a Jordan triple system with respect to triple circle multiplication. In other words, show that the two axioms for Jordan triple systems in Table 4 are satisfied if abc denotes

$$(ab^*c + cb^*a)/2 \quad (a, b \text{ and } c \text{ denote matrices}).$$

(Use the notation $\{abc\}$ for $(ab^*c + cb^*a)/2$)

(6) Let us write $\delta_{a,b}$ for the linear process

$$\delta_{a,b}(x) = abx$$

in a Lie triple system. Show that $\delta_{a,b}$ is a derivation of the Lie triple system by using the axioms for Lie triple systems in Table 4. (Use the notation $[abc]$ for the triple product in any Lie triple system, so that, for example, $\delta_{a,b}(x)$ is denoted by $[abx]$)

(7) Let us write $\delta_{a,b}$ for the linear process

$$\delta_{a,b}(x) = abx - bax$$

in a Jordan triple system. Show that $\delta_{a,b}$ is a derivation of the Jordan triple system by using the axioms for Jordan triple systems in Table 4. (Use the notation $\{abc\}$ for the triple product in any Jordan triple system, so that, for example, $\delta_{a,b}(x) = \{abx\} - \{bax\}$)

(8) On the Jordan algebra $M_n(\mathbf{C})$ with the circle product $a \circ b = ab + ba$, define a triple product

$$\{abc\} = (a \circ b) \circ c + (c \circ b) \circ a - (a \circ c) \circ b.$$

Show that $M_n(\mathbf{C})$ is a Jordan triple system with this triple product. Hint: show that $\{abc\} = 2abc + 2cba$

(9) On the vector space $M_n(\mathbf{C})$, define a triple product $\langle abc\rangle = abc$ (matrix multiplication without the adjoint in the middle). Formulate the definition of a derivation of the resulting triple system, and state and prove a result corresponding to Proposition 2.1. Is this triple system associative?

(10) In an associative algebra, define a triple product $\langle abc\rangle$ to be abc. Show that the algebra becomes an associative triple system with this triple product.

(11) In an associative triple system with triple product denoted $\langle abc\rangle$, define a binary product ab to be $\langle aub\rangle$, where u is a fixed element. Show that the triple system becomes an associative algebra with this product. Suppose further that $\langle auu\rangle = \langle uua\rangle = a$ for all a. Show that we get a unital involutive algebra with involution $a^\sharp = \langle uau\rangle$.

(12) In a Lie algebra with product denoted by $[a, b]$, define a triple product $[abc]$ to be $[[a, b], c]$. Show that the Lie algebra becomes a Lie triple system with this triple product. ([24] ch. 6, ex. 1, p. 43)

(13) Let A be an algebra (associative, Lie, or Jordan; it doesn't matter). Show that the set $\mathcal{D} := \mathrm{Der}\,(A)$ of all derivations of A is a Lie subalgebra of $\mathrm{End}\,(A)$. That is, $\mathcal{D}$ is a linear subspace of the vector space of linear transformations on A, and if $D_1, D_2 \in \mathcal{D}$, then $D_1D_2 - D_2D_1 \in \mathcal{D}$.

(14) Let A be a triple system (associative, Lie, or Jordan; it doesn't matter). Show that the set $\mathcal{D} := \mathrm{Der}\,(A)$ of derivations of A is a Lie subalgebra of $\mathrm{End}\,(A)$. That is, $\mathcal{D}$ is a linear subspace of the vector space of linear transformations on A, and if $D_1, D_2 \in \mathcal{D}$, then $D_1D_2 - D_2D_1 \in \mathcal{D}$.

2.4. *Supplemental exercises (Gradus ad Parnassum)—Algebras and triple systems*

(1) In an arbitrary Jordan triple system, with triple product denoted by $\{abc\}$, define a triple product by

$$[abc] = \{abc\} - \{bac\}.$$

Show that the Jordan triple system becomes a Lie triple system with this new triple product.
([24] ch. 11, Th. 1, p. 108)

(2) In an arbitrary associative triple system, with triple product denoted by $\langle abc\rangle$, define a triple product by

$$[xyz] = \langle xyz\rangle - \langle yxz\rangle - \langle zxy\rangle + \langle zyx\rangle.$$

Show that the associative triple system becomes a Lie triple system with this new triple product.
([24] ch. 6, ex. 3, p. 43)

(3) In an arbitrary Jordan algebra, with product denoted by xy, define a triple product by $[xyz] = x(yz) - y(xz)$. Show that the Jordan algebra becomes a Lie triple system with this new triple product.
([24] ch. 6, ex. 4, p. 43)

(4) In an arbitrary Jordan triple system, with triple product denoted by $\{abc\}$, fix an element y and define a binary product by

$$ab = \{ayb\}.$$

Show that the Jordan triple system becomes a Jordan algebra with this (binary) product.
([24] ch. 10, Th. 1, p. 94—using different language; see also[41] Prop. 19.7, p. 317)

(5) In an arbitrary Jordan algebra with multiplication denoted by ab, define a triple product

$$\{abc\} = (ab)c + (cb)a - (ac)b.$$

Show that the Jordan algebra becomes a Jordan triple system with this triple product. ([24] ch. 10, p. 93—using different language; see also[41] Cor. 19.10, p. 320)

(6) Show that every Lie triple system, with triple product denoted $[abc]$ is a subspace of some Lie algebra, with product denoted $[a, b]$, such that $[abc] = [[a, b], c]$.
([24] ch. 6, Th. 1, p. 45)

(7) Find out what a semisimple associative algebra is and prove that every derivation of a finite dimensional semisimple associative algebra is inner, that is, of the form $x \mapsto ax - xa$ for some fixed a in the algebra.
([20] Theorem 2.2)

(8) Find out what a semisimple Lie algebra is and prove that every derivation of a finite dimensional semisimple Lie algebra is inner, that is, of the form $x \mapsto [a, x]$ for some fixed a in the algebra.
([24] ch. 5, Th. 2, p. 42; see also[20] Theorem 2.1)

(9) Find out what a semisimple Jordan algebra is and prove that every derivation of a finite dimensional semisimple Jordan algebra is inner, that is, of the form $x \mapsto \sum_{i=1}^{n}(a_i(b_i x) - b_i(a_i x))$ for some fixed elements $a_1, \ldots, a_n$ and $b_1, \ldots, b_n$ in the algebra.[42] p. 320 and[43] p. 270

(10) In an associative triple system with triple product $\langle xyz \rangle$, show that you get a Jordan triple system with the triple product

$$\{xyz\} = \langle xyz \rangle + \langle zyx \rangle.$$

Then use Theorem 2.3 to prove Theorem 2.1.

(11) Find out what a semisimple associative triple system is and prove that every derivation of a finite dimensional semisimple associative triple system is inner (also find out what inner means in this context). ([35])

(12) Find out what a semisimple Lie triple system is and prove that every derivation of a finite dimensional semisimple Lie triple system is inner, that is, of the form $x \mapsto \sum_{i=1}^{n} [a_i b_i x]$ for some fixed elements $a_1, \ldots, a_n$ and $b_1, \ldots, b_n$ in the Lie triple system.
([24] ch. 6, Th.10, p. 57)

(13) Find out what a semisimple Jordan triple system is and prove that every derivation of a finite dimensional semisimple Jordan triple system is inner, that is, of the form $x \mapsto \sum_{i=1}^{n} (\{a_i b_i x\} - \{b_i a_i x\})$ for some fixed elements $a_1, \ldots, a_n$ and $b_1, \ldots, b_n$ in the Jordan triple system.
([24] ch. 11, Th. 8, p.123 and Cor. 2, p. 124)

3. Derivations on operator algebras and operator triple systems

The theory of derivations on operator algebras is an important and well-investigated part of the general theory of operator algebras, with application in mathematical physics. Investigation of unbounded derivations were motivated mainly by the needs of mathematical physics, in particular by the problem of constructing the dynamics in quantum statistical mechanics. We do not consider unbounded derivations *per se* in this paper, cf. [44,45].

In this section we shall review the literature about derivations on associative and non associative operator algebras and operator triple systems, with the main focus being on some new results concerning the notions of ternary weak amenability and triple derivations on von Neumann algebras.

There are two basic questions concerning derivations of Banach algebras and Banach triple systems, namely, whether they are automatically continuous, and to what extent are they all inner, in the appropriate context. The automatic continuity of various algebraic mappings plays important roles in the general theory of Banach algebras and in particular in operator algebra theory. These questions are significant in the case of a derivation of a space into itself, or into an appropriate module.

We shall consider derivations primarily in the following contexts:

(i) C*-ALGEBRAS (more generally, associative Banach algebras)
(ii) JC*-ALGEBRAS (more generally, Jordan Banach algebras)
(iii) JC*-TRIPLES (more generally, Banach Jordan triples)

One could and should also consider:

(i′) Banach associative triple systems
(ii′) Banach Lie algebras
(iii′) Banach Lie triple systems

3.1. *C*-algebras*

In the context of associative algebras, a derivation is a linear mapping satisfying the property $D(ab) = a \cdot Db + Da \cdot b$, and an inner derivation is a derivation of the form ad $\mathrm{x}(a) = x \cdot a - a \cdot x$.

The main automatic continuity results for C^*-algebras are due to Kaplansky (commutative and type I cases), Sakai, and Ringrose[17]. The main inner derivation results are due primarily to Sakai, Kadison, Connes[18], and Haagerup[19]. We won't define the terms amenable and nuclear here. We refer to the excellent notes by Runde[46] for these definitions and much other information. Weak amenability will be defined shortly, as it is one of the main objects of our discussion.

Theorem 3.1 (Sakai 1960[14]). *Every derivation from a C^*-algebra into itself is continuous.*

[17]**John Ringrose** (b. 1932) John Ringrose is a leading world expert on non-self-adjoint operators and operator algebras. He has written a number of influential texts including Compact non-self-adjoint operators (1971) and, with R V Kadison, Fundamentals of the theory of operator algebras in four volumes published in 1983, 1986, 1991 and 1992

[18]**Alain Connes** (b. 1947) Alain Connes is the leading specialist on operator algebras. In his early work on von Neumann algebras in the 1970s, he succeeded in obtaining the almost complete classification of injective factors. Following this he made contributions in operator K-theory and index theory, which culminated in the Baum-Connes conjecture. He also introduced cyclic cohomology in the early 1980s as a first step in the study of noncommutative differential geometry. Connes has applied his work in areas of mathematics and theoretical physics, including number theory, differential geometry and particle physics

[19]**Uffe Haagerup** (b. 1950) Haagerup's research is in operator theory, and covers many subareas in the subject which are currently very active - random matrices, free probability, C*-algebras and applications to mathematical physics

Theorem 3.2 (Ringrose 1972[47]). *Every derivation from a C^*-algebra into a Banach A-**bimodule** is continuous.*

Theorem 3.3 (Sakai[28], Kadison[29]). *Every derivation of a C^*-algebra is of the form $x \mapsto ax - xa$ for some a in the weak closere of the C^*-algebra*

Theorem 3.4 (Connes 1976[48]). *Every amenable C^*-algebra is nuclear.*

Theorem 3.5 (Haagerup 1983[49]). *Every nuclear C^*-algebra is amenable.*

Theorem 3.6 (Haagerup 1983[49]). *Every C^*-algebra is weakly amenable.*

3.2. *Algebras of unbounded operators*

3.2.1. *Murray-von Neumann algebras*

Physics considerations demand that the Hamiltonian of a quantum system will, in general, correspond to an unbounded operator on a Hilbert space H. These unbounded operators will not lie in a von Neumann algebra, but they may be affiliated with the von Neumann algebra corresponding to the quantum system. In general, unbounded operators do not behave well with respect to addition and multiplication. However, as noted by Murray and von Neumann[50], for the finite von Neumann algebras, their families of affiliated operators do form a *-algebra. Thus, it natural to study derivations of algebras that include such unbounded operators.

A closed densely defined operator T on a Hilbert space H is affiliated with a von Neumann algebra R when $UT = TU$ for each unitary operator U in R', the commutant of R. If operators S and T are affiliated with R, then S + T and ST are densely defined, preclosed and their closures are affiliated with R. Such algebras are referred to as **Murray-von Neumann algebras**.

If R is a finite von Neumann algebra, we denote by $A_f(R)$ its associated Murray-von Neumann algebra. It is natural to conjecture that every derivation of $A_f(R)$ should be inner. In[51] Theorem 5 it is proved that extended derivations of $A_f(R)$ (those that map R into R) are inner. In[51] Theorem 12, it is proved that each derivation of $A_f(R)$ with R a factor of type II_1 that maps $A_f(R)$ into R is 0. In other words, by requiring that the range of the derivation is in R, the bounded part of $A_f(R)$, allows a noncommutative unbounded version of the well-known Singer-Wermer conjecture. This is extended to the general von Neumann algebra of type II_1 in[52].

We turn our attention now to commutators in $A_f(N)$, where N is a finite von Neumann algebra. The following theorem is related to the work of Kadison and Liu described above.

Theorem 3.7 (Theorem 4.13 of[53]). *Let N be II_1-factor. If the element b of N is a commutator of self-adjoint elements in $A_f(N)$, then b has trace zero. In particular, the scalar operator $i1 \in N$ is not a commutator of self-adjoint elements in $A_f(N)$.*

This has been complemented in[54] as follows.

Theorem 3.8 (Theorem 4 of[54]). *Let $a, b \in A_f(N)$ where N is a II_1-factor. Assume that either a, b belong to the Haagerup-Schultz algebra (a *-subalgebra of $A_f(N)$, see[55]) or if $ab \in A_f(N)$ is conjugate to a self-adjoint element. If $[a, b] = \lambda 1$, then $\lambda = 0$.*

As for sums of commutators, we have the following theorem.

Theorem 3.9 (Theorems 5 and 6 of[54]). *Let N be a II_1-factor. There exist $a, b, c, d \in A_f(N)$ such that $1 = [a, b] + [c, d]$. Every element of $A_f(N)$ is the sum of two commutators.*

3.2.2. *Algebras of measurable operators*

Noncommutative integration theory was initiated by Segal ([56]), who considered new classes of algebras of unbounded operators, in particular the algebra $S(M)$ of all measurable operators affiliated with a von Neumann algebra M. A study of the derivations on the algebra $S(M)$ was initiated by Ayupov[20] in 2000. For example, in the commutative case $M = L^\infty(\Omega, \Sigma, \mu)$, $S(M)$ is isomorphic to the algebra $L^0(\Omega)$ of all complex measurable functions and it is shown in[57] that $L^0(0, 1)$ admits nonzero derivations which are discontinuous in the measure topology. The study of derivations on various subalgebras of the algebra $LS(M)$ of all locally measurable operators in the general semifinite case was initiated in[58,59], with the most complete results obtained in the type I case,[60]. See the survey[61] for more detail up to 2010.

It was shown in[60] (and in[62] for separable predual) that if M is properly infinite and of type I, then every derivation of $LS(M)$ is continuous in the

[20]**Shavkat Ayupov** (b. 1952) is the Director of the Institute of Mathematics, National University of Uzbekistan, well-known for his work in operator algebras and quantum probability, as well as in the theory of Leibniz algebras

local measure topology $t(M)$ on $LS(M)$. The same holds for M of type III as shown in[61] and for type II_∞ in[63].

The following is the main theorem of[64] and illustrates the state of the art.

Theorem 3.10. *Let M be any von Neumann algebra. Every derivation on the *-algebra $LS(M)$ continuous with respect to the topology $t(M)$ is inner.*

Corollary 3.1. *If M is a properly infinite von Neumann algebra, then every derivation on $LS(M)$ is inner.*

3.3. *A bridge to Jordan algebras*

A *Jordan derivation* from a Banach algebra A into a Banach A-module is a linear map D satisfying $D(a^2) = aD(a) + D(a)a$, $(a \in A)$, or equivalently, $D(ab + ba) = aD(b) + D(b)a + D(a)b + bD(a)$, $(a, b \in A)$. Sinclair proved in 1970[27] that a bounded Jordan derivation from a semisimple Banach algebra to itself is a derivation, although this result fails for derivations of semisimple Banach algebras into a Banach bi-module.

Nevertheless, a celebrated result of B.E. Johnson in 1996[65] states that every bounded Jordan derivation from a C*-algebra A to a Banach A-bimodule is an associative derivation. In view of the intense interest in automatic continuity problems in the past half century, it was natural to consider the following problem.

Problem 3.1. *Is every Jordan derivation from a C*-algebra A to a Banach A-bimodule automatically continuous (and hence a derivation, by Johnson's theorem)?*

In 2004, J. Alaminos, M. Brešar and A.R. Villena[66] gave a positive answer to the above problem for some classes of C*-algebras including the class of abelian C*-algebras. Combining a theorem of Cuntz from 1976[67] with the theorem just quoted yields

Theorem 3.11. *Every Jordan derivation from a C*-algebra A to a Banach A-module is continuous.*

In the same way, using the solution in 1996 by Hejazian-Niknam[68] in the commutative case we have

Theorem 3.12. *Every Jordan derivation from a C*-algebra A to a <u>Jordan</u> Banach A-module is continuous.*

(Jordan module will be defined below)

These last two results are among the consequences of some work of Peralta and Russo[3], described below, on automatic continuity of derivations into Jordan triple modules.

3.4. *JC*-algebras*

In the context of Jordan algebras and Jordan modules, a derivation is a linear mapping satisfying the property $D(a \circ b) = a \circ Db + Da \circ b$, and an inner derivation is a derivation of the form $\sum_i [L(x_i)L(a_i) - L(a_i)L(x_i)]$, $(x_i \in M, a_i \in A)$, that is, $b \mapsto \sum_i [x_i \circ (a_i \circ b) - a_i \circ (x_i \circ b)]$.

The main automatic continuity results for JC^*-algebras are due to Upmeier[21], Hejazian-Niknam, and Alaminos-Bresar-Villena. The main inner derivation results are due to Jacobson and Upmeier. We have already mentioned the results of Hejazian-Niknam and Alaminos-Bresar-Villena in this context.

Theorem 3.13 (Upmeier 1980[69]). *Every derivation of a reversible JC^*-algebra extends to a derivation of its enveloping C^*-algebra.*

In particular, this theorem improves Theorem 1.7 in the case of a C^*-algebra.

Theorem 3.14 (Jacobson 1951[42]). *Every derivation of a finite dimensional semisimple Jordan algebra into a (Jordan) module is inner.*

The proof of this theorem uses the theory of Lie algebras and Lie triple systems.

Because of the structure theory for real Jordan operator algebras, the following theorem completely answers the question of when all derivations are inner on a JBW-algebra.

Theorem 3.15 (Upmeier 1980 (Theorem 3.10 in[69])). *1. Purely exceptional JBW-algebras have the inner derivation property, that is, every derivation is inner*
2. Reversible JBW-algebras have the inner derivation property
3. $\oplus L^\infty(S_j, U_j)$ has the inner derivation property if and only if $\sup_j \dim U_j < \infty$ (U_j spin factors).

---

[21]**Harald Upmeier** (b. 1950), Professor at University of Marburg, Germany, pioneer of topological Jordan structures

3.5. *Lie derivations of operator algebras and Banach algebras*

In this subsection, we record some results on derivations of the Lie algebra structure of operator algebras, including triple structure, which is one theme of this survey paper.

Theorem 3.16 (Miers 1973[70]). *If M is a von Neumann algebra, $[M,M]$ the Lie algebra linearly generated by*

$$\{[X,Y] = XY - YX : X, Y \in M\}$$

and $L : [M,M] \to M$ a Lie derivation, i.e., L is linear and

$$L[X,Y] = [LX,Y] + [X,LY],$$

then L has an extension $D : M \to M$ that is a derivation of the associative algebra.

The proof involves matrix-like computations. Using Theorem 1.9 leads to

Theorem 3.17 (Miers 1973[70]). *If $L : M \to M$ is a Lie derivation, then $L = D + \lambda$, where D is an associative derivation and λ is a linear map into the center of M vanishing on $[M,M]$.*

A Lie triple derivation of an associative algebra M is a linear map $L : M \to M$ such that

$$L[[X,Y],Z] = [[L(X),Y],Z] + [[X,L(Y)],Z] + [[X,Y],L(Z)]$$

for all $X, Y, Z \in M$.

Theorem 3.18 (Miers 1978[71]). *If M is a von Neumann algebra with no central Abelian summands and L is a Lie triple derivation, then there exists an operator $A \in M$ such that $L(X) = [A,X] + \lambda(X)$ where λ, from M into Z_M, is a linear map which annihilates brackets of operators in M.*

Notice that no assumption of continuity was made in these theorems. Continuous Lie derivations of associative Banach algebras have also been studied outside the operator algebra context, as evidenced in the following.

Theorem 3.19 (Johnson 1996[65]). *Every continuous Lie derivation of a symmetrically amenable Banach algebra A into a Banach A-module X is the sum of an associative derivation and a "trivial" derivation, that is, a linear map which vanishes on commutators and maps into the "center" of the module. The same holds if A is a C^*-algebra.*

As shown in[72] and[73], the continuity assumption can be dropped when $X = A$ and A is a C*-algebra or a semisimple symmetrically amenable Banach algebra.

According to a review by Garth Dales[74]: "It remains an open question whether an analogous result (automatic continuity) for Lie derivations from A into a Banach A-bimodule holds when A is an arbitrary C*-algebra and when A is an arbitrary symmetrically amenable Banach algebra. It is also an interesting open question whether or not every Lie derivation on a semisimple Banach algebra to itself has this form."

As pointed out to the author by Martin Mathieu, a negative answer has been given, by Read[75], to the last question raised by Dales in his review.

A search for "Lie derivation" in the "review text" option in MathSciNet turns up 172 entries. A good number of these concern Lie derivations on nonselfadjoint operator algebras, for example nest algebras, reflexive algebras, CSL algebras, UHF-algebras[22]. Several others are concerned with von Neumann algebras. We quote a few of them.

Theorem 3.20 (Ji and Qi 2011 Corollary 2.1 in[79]). *Let N be an arbitrary nest on a Hilbert space H of dimension greater than 2, and $AlgN$ be the associated nest algebra. Let δ be a linear mapping on $AlgN$ which satisfies the Lie derivation property on commutators $[a,b]$ with $ab = 0$. Then there exists an operator $r \in AlgN$ and a linear center valued map τ vainshing on such commutators such that $\delta(a) = ra - ar + \tau(a)$.*

Theorem 3.21 (Zhuraev 2011[80]). *For a von Neumann algebra M on a Hilbert space H denote by $S(M)$ the unital *-algebra of all measurable operators affiliated with M. If M is a homogeneous von Neumann algebra of type I_n, then very Lie derivation on $S(M)$ can be uniquely decomposed into the sum of a derivation and a centre-valued trace.*

In any algebra A set $p_1(x) = x$ and for $n \geq 2$,

$$p_n(x_1, \dots, x_n) = [p_{n-1}(x_1, \dots, x_{n-1}), x_n].$$

Thus $p_2(x_1, x_2) = [x_1, x_2]$, $p_3(x_1, x_2, x_3) = [[x_1, x_2], x_3]$. A map $\varphi : A \to A$ is called a multiplicative Lie n-derivation if

$$\varphi(p_n(x_1, \dots, x_n)) = \sum_i^n p_n(x_1, \dots, \varphi(x_i), x_{i+1}, \dots, x_n).$$

[22]It is worth mentioning in connection with section 5, that there are corresponding results for the cohomology of nonselfadjoint operator algebras. See references [14]-[17],[38],[39] in[76] as well as the recent papers[77] and[78]

Theorem 3.22 (Abdullaev 1992[81]). *Every n-Lie derivation L on some von Neumann algebra M (or on its skew-adjoint part) can be decomposed as $L = D + E$, where D is an ordinary derivation on M and E is a linear map from M into its center which annihilates the Lie products.*

3.6. *JC*-triples*

In the context of Jordan triple systems, a derivation is a linear or conjugate linear mapping satisfying the property

$$D\{a, b, c\} = \{Da.b, c\} + \{a, Db, c\} + \{a, b, Dc\},$$

and an inner derivation is a derivation of the form $\sum_i [L(x_i, a_i) - L(a_i, x_i)]$ $(x_i \in M, a_i \in A)$, that is, $b \mapsto \sum_i [\{x_i, a_i, b\} - \{a_i, x_i, b\}]$.

For reasons related to the definition of Jordan triple module in the complex case, and which are discussed in both[2] and[3], although a derivation from a Jordan triple system to itself is linear, nevertheless, in the complex case, a derivation of a Jordan triple system into a Jordan triple module (other than the triple itself) is declared to be a conjugate linear map.

The main automatic continuity results for JC^*-triples, where the triple product is given by $\{x, y, z\} = (xy^*z + zy^*x)/2$ are due to Barton[23] and Friedman[24]; and Peralta[25] and Russo[26], and the main inner derivation results are due to Ho-Martinez-Peralta-Russo, Meyberg, Kühn-Rosendahl, and Ho-Peralta-Russo.

Before stating these results, we must acknowledge the pioneering efforts of Harris[27] to the study of Jordan operator triple systems, especially for the papers[82] and[83] devoted respectively to infinite dimensional holomorphy and spectral and ideal theory.

Theorem 3.23 (Barton-Friedman 1990[84]). *Every derivation of a JB^*-triple is continuous.*

---

[23]**Tom Barton** (b. 1955) Tom Barton is Senior Director for Architecture, Integration and CISO at the University of Chicago. He had similar assignments at the University of Memphis, where he was a member of the mathematics faculty before turning to administration

[24]**Yaakov Friedman** (b. 1948) Yaakov Friedman is director of research at Jerusalem College of Technology

[25]**Antonio Peralta** (b. 1974), Professor at University of Granada, Spain

[26]**Bernard Russo** (b. 1939), Professor Emeritus at University of California, Irvine

[27]**Lawrence A. Harris** (PhD 1969), Professor at University of Kentucky

Theorem 3.24 (Peralta-Russo 2010[3]). *Necessary and sufficient conditions under which a derivation of a JB*-triple into a Jordan triple module is continuous.*

(JB*-triple and Jordan triple module are defined below)

The proofs of the following two theorems rely heavily on the theories of Lie algebras and Lie triple systems.

Theorem 3.25 (Meyberg 1972[24]). *Every derivation of a finite dimensional semisimple Jordan triple system is inner.*

Theorem 3.26 (Kühn-Rosendahl 1978[85]). *Every derivation of a finite dimensional semisimple Jordan triple system into a Jordan triple module is inner.*

The proofs of the following two theorems rely on operator theory and functional analysis.

Theorem 3.27 (Ho-Martínez-Peralta-Russo 2002 (Th. 2 in[1])**).** *Cartan factors of type $I_{n,n}$, II (even or ∞), and III have the inner derivation property.*

Theorem 3.28 (Ho-Martínez-Peralta-Russo 2002[1]). *Infinite dimensional Cartan factors of type $I_{m,n}, m \neq n$, and IV do not have the inner derivation property.*

3.6.1. *Ternary weak amenability*

Recall that a Banach algebra A is said to be *amenable* if every bounded derivation of A into a dual A-module is inner, and *weakly amenable* if every (bounded) derivation from A to A^* is inner. A Jordan Banach triple E is said to be *weakly amenable* or *ternary weakly amenable* if every continuous triple derivation from E into its dual space is necessarily inner. The main results of[2] are:

(1) Commutative C^*-algebras are ternary weakly amenable.
(2) Commutative JB^*-triples are approximately weakly amenable, that is, every derivation into its dual is approximated in norm by inner derivations.
(3) $B(H), K(H)$ are ternary weakly amenable if and only if they are finite dimensional.
(4) Cartan factors of type $I_{m,n}$ of finite rank with $m \neq n$, and of type IV are ternary weakly amenable if and only if they are finite dimensional.

We provide a taste of the details of one of the results of[2].

Lemma 3.1. *The C^*-algebra $A = K(H)$ of all compact operators on an infinite dimensional Hilbert space H is not Jordan weakly amenable, that is, not every Jordan derivation into the dual is inner.*

Proof: We shall identify A^* with the trace-class operators on H. Supposing that A were Jordan weakly amenable, let $\psi \in A^*$ be arbitrary. Then D_ψ ($= \operatorname{ad}\psi$) is an associative derivation and hence a Jordan derivation, so by assumption would be an inner Jordan derivation. Thus there would exist $\varphi_j \in A^*$ and $b_j \in A$ such that

$$D_\psi(x) = \sum_{j=1}^{n} [\varphi_j \circ (b_j \circ x) - b_j \circ (\varphi_j \circ x)]$$

for all $x \in A$. For $x, y \in A$, a direct calculation yields

$$\psi(xy - yx) = -\frac{1}{4}\left(\sum_{j=1}^{n} b_j\varphi_j - \varphi_j b_j\right)(xy - yx).$$

It is known (Pearcy-Topping 1971[86]) that every compact operator on a separable (which we may assume without loss of generality) infinite dimensional Hilbert space is a finite sum of commutators of compact operators, and it follows that the trace-class operator

$$\psi = -\frac{1}{4}\left(\sum_{j=1}^{n} b_j\varphi_j - \varphi_j b_j\right)$$

is a finite sum of commutators of compact and trace-class operators, and hence has trace zero. This is a contradiction, since ψ was arbitrary, completing the proof.

3.6.2. *Automatic continuity*

We shall now give some details about the paper[3] which has been mentioned several times already. This involves Jordan triples and Jordan triple modules.

A complex (resp., real) *Jordan triple* is a complex (resp., real) vector space E equipped with a non-trivial triple product $(x, y, z) \mapsto \{xyz\}$, from $E \times E \times E$ to E, which is bilinear and symmetric in the outer variables and conjugate linear (resp., linear) in the middle one, satisfying the so-called *"Jordan Identity"*:

$$L(a,b)L(x,y) - L(x,y)L(a,b) = L(L(a,b)x,y) - L(x,L(b,a)y), \qquad (1)$$

for all a, b, x, y in E, where $L(x,y)z := \{xyz\}$.

The Jordan identity can also be written in the form

$$\{ab\{xyz\}\} - \{xy\{abz\}\} = \{\{abx\}yz\} - \{x\{bay\}z\},$$

which is called the "main identity."

A JB*-algebra is a complex Jordan Banach algebra A equipped with an algebra involution * satisfying $\|\{a,a^*,a\}\| = \|a\|^3$, for all $a \in A$. (Recall that $\{a,a^*,a\} = 2(a\circ a^*)\circ a - a^2\circ a^*)$. A (complex) *JB*-triple* is a complex Jordan Banach triple E satisfying the following axioms:

(a) For each a in E the map $L(a,a)$ is an hermitian operator on E with non negative spectrum.
(b) $\|\{a,a,a\}\| = \|a\|^3$ for all a in A.

Every C*-algebra (resp., every JB*-algebra) is a JB*-triple with respect to the product $\{abc\} = \frac{1}{2}\,(ab^*c + cb^*a)$

$$(\text{resp., } \{abc\} := (a\circ b^*)\circ c + (c\circ b^*)\circ a - (a\circ c)\circ b^*).$$

In order to motivate the definition of triple module, we first recall two basic definitions. If A is an associative algebra, an *A-bimodule* is a vector space X, equipped with two bilinear products $(a,x)\mapsto ax$ and $(a,x)\mapsto xa$ from $A\times X$ to X satisfying the following axioms:

$$a(bx) = (ab)x,\ \ a(xb) = (ax)b \text{ and } (xa)b = x(ab),$$

for every $a, b\in A$ and $x\in X$. If J is a Jordan algebra, a *Jordan J-module* is a vector space X, equipped with two bilinear products $(a,x)\mapsto a\circ x$ and $(x,a)\mapsto x\circ a$, from $J\times X$ to X, satisfying:

$$a\circ x = x\circ a,$$
$$a^2\circ(x\circ a) = (a^2\circ x)\circ a$$
$$2((x\circ a)\circ b)\circ a + x\circ(a^2\circ b) = 2(x\circ a)\circ(a\circ b) + (x\circ b)\circ a^2,$$

for every $a, b\in J$ and $x\in X$.

If E is a complex Jordan triple, a *Jordan triple E-module* (also called *triple E-module*) is a vector space X equipped with three mappings

$$\{.,.,.\}_1 : X\times E\times E\to X$$

$$\{.,.,.\}_2 : E \times X \times E \to X$$
$$\{.,.,.\}_3 : E \times E \times X \to X$$

satisfying:

(1) $\{x, a, b\}_1$ is linear in a and x and conjugate linear in b, $\{abx\}_3$ is linear in b and x and conjugate linear in a and $\{a, x, b\}_2$ is conjugate linear in a, b, x.
(2) $\{x, b, a\}_1 = \{a, b, x\}_3$, and $\{a, x, b\}_2 = \{b, x, a\}_2$ for every $a, b \in E$ and $x \in X$.
(3) Denoting by $\{.,.,.\}$ any of the products $\{.,.,.\}_1$, $\{.,.,.\}_2$ and $\{.,.,.\}_3$, the main identity for Jordan triple modules

$$\{a, b, \{c, d, e\}\} = \{\{a, b, c\}, d, e\} - \{c, \{b, a, d\}, e\} + \{c, d, \{a, b, e\}\}$$

holds whenever one of the elements a, b, c, d, e is in X and the rest are in E.

It is a little bit laborious to check that the dual space, E^*, of a complex (resp., real) Jordan Banach triple E is a complex (resp., real) triple E-module with respect to the products:

$$\{a, b, \varphi\}(x) = \{\varphi, b, a\}(x) := \varphi\{b, a, x\} \tag{2}$$
$$\{a, \varphi, b\}(x) := \overline{\varphi\{a, x, b\}}, \forall \varphi \in E^*, a, b, x \in E. \tag{3}$$

For each submodule S of a triple E-module X, we define its *quadratic annihilator*, $\text{Ann}_E(S)$, as the set $\{a \in E : Q(a)(S) = \{a, S, a\} = 0\}$.

Theorem 3.29. *Let E be a complex JB*-triple, X a Banach triple E-module, and let $\delta : E \to X$ be a triple derivation. Then δ is continuous if and only if $Ann_E(\sigma_X(\delta))$ is a (norm-closed) linear subspace of E and*

$$\{Ann_E(\sigma_X(\delta)), Ann_E(\sigma_X(\delta)), \sigma_X(\delta)\} = 0.$$

Corollary 3.2. *Let E be a real or complex JB*-triple. Then*

(a) Every derivation $\delta : E \to E$ is continuous. (Barton-Friedman)
(b) Every derivation $\delta : E \to E^*$ is continuous (motivated ternary weak amenability). In particular,
(c) Automatic continuity of derivation of JB-algebra into a Jordan module (Hejazian-Niknam)*
(d) Automatic continuity of derivation of C-algebra into a module (Ringrose)*
(e) Automatic continuity of Jordan derivation of C-algebra into a Jordan module (Hejazian-Niknam)*

3.7. *Triple derivations on von Neumann algebras*

As noted in subsection 3.1, building on earlier work of Kadison[29], Sakai[28] proved that very derivation of a von Neumann algebra into itself is inner. Building on earlier work of Bunce and Paschke[87], Haagerup, on his way to proving that every C*-algebra is weakly amenable, showed in[49] that every derivation of a von Neumann algebra into its predual is inner. Thus the first Hochschild cohomology groups $H^1(M, M)$ and $H^1(M, M_*)$ vanish for any von Neumann algebra M.

It is also known that every triple derivation of a von Neumann algebra into itself is an inner triple derivation (Theorem 3.27), and every Jordan derivation of a von Neumann algebra into itself is an inner Jordan derivation (Theorem 3.15). Triple derivations and inner triple derivations into a triple module are defined and discussed in subsection 3.6.2.[28] However, in this subsection we shall only be concerned with these concepts for von Neumann algebras, with their associated Jordan structure.

Two consequences of[2] are that finite dimensional von Neumann algebras and abelian von Neumann algebras have the property that every triple derivation into the predual is an inner triple derivation, analogous to the Haagerup result. This property was called **normal ternary weak amenability** in[88] where it is shown that it rarely holds in a general von Neumann algebra, but that it comes close. The main results of[88] are the following two theorems. The first one gives a cohomological characterization of finite factors and the second gives a "zero-one" law for factors.

Theorem 3.30. *Let M be a von Neumann algebra.*

(a) *If every triple derivation of M into M_* is approximated in norm by inner triple derivations, then M is finite.*
(b) *If M is a finite von Neumann algebra acting on a separable Hilbert space or if M is a finite factor, then every triple derivation of M into M_* is approximated in norm by inner triple derivations.*

Theorem 3.31. *If M is a properly infinite factor, then the real vector space of triple derivations of M into M_*, modulo the norm closure of the inner triple derivations, has dimension 1.*

[28]In this subsection, we use the terms 'triple' and 'ternary' interchangeably, while mindful that in some quarters 'triple' means 'Jordan triple' and 'ternary' refers to an associative triple setting, such as a TRO (ternary ring of operators)

Inner triple derivations on a von Neumann algebra M into its predual M_* are closely related to the span of commutators of normal functionals with elements of M, denoted by $[M_*, M]$. As noted earlier, for a finite factor of type I_n, all triple derivations into the predual are inner triple derivations. A consequence of[88] Proposition 4.1 and the work of Dykema and Kalton[89] is that no factor of type II_1 is normally ternary weakly amenable ([88] Corollary 4.3), even though each triple derivation into the predual is a norm limit of inner triple derivations by Theorem 3.30.

It is also shown in[88] that a finite countably decomposable von Neumann algebra M of type I_n is normally ternary weakly amenable if and only if every element of M_* of central trace zero belongs to $[M_*, M]$. In particular, for such an algebra M with a faithful normal finite trace, if every element of M_* of trace zero belongs to $[M_*, M]$ then M is normally ternary weakly amenable.

3.8. *Local derivations on JB^*-triples*

Linear maps which agree with a derivation at each point are called local derivations. These have been studied in the Banach setting by Kadison[90], Johnson[91], Ajupov et. al[92],[93],[94], among others. Kadison proved that a continuous local derivation from a von Neumann algebra into a Banach module is a derivation.

Let X and Y be Banach spaces. According to the terminology employed in the literature (see, for example,[95]), a subset $\mathcal{D}$ of the Banach space $B(X,Y)$, of all bounded linear operators from X into Y, is called *algebraically reflexive* in $B(X,Y)$ when it satisfies the property:

$$T \in B(X,Y) \text{ with } T(x) \in \mathcal{D}(x),\ \forall x \in X \Rightarrow T \in \mathcal{D}. \tag{4}$$

Algebraic reflexivity of $\mathcal{D}$ in the space $L(X,Y)$, of all linear mappings from X into Y, a stronger version of the above property not requiring continuity of T, is defined by:

$$T \in L(X,Y) \text{ with } T(x) \in \mathcal{D}(x),\ \forall x \in X \Rightarrow T \in \mathcal{D}. \tag{5}$$

In 1990, Kadison proved that (4) holds if $\mathcal{D}$ is the set $\mathrm{Der}(M,X)$ of all (associative) derivations on a von Neumann algebra M into a dual M-bimodule X[90]. Johnson extended Kadison's result by establishing that the set $\mathcal{D} = \mathrm{Der}(A,X)$, of all (associative) derivations from a C*-algebra A into a Banach A-bimodule X satisfies (5)[91].

Michael Mackey gave a talk on this topic at the conference in honor of Cho-Ho Chu's 65th birthday in May 2012 in Hong Kong. He proved that every continuous local derivation on a JBW^*-triple is a derivation (see[96]), and he suggested some problems, among them whether every local derivation on a JB^*-triple into itself or into a Banach module is automatically continuous, and if so, whether it is a derivation. There are other problems in this area, some involving nonlinear maps. Many of these have now been answered.

Algebraic reflexivity of the set of triple derivations on a C*-algebra and on a JB*-triple have been studied in[96–98] and[99]. More precisely, Mackey proves in[96] that the set $\mathcal{D} = \text{Der}_t(M)$, of all triple derivations on a JBW*-triple M satisfies (4). The result has been supplemented in[98], where Burgos, Fernndez-Polo and Peralta prove that for each JB*-triple E, the set $\mathcal{D} = \text{Der}_t(E)$ of all triple derivations on E satisfies (5). In what follows, *algebraic reflexivity* will refer to the stronger version (5) which does not assume the continuity of T.

In[100], Brešar and Šemrl proved that the set of all (algebra) automorphisms of $B(H)$ is algebraically reflexive whenever H is a separable, infinite-dimensional Hilbert space. Given a Banach space X. A linear mapping $T : X \to X$ satisfying the hypothesis at (5) for $\mathcal{D} = \text{Aut}(X)$, the set of automorphisms on X, is called a *local automorphism*. Larson and Sourour showed in[101] that for every infinite dimensional Banach space X, every surjective local automorphism T on the Banach algebra $B(X)$, of all bounded linear operators on X, is an automorphism.

Motivated by the results of Šemrl in[102], references witness a growing interest in a subtle version of algebraic reflexivity called *algebraic 2-reflexivity* (cf.[103–110] and[111]). A subset $\mathcal{D}$ of the set $\mathcal{M}(X,Y) = Y^X$, of all mappings from X into Y, is called *algebraically 2-reflexive* when the following property holds: for each mapping T in $\mathcal{M}(X,Y)$ such that for each $a, b \in X$, there exists $S = S_{a,b} \in \mathcal{D}$ (depending on a and b), with $T(a) = S_{a,b}(a)$ and $T(b) = S_{a,b}(b)$, then T lies in $\mathcal{D}$. A mapping $T : X \to Y$ satisfying the property that for each $a, b \in X$, there exists $S = S_{a,b} \in \mathcal{D}$ (depending on a and b), with $T(a) = S_{a,b}(a)$ and $T(b) = S_{a,b}(b)$ will be called a 2-local $\mathcal{D}$-mapping.

Šemrl establishes in[102] that for every infinite-dimensional separable Hilbert space H, the sets $\text{Aut}(B(H))$ and $\text{Der}(B(H))$, of all (algebra) automorphisms and associative derivations on $B(H)$, respectively, are algebraically 2-reflexive in $\mathcal{M}(B(H)) = \mathcal{M}(B(H), B(H))$. Ayupov and Kudaybergenov proved in[103] that the same statement remains true for general Hilbert spaces (see[112] for the finite dimensional case). Actually, the set $\text{Hom}(A)$, of all homomorphisms on a general C*-algebra A, is algebraically 2-reflexive in the Banach algebra $B(A)$, of all bounded linear operators on A, and the set *-Hom(A), of all *-homomorphisms on A, is algebraically 2-reflexive in the space $L(A)$, of all linear operators on A (cf.[113]).

In recent contributions, Burgos, Fernández-Polo and Peralta prove that the set Hom(M) (respectively, $\text{Hom}_t(M)$), of all homomorphisms (respectively, triple homomorphisms) on a von Neumann algebra (respectively, on a JBW*-triple) M, is an algebraically 2-reflexive subset of $\mathcal{M}(M)$ (cf.[105],[106], respectively), while Ayupov and Kudaybergenov establish that the set Der(M) of all derivations on M is algebraically 2-reflexive in $\mathcal{M}(M)$ (see[104]).

In[114] the set $\text{Der}_t(A)$ of all triple derivations on a C*-algebra A is considered. We recall that every C*-algebra A can be equipped with a ternary product of the form

$$\{a,b,c\} = \frac{1}{2}(ab^*c + cb^*a).$$

When A is equipped with this product it becomes a JB*-triple in the sense of[115]. A linear mapping $\delta : A \to A$ is said to be a *triple derivation* when it satisfies the (triple) Leibnitz rule:

$$\delta\{a,b,c\} = \{\delta(a),b,c\} + \{a,\delta(b),c\} + \{a,b,\delta(c)\}.$$

According to the standard notation, 2-local $\text{Der}_t(A)$-mappings from A into A are called *2-local triple derivations.*

The main result of[114] is the following theorem.

Theorem 3.32. *Every (not necessarily linear nor continuous) 2-local triple derivation on an arbitrary von Neumann algebra M is a triple derivation (hence linear and continuous), equivalently, $Der_t(M)$ is algebraically 2-reflexive in $\mathcal{M}(M)$.*

Two recent extensive surveys of the material in this subsection are: S. Ayupov, K. Kudaybergenov, A. M. Peralta, "A survey on local and 2-local derivatins on C*-algebras and von Neumann algebras," and S. Ayupov, K. Kudaybergenov, "Derivations, local, and 2-local derivations on algebras of measurable operators", both of which are to appear in a forthcoming volume of Contemporary Mathematics.

Part II
Cohomology

4. Cohomology of finite dimensional algebras

4.1. *Associative algebras*

The starting point for the cohomology theory of algebras is the paper of Hochschild from 1945[116]. The standard reference of the theory is[4]. Two other useful references are due to Weibel ([117],[118]). We review here the definitions of the cohomology groups $H^n(M,X)$ and the interpretation of them in the cases $n=1,2$.

Let M be an associative algebra and X a two-sided M-module. For $n \geq 1$, let $L^n(M,X)$ be the linear space of all n-linear maps from M to X ($L^0(M,X)=X$). The coboundary operator is the linear mapping $\partial : L^n \to L^{n+1}$ (for $n \geq 1$) defined by

$$\begin{aligned}\partial\phi(a_1,\cdots,a_{n+1}) &= a_1\phi(a_2,\cdots,a_{n+1}) \\ &+ \sum(-1)^j\phi(a_1,\cdots,a_{j-1},a_ja_{j+1},\cdots,a_{n+1}) \\ &+ (-1)^{n+1}\phi(a_1,\cdots,a_n)a_{n+1}.\end{aligned}$$

For $n=0$, $\partial : X \to L(M,X)$ is defined by $\partial x(a) = ax - xa$. Since $\partial\circ\partial = 0$, $\mathrm{Im}(\partial : L^{n-1} \to L^n) \subset \ker(\partial : L^n \to L^{n+1})$, and $H^n(M,X) = \ker\partial/\mathrm{Im}\partial$ is a vector space.

For $n=1$, $\ker\partial = \{\phi : M \to X : a_1\phi(a_2) - \phi(a_1a_2) + \phi(a_1)a_2 = 0\}$ which is the space of derivations from M to X; and $\partial : X \to L(M,X)$ is given by $\partial x(a) = ax - xa$ so that $\mathrm{Im}\,\partial$ is the space of inner derivations

Thus $H^1(M,X)$ measures how close derivations are to inner derivations.

An associative algebra B is an extension of associative algebra A if there is a homomorphism σ of B onto A. The extension splits if $B = \ker\sigma \oplus A^*$ where A^* is an algebra isomorphic to A, and is singular if $(\ker\sigma)^2 = 0$.

Proposition 4.1. *There is a one to one correspondence between isomorphism classes of singular extensions of A and $H^2(A,A)$*

4.2. *Lie algebras*

Shortly after the introduction of the cohomology for associative algebras was introduced, there appeared a corresponding theory for Lie algebras. We follow[25] for the definitions and initial results. Applications can be found in[5] and[6].

If L is a Lie algebra, then an L-module is a vector space M and a mapping of $M \times L$ into M, $(m, x) \mapsto mx$, satisfying $(m_1 + m_2)x = m_1x + m_2x$, $\alpha(mx) = (\alpha m)x = m(\alpha x)$, and $m[x_1, x_2] = (mx_1)x_2 - (mx_2)x_1$.

Let L be a Lie algebra, M an L-module. If $i \geq 1$, an *i-dimensional M-cochain* for L is a skew symmetric i-linear mapping f of $L \times L \times \cdots \times L$ into M. Skew symmetric means that if two arguments in $f(x_1, \cdots, x_i)$ are interchanged, the value of f changes sign. A 0-dimensional cochain is a constant function from L to M.

The coboundary operator δ (for $i \geq 1$) is:

$$\delta(f)(x_1, \cdots, x_{i+1}) = \sum_{q=1}^{i+1} (-1)^{i+1} f(x_1, \cdots, \hat{x}_q, \cdots, x_{i+1})x_q + \sum_{q<r=2}^{i+1} (-1)^{r+q} f(x_1, \cdots, \hat{x}_q, \cdots, \hat{x}_r, \cdots, x_{i+1}, [x_q, x_r])$$

and for $i = 0$, $\delta(f)(x) = ux$ (module action), if f is the constant $u \in M$.

One verifies that $\delta^2 = 0$ giving rise to cohomology groups

$$H^i(L, M) = Z^i(L, M)/B^i(L, M).$$

If $i = 0$ we take $B^i = 0$ and

$$H^0(L, M) = Z^0(L, M) = \{u \in M : ux = 0, \forall x \in L\}.$$

Theorem 4.1 (Whitehead lemmas). *If L is a finite dimensional semisimple Lie algebra over a field of characteristic 0, then*

$$H^1(L, M) = H^2(L, M) = 0$$

for every finite dimensional module M of L.

Theorem 4.2 (Whitehead). *If L is a finite dimensional semisimple Lie algebra over a field of characteristic 0, then $H^i(L, M) = 0$ ($\forall i \geq 0$) for every finite dimensional <u>irreducible</u> module M of L such that $ML \neq 0$.*

4.3. *Jordan algebras*

The cohomology theory for Jordan algebras is less well developed than for associative and Lie algebras. A starting point would seem to be the papers of Gerstenhaber in 1964[119] and Glassman in 1970[120], which concern arbitrary nonassociative algebras. A study focussed primarily on Jordan algebras is[121].

Let A be an algebra defined by a set of identities and let M be an A-module. A singular extension of length 2 is, by definition, a null extension of A by M. A null extension is a short exact sequence

$$0 \to M \overset{\alpha}{\to} E \overset{\beta}{\to} A \to 0$$

where, provisionally, M is an algebra (rather than an A-module) such that $M^2 = 0$. If $n > 2$, a singular extension of length n is an exact sequence of bimodules

$$0 \to M \to M_{n-1} \to \cdots \to M_2 \to E \to A \to 0.$$

Morphisms, equivalences, addition, and scalar multiplication of equivalence classes of singular extensions can be defined. Then for $n \geq 2$, $H^n(A, M) :=$ equivalence classes of singular extensions of length n. These definitions are equivalent to the classical ones in the associative and Lie cases.

It is shown in[119] (using generalized projective resolutions) that if $0 \to M' \to M \to M'' \to 0$ is an exact sequence of A-bimodules, then there are natural homomorphisms δ^n so that the long sequence

$$0 \to \text{Der}\,(A, M') \to \text{Der}\,(A, M) \to \text{Der}\,(A, M'') \overset{\delta^1}{\to}$$

$$H^2(A, M') \to H^2(A, M) \to H^2(A, M'') \overset{\delta^2}{\to}$$

$$H^3(A, M') \to H^3(A, M) \to H^3(A, M'') \to$$

$$\cdots \to H^n(A, M'') \overset{\delta^n}{\to} H^{n+1}(A, M') \to \cdots$$

is exact. In particular,

$$H^n(A, M) = \ker \delta^n / \text{im}\, \delta^{n-1} \quad (n \geq 2).$$

What about $H^0(A, M)$ and $H^1(A, M)$ and Jordan algebras? For this we turn first to Glassman's thesis of 1968, embodied in[121] and[120]. Given an algebra A, consider the functor $\mathcal{T}$ from the category $\mathcal{C}$ of A-bimodules and A-homomorphisms to the category $\mathcal{V}$ of vector spaces and linear maps:

$$M \in \mathcal{C} \mapsto \text{Der}\,(A, M) \in \mathcal{V}, \quad \eta \mapsto \tilde{\eta}$$

where $\eta \in \text{Hom}_A(M_1, M_2)$ and $\tilde{\eta} \in \text{Hom}(\text{Der}\,(A, M_1), \text{Der}\,(A, M_2))$ is given by $\tilde{\eta} = \eta \circ D$. $\left(A \overset{D}{\to} M_1 \overset{\eta}{\to} M_2\right)$

An <u>inner derivation functor</u> is a subfunctor $\mathcal{J}$ which respects epimorphisms, that is,

$$M \in \mathcal{C} \mapsto \mathcal{J}(A, M) \subset \text{Der}\,(A, M) \in \mathcal{V}$$

and if $\eta \in \text{Hom}_A(M_1, M_2)$ is onto, then so is $\mathcal{J}(A, \eta) := \tilde{\eta}|\mathcal{J}(A, M)$.

Relative to the choice of $\mathcal{J}$ one defines $H^1(A, M) = \mathrm{Der}\,(A, M)/\mathcal{J}(A, M)$. (The definition of $H^0(A, M)$ is more involved so is omitted here). Glassman then proves that

$$0 \to H^0(A, M') \to H^0(A, M) \to H^0(A, M'') \overset{\delta^0}{\to}$$

$$H^1(A, M') \to H^1(A, M) \to H^1(A, M'') \overset{\delta^1}{\to}$$

$$H^2(A, M') \to H^2(A, M) \to H^2(A, M'') \overset{\delta^2}{\to}$$

$$H^3(A, M') \to H^3(A, M) \to H^3(A, M'') \to$$

$$\cdots \to H^n(A, M'') \overset{\delta^n}{\to} H^{n+1}(A, M') \to \cdots$$

is exact.

We mention just one other result from[121]. In a Jordan algebra, recall that $\{xby\} = (xb)y + (by)x - (xy)b$. The b-homotope of J, written $J^{(b)}$ is the Jordan algebra structure on the vector space J given by the multiplication $x \cdot_b y = \{xby\}$. If M is a bimodule for J and b is invertible, the corresponding bimodule $M^{(b)}$ for $J^{(b)}$ is the vector space M with action

$$a \cdot_b m = m \cdot_b a = \{abm\}.$$

Lemma 4.1. $H^n(J, M) \sim H^n(J^{(b)}, M^{(b)})$.

We next record the Jordan analogs of the first and second Whitehead lemmas as described in[122]. The first one has already been mentioned in Theorem 3.14.

Theorem 4.3 (Jordan analog of first Whitehead lemma[123]). *Let J be a finite dimensional semisimple Jordan algebra over a field of characteristic 0 and let M be a J-module. Let f be a linear mapping of J into M such that $f(ab) = f(a)b + af(b)$. Then there exist $v_i \in M, b_i \in J$ such that*

$$f(a) = \sum_i ((v_i a)b - v_i(ab_i)).$$

Theorem 4.4 (Jordan analog of second Whitehead lemma[124]). *Let J be a finite dimensional separable[29] Jordan algebra and let M be a J-module. Let f be a bilinear mapping of $J \times J$ into M such that*

$$f(a, b) = f(b, a)$$

[29]See the footnote to Theorem 4.14

$$f(a^2, ab) + f(a, b)a^2 + f(a, a)ab = f(a^2b, a) + f(a^2, b)a + (f(a, a)b)a.$$

Then there exist a linear mapping g from J into M such that

$$f(a, b) = g(ab) - g(b)a - g(a)b.$$

Two proofs of Theorem 4.4 are given in[42]. One of them, which uses the classification of finite dimensional Jordan algebras, is outlined in the next subsection (see Corollary 4.1). The other proof uses Lie algebras and is contained in[42] pp. 324–336 and will not be discussed here.

4.3.1. *Jordan classification approach*

In this subsection, as just indicated, we shall give a detailed outline of one proof of Theorem 4.4, following[42]. This will serve as a model, or road map, for a proposed infinite dimensional generalization.

Step 1 Extensions of algebras and factor sets ([42] pp.91-92)

Let A and M be algebras (associative, Jordan, Lie, alternative). An **extension of** A **by** M is a short exact sequence $0 \to M \overset{\alpha}{\to} E \overset{\beta}{\to} A \to 0$. Thus α is an injective homomorphism and β is a surjective homorphism. Extensions $0 \to M \overset{\alpha}{\to} E \overset{\beta}{\to} A \to 0$ and $0 \to M \overset{\alpha'}{\to} E' \overset{\beta'}{\to} A \to 0$ are **equivalent** if there exists a homomorphism $\gamma : E \to E'$ such that $\alpha' = \gamma \circ \alpha$ and $\beta = \beta' \circ \gamma$. Thus γ is an isomorphism of E onto E'.

An extension $0 \to M \overset{\alpha}{\to} E \overset{\beta}{\to} A \to 0$ is **split** (or **inessential**) if there exists a homomorphism $\delta : A \to E$ with $\beta \circ \delta = 1_A$. Therefore, $E = \delta(A) \oplus \alpha(M)$ as vector spaces and $\delta(A)$ is a subalgebra of E isomorphic to A. An extension $0 \to M \overset{\alpha}{\to} E \overset{\beta}{\to} A \to 0$ is **null** if $M^2 = 0$, that is, all products in M are zero.

Let $0 \to M \overset{\alpha}{\to} E \overset{\beta}{\to} A \to 0$ be any extension and identify M with $\alpha(M) \subset E$. Then we may write $E = M \oplus \delta(A)$, a vector space direct sum, for some linear map $\delta : A \to E$ such that $\beta \circ \delta = 1_A$. E is an E-module and since M is an ideal in E, it is a submodule. If $M^2 = 0$ then M is an E/M-module via $(e+M)\cdot u = eu$, $u \cdot (e+M) = ue$. Now the isomorphism $\beta(e) \to e+M$ of A with E/M completes the proof of the following lemma.

Lemma 4.2. *M is a A-module under the module actions $a \cdot u = \delta(a)u$, $u \cdot a = u\delta(a)$ for $a \in A$ and $u \in M \subset E$. (Multiplication in E)*

Although the next theorem is proved in[42] simultaneously for associative, Lie, and Jordan algebras, we shall now restrict to Jordan algebras.

Definition 4.1. Let M be a Jordan A-module. A bilinear map h, from $A \times A$ to M, is a (Jordan) **2-cocycle** if it is symmetric and satisfies

$$(h(a,a)\cdot b)\cdot a + h(a^2,b)\cdot a + h(a^2b,a) = a^2\cdot h(b,a) + h(a,a)\cdot(ba) + h(a^2,ba).$$

A map $h : A \times A \to M$ is a (Jordan) **2-coboundary** if it is of the form

$$h(a,b) = \mu(ab) - a\cdot\mu(b) - \mu(a)\cdot b$$

for some linear map $\mu : A \to M$. The vector space of all 2-cocycles modulo 2-coboundaries is denoted $H^2(A,M)$.

Theorem 4.5 (THEOREM 12, p.94). *Let M be a Jordan A-module. Then there is a bijection of $H^2(A,M)$ and the set of equivalence classes of null extensions of A by M such that the associated bimodule structure on M given by Lemma 4.2 is the given one. In this correspondence the equivalence class of 0 in $H^2(A,M)$ corresponds to the isomorphism class of inessential extensions.*

Step 2 Peirce decomposition; connected idempotents ([42] pp.117–124).

Let e be an idempotent in a Jordan algebra J. The **Peirce decomposition** of J with respect to e is $J = J_0(e) \oplus J_1(e) \oplus J_{\frac{1}{2}}(e)$, where

$$J_i(e) = \{x_i \in J : x_i e = i x_i\}.$$

Lemma 4.3 (Lemma 1, p.119). *If e is an idempotent, then*

- $J_0^2 \subset J_0,\ J_1^2 \subset J_1,\ J_0J_1 = 0$
- $J_{\frac{1}{2}}^2 \subset J_0 + J_1,\ J_{\frac{1}{2}}(J_0+J_1) \subset J_{\frac{1}{2}}$

Suppose J has an identity element and $1 = e_1 + \ldots + e_n$ where $e_i^2 = e_i$ and $e_ie_j = 0$ $(i \neq j)$. The **Peirce decomposition** of J with respect to the idempotents e_i (orthogonal with sum 1) is $J = \sum_{i\le j} \oplus J_{ij}$, where

$$J_{ii} = \{x_{ii} \in J : x_{ii}e_i = x_{ii}\} = J_1(e_i)$$
$$J_{ij} = J_{\frac{1}{2}}(e_i) \cap J_{\frac{1}{2}}(e_j) = \{x_{ij} : x_{ij}e_i = \frac{1}{2}x_{ij} = x_{ij}e_j\}.$$

Lemma 4.4 (Lemma 2, p.120). *If e_i are orthogonal idempotents with sum 1, then*

- $J_{ii}^2 \subset J_{ii},\ J_{ij}J_{ii} \subset J_{ij},\ J_{ij}^2 \subset J_{ii} + J_{jj},\ J_{ii}J_{jj} = 0\ (i \neq j)$

- $J_{ij}J_{jk} \subset J_{ik}$, $J_{ij}J_{kk} = 0$, $J_{ij}J_{kl} = 0$ *if* i, j, k, l *are distinct.*

Let e_1 and e_2 be orthogonal idempotents so that we have

$$J_1(e_1 + e_2) = J_{11} + J_{22} + J_{12} = J_1(e_1) + J_1(e_2) + J_{12}.$$

The orthogonal idempotents e_1 and e_2 are **connected** if there exists u_{12} in J_{12} which is invertible in $J_1(e_1 + e_2)$. They are **strongly connected** if here exists $u_{12} \in J_{12}$ with $u_{12}^2 = e_1 + e_2$.

Step 3 Jordan Matrix Algebras ([42] pp.125–131)

Let D be a unital algebra with involution $j(d) = \overline{d}$. Then D_n denotes the algebra of n by n matrices with entries from D, with involution

$$X \mapsto X^J = \overline{X}^t,$$

and $H(D_n) \subset D_n^+$ denotes the subalgebra of symmetric elements. Algebras of the form $H(D_n)$ which are Jordan algebras are called **Jordan matrix algebras.**

Let e_{ij} be the usual matrix units in D_n and if $x \in D$, we identify x with the diagonal matrix in D_n all of whose diagonal entries are x. Then xe_{ij} is the matrix with x in the (i, j)-position and zeros elsewhere. We set

$$x[ij] = xe_{ij} + (xe_{ij})^J.$$

Theorems 4.6 and 4.7 are not used in this paper, but are stated here for their intrinsic interest.

Theorem 4.6 (THEOREM 1, p.127). *For* $n \geq 3$, $H(D_n)$ *is a Jordan algebra if and only if* D *is associative, or* $n = 3$ *and* D *is alternative with symmetric elements in the nucleus.*

Theorem 4.7 (THEOREM 2, p.129). *For* $n \geq 3$, *let* $H = H(D_n)$ *be a Jordan matrix algebra. There are one to one correspondences*

$$\{\text{*-subalgebras } E \text{ of } D\} \leftrightarrow \{\text{*-subalgebras } H \cap E_n \text{ containing } 1[ij]\}$$
$$\{\text{*-ideals } E \text{ of } D\} \leftrightarrow \{\text{*-ideals } H \cap E_n\}.$$

In each case, the map $E \mapsto H \cap E_n$ *is a lattice isomorphism, and in the second case,* $(H \cap E_n)^2 = 0$ *if and only if* $E^2 = 0$.

Theorem 4.8 (THEOREM 3, p.130). *For* $n \geq 3$, *let* $H(D_n)$ *and* $H(E_n)$ *be Jordan matrix algebras, and suppose* $\eta : D \to E$ *is a unital *-homomorphism (into). Then the restriction* σ *to* $H(D_n)$ *of the mapping*

*$(d_{ij}) \mapsto (\eta(d_{ij}))$ of D_n is a *-homomorphism of $H(D_n)$ into $H(E_n)$ such that $\sigma(1[ij]]) = 1[ij]$.*

*Conversely, if σ is a *-homomorphism of $H(D_n)$ into $H(E_n)$ such that $\sigma(1[ij]]) = 1[ij]$, then there exists a unital *-homomorphism η of D into E such that σ is the restriction of the mapping $(d_{ij}) \mapsto (\eta(d_{ij}))$.*

Step 4 Coordinatization Theorems ([42] pp.132–138)

Remark 4.1. If $J = H(D_n)$ and we write

$$x[ij] = xe_{ij} + (xe_{ij})^J, \ e_i = \frac{1}{2}[ii] = e_{ii},$$

then the e_i are orthogonal idempotents with sum 1, and e_1 and e_j are strongly connected.

Theorem 4.9 (THEOREM 5, p.133; strong coordinatization theorem). *Let J be a Jordan algebra whose identity is the sum of $n \geq 3$ strongly connected idempotents. Then J is isomorphic to a Jordan matrix algebra $H(D_n)$.*

More precisely, if $1 = \sum_1^n e_i$ and we have u_{1j} with $e_1 u_{1j} = \frac{1}{2}u_{1j} = e_j u_{1j}$ and $u_{1j}^2 = e_1 + e_j$, then there is a Jordan matrix algebra $H(D_n)$ and an isomorphism η of J onto $H(D_n)$ such that $\eta(e_i) = \frac{1}{2}[ii]$ and $\eta(u_{1j}) = 1[1j]$.

Step 5 Lifting of idempotents ([42] pp.148–151)

An algebra is **nil** if every element is nilpotent.

Lemma 4.5 (Lemma 5, p.150). *Let J be a Jordan algebra with 1, N a nil ideal, e_1 and e_2 nonzero orthogonal idempotents in J. Then e_1 and e_2 are connected (strongly connected) if and only if $\overline{e}_1 = e_1 + N$ and $\overline{e}_2 = e_2 + N$ are connected (strongly connected) in $\overline{J} = J/N$.*

Theorem 4.10 (THEOREM 10, p.151). *Let J be a Jordan algebra with 1, N a nil ideal such that $\overline{J} = J/N$ is isomorphic to a Jordan matrix algebra $H(\overline{D}_n)$ of order $n \geq 3$. Then J is isomorphic to a Jordan matrix algebra $H(D_n)$ where the ideal in $H(D_n)$ corresponding to N has the form $M_n \cap H(D_n)$ where M is an ideal in D and $\overline{D}$ is isomorphic to D/M as algebras with involution.*

Step 6 Solvable ideals and the radical ([42] pp.190–196)

A subspace B if a Jordan algebra J is an **associator ideal** if any associator $[a_1, a_2, a_3] := (a_1 a_2)a_3 - a_1(a_2 a_3) \in B$ whenever any one of $a_i \in B$.

Theorem 4.11 (THEOREM 1, p.191). *Let J be a Jordan algebra, B an ideal of finite codimension m. Define*

(1) $B^{[0]} = B, \quad B^{[k]} = B^{[k-1]}B + (B^{[k-1]}B)J$
(2) $B^{(0)} = B, \quad B^{(k)} = B^{(k-1)}B^{(k-1)} + (B^{(k-1)}B^{(k-1)})J.$

Then $B^{[k]}$ and $B^{(k)}$ are ideals and $B^{[m+1]} \cup B^{(m+1)} \subset B^2$.

The powers J^{2^k} of a Jordan algebra J are $J^{2^0} = J$ and $J^{2^k} = (J^{2^{k-1}})^2$. J^2 is an ideal and J^{2^k} is a subalgebra. J is **solvable** if there exists an integer N with $J^{2^N} = 0$. A solvable Jordan algebra is nil. If J is finite dimensional (more generally, if J satisfies the maximum condition for ideals), then it contains a solvable ideal R, the **radical** (= rad J), such that R contains every solvable ideal of J. J is **semisimple** if $R = 0$ (J has no nonzero solvable ideals). Note that $J/\text{rad}\, J$ is semisimple.

The next two theorems, due to Albert, are included for their intrinsic interest. An algebra is **nilpotent** if there exists an integer N such that every product (in any association) of N (distinct) elements is zero. A nilpotent Jordan algebra is solvable.

Theorem 4.12 (COROLLARY 1, p.195). *A finite dimensional solvable Jordan algebra is nilpotent.*

Theorem 4.13 (THEOREM 3, p.196). *Any finite dimensional nil algebra is solvable.*

Step 7 The theorem of Albert-Penico-Taft ([42] pp.287–292)

Theorem 4.14 (Theorem 13, p. 292). *Let E be a finite dimensional Jordan algebra, M an ideal in E such that E/M is separable*[30]*. Then there is a subalgebra R of E such that $E = R \oplus M$ as vector spaces. In other words, every finite dimensional extension of a separable Jordan algebra splits.*

By Theorem 4.5, we have vanishing of the second cohomology group and a proof of Theorem 4.4.

Corollary 4.1 (Corollary, p 292). *If M is a finite dimensional bimodule for J, a separable Jordan algebra, then $H^2(J, M) = 0$.*

[30]Separable, in this context, means that the algebra remains semisimple with respect to all extensions of the ground field. For algebraically closed fields, this is the same as being semisimple

Reduction I, p.288 If Theorem 4.14 is true for solvable M, then it is true for arbitrary M.

Reduction II, p.288 If Theorem 4.14 is true for solvable M with $M^2 = 0$, then it is true for arbitrary solvable M.

Reduction III, p.288 If Theorem 4.14 is true for algebras E with a unit, then it is true for arbitrary E.

Reduction IV, p.288-289 If Theorem 4.14 is true for algebras E over an algebraically closed field, then it is true for arbitrary E

Reduction V, p.289-290 If Theorem 4.14 is true for algebras E with E/M simple, then it is true for arbitrary E

The proof is now completed by appealing to the classification of unital simple Jordan algebras over an algebraically closed field in order to prove

Proposition 4.2 (Lemmas 2-4, pp. 290–291). *Let E be a finite dimensional Jordan algebra with unit over an algebraically closed field and let M be an ideal in E such that $M^2 = 0$ and E/M is separable. Then E contains a subalgebra isomorphic to E/M.*

For later use, we indicate the proof of Reduction IV when the underlying field is the real numbers. If we complexify the short exact sequence

$$0 \to M \to E \to J = E/M \to 0, \tag{1}$$

and this complexified sequence splits, then so does the original one. To show this, one uses the equivalence given by Theorem 4.5. as follows:

Let $h : J \times J \to M$ be a Jordan 2-cocycle (over the real field). We need to show that there is a linear map $\mu : J \to M$ such that

$$h(a, b) = \mu(ab) - \mu(a) \cdot b - \mu(b) \cdot a. \tag{2}$$

Let $u_1, \ldots, u_n$ be a basis for J over $\mathbf{R}$ and $v_1, \ldots, v_r$ a basis for M over $\mathbf{R}$. Then (2) holds if and only if it holds for $a, b \in \{u_1, \ldots, u_n\}$. With μ provisionally defined by $\mu(u_i) = \sum_p \mu_{ip} v_p$, define the quantities η_{ijq}, γ_{ijk}, δ_{piq} by the formulas

$$h(u_i, u_j) = \sum_q \eta_{ijq} v_q, \quad u_i u_j = \sum_k \gamma_{ijk} u_k; \quad v_p \cdot u_i = \sum_q \delta_{piq} v_q.$$

Then (2) for $a = u_i$ and $b = u_j$ is equivalent to the set of linear equations

$$\eta_{ijk} = \sum_k \gamma_{ijk} \mu_{kq} - \sum_p \mu_{ip} \delta_{pjq} - \sum_p \mu_{jp} \delta_{piq}. \tag{3}$$

for the μ's in $\mathbf{R}$.

The solvability of (3) for the μ's in $\mathbf{R}$ is a necessary and sufficient condition that the extension splits. If we complexity the sequence (1), and extend the maps α, β, and h, we still have $(M^{\mathbf{C}})^2 = 0$, $J^{\mathbf{C}}$ is separable and $h^{\mathbf{C}}$ is a 2-cocycle. The bases $\{u_i\}$ and $\{v_j\}$ remain bases over the complex field. Assuming the theorem holds in the algebraically closed case, the equations (3) have a solution $\{\mu_{ip}\}$ in $\mathbf{C}$. Since these are linear equations with coefficients in $\mathbf{R}$, it follows that they have a solution in $\mathbf{R}$.

4.3.2. *Cohomology groups of Jordan algebras and triples, after McCrimmon*

A study of low dimensional cohomology for quadratic Jordan algebras is given in[125]. Since quadratic Jordan algebras (which we do not define here as they coincide with "linear" Jordan algebras over characteristic 0 fields) can be considered a bridge from Jordan algebras to Jordan triple systems, this would seem to be a good place to look for exploring cohomology theory for Jordan triples. Indeed, this is hinted at in[126]. Let us review some aspects of[125,126]. Although[125] is about Jordan algebras, the concepts are phrased in terms of the associated triple product $\{abc\} = (ab)c + (cb)a - (ac)b$.

Let J be a unital (quadratic) Jordan algebra and $\mathcal{M}$ a unital J-module. A linear map $d : J \to \mathcal{M}$ is a *derivation* if $d(1) = 0$ and $d\{xyx\} = \{xd(y)x\} + 2\{d(x)yx\}$. An *inner derivation* is a sum of derivations of the form $d_{x,m}(y) = \{xmy\} - \{mxy\}$. We shall use the notation $\text{Inder}\,(J, \mathcal{M}) \subset \text{Der}\,(J, \mathcal{M})$, $H^1(J, \mathcal{M}) = \text{Der}\,(J, \mathcal{M})/\text{Inder}\,(J, \mathcal{M})$.

We can express H^1 (and define H^2) in terms of cocycles and coboundaries.

Definition 4.2. (n-cochains, $n = 0, 1, 2, 3$)

- $C^0(J, \mathcal{M}) = J \otimes \mathcal{M}$
- $C^1(J, \mathcal{M}) =$ 1-cochains $= \{f : J \to \mathcal{M} : f \text{ linear}, f(1) = 0\}$
- $C^2(J, \mathcal{M}) = \{q : J \times J \to \mathcal{M} : q(x; y) \text{ quadratic in } x; \text{linear in } y, q(1; y) \equiv 0\}$
- $C^3(J, \mathcal{M}) = C^3_1(J, \mathcal{M}) \oplus C^3_2(J, \mathcal{M})$, where

$$C^3_1(J, \mathcal{M}) = \{q : J^3 \to \mathcal{M} : q \text{ quardic in } x;\ \text{quadratic in } y;\ \text{linear in } z\}$$
$$C^3_2(J, \mathcal{M}) = \{q : J^3 \to \mathcal{M} : q \text{ cubic in } x;\ \text{linear in } y, z\}.$$

Definition 4.3. (coboundary maps)

- $\partial^0 : C^0 \to C^1$: $\partial^0(x \otimes m)(y) = \{x\}my - \{m\}xy$
- $\partial^1 : C^1 \to C^2$: $\partial^1 f(x, y) = f\{x\}yx - 2\{x\{y\}f(x)\} - \{x\{f(y)\}x\}$

- $\partial^2 = \partial_1^2 \oplus \partial_2^2 : C^2 \to C^3$, where $\partial_1^2 : C^2 \to C_1^3$ and $\partial_2^2 : C^2 \to C_2^3$ are given by

$$\begin{aligned}\partial_1^2 q(x;y;z) &= q(\{x\}yx;z) + \{\{x\}yx\{z\}q(x;y)\}\\ &\quad - q(x;\{y\{\{x\}zx\}y\}) - \{x\{q(y;\{x\}zx)\}x\}\\ &\quad - \{x\{\{y\{q(x;z)\}y\}\}x\},\\ \partial_2^2 q(x;y;z) &= q(x,\{x\}zx;y) + 2\{x\{y\}q(x;z)\}\\ &\quad - 2q(x;\{y\}xz) - \{x\{q(y,z;x)\}x\},\end{aligned}$$

  and where $q(x,y;z) = (q(x+y,z) - q(x,z) - q(y,z))/2$ denotes the bilinearization of $q(x;z)$ in the first variable.

Definition 4.4. (cocycles and coboundaries)

- $B^1(J,\mathcal{M}) = \operatorname{Im}\partial^0 = \{\sum(\{x_i\}m_i\cdot - \{m_i\}x_i\cdot)\}$ (=Inder $(J,\mathcal{M})$)
- $Z^1(J,\mathcal{M}) = \ker\partial^1 (= \operatorname{Der}(J,\mathcal{M}))$
- $B^2(J,\mathcal{M}) = \operatorname{Im}\partial^1$
- $Z^2(J,\mathcal{M}) = \ker\partial^2\ (= \ker\partial_1^2 \oplus \ker\partial_2^2)$.

Then $H^n(J,\mathcal{M}) = Z^n(J,\mathcal{M})/B^n(J,\mathcal{M})$, $n = 1,2$.

The paper[125], which is mostly concerned with representation theory, proves a version of Theorem 4.5 for the cohomology groups just defined, and the linearity of the functor H^n:

$$H^n(J,\oplus_i M_i) = \oplus_i H^n(J,M_i), \quad n = 1,2.$$

The paper[126], which is mostly concerned with compatibility of tripotents in Jordan triple systems, proves versions of the linearity of the functor H^n corresponding to the Jordan triple structure.

5. Cohomology of operator algebras

We are now going to summarize the main points of this theory using the two survey articles[76],[127] as our primary guide, and using the notation found there.

5.1. *Continuous Hochschild cohomology*

As we saw in section 4, Hochschild cohomology involves an associative algebra A and A-bimodules X and gives rise to

- n-cochains $L^n(A,X)$,

- coboundary operators Δ_n,
- n-coboundaries B^n,
- n-cocycles Z^n and
- cohomology groups $H^n(A, X)$.

If A is a Banach algebra and X is a Banach A-bimodule (that is, a Banach space with module actions jointly continuous) we have the continuous versions of the above concepts L^n_c , B^n_c , Z^n_c , $H^n_c(A, X)$. Warning: B^n_c is not always closed, so H^n_c is still only a vector space.

Let A be a C*-algebra of operators acting on a Hilbert space H and let X be a dual normal A-module (X is a dual space and the module actions are separately ultra weakly-weak*-continuous). Further, we now have

- normal n-cochains $L^n_w(A, X)$, that is, bounded and separately weakly continuous n-cochains
- coboundary operators Δ_n,
- normal n-coboundaries B^n_w,
- normal n-cocycles Z^n_w and
- normal cohomology groups $H^n_w(A, X)$.

Thus, for a C*-algebra acting on a Hilbert space we have three possible cohomology theories:

- the purely algebraic Hochschild theory H^n
- the bounded theory H^n_c
- the normal theory H^n_w

Theorem 5.1 (Johnson,Kadison,Ringrose 1972[128]).

$$H^n_w(A, X) \sim H^n_w(R, X) \ (R = \textit{ultraweak closure of } A).$$

Theorem 5.2 (Johnson,Kadison,Ringrose 1972[128]).

$$H^n_w(A, X) \sim H^c_w(A, X).$$

By Theorems 5.1 and 5.2, all four cohomology groups

$$H^n_w(A, X) \ , \ H^n_w(R, X) \ , \ H^n_c(R, X) \ , \ H^n_w(R, X)$$

are pairwise isomorphic.

Theorem 5.3 (Johnson,Kadison,Ringrose 1972[128]).

$$H^n_c(R, X) = 0 \ \forall n \geq 1 \ (R = \textit{hyperfinite von Neumann algebra}).$$

Theorem 5.4 (Connes 1978[129]). *If R is a von Neumann algebra with a separable predual, and $H_c^1(R, X) = 0$ for every dual normal R-bimodule X, then R is hyperfinite.*

At this point, there were two outstanding problems of special interest;

Problem 5.1. $H_c^n(R, R) = 0\ \forall n \geq 1$ *for every von Neumann algebra R?*

Problem 5.2. $H_c^n(R, B(H)) = 0\ \forall n \geq 1$ *for every von Neumann algebra R acting on a Hilbert space H?*

In the words of John Ringrose, "The main obstacle to advance was a paucity of information about the general bounded linear (or multilinear) mapping between operator algebras. The major breakthrough, leading to most of the recent advances, came through the development of a rather detailed theory of completely bounded mappings." This development occurred about a decade later and will be described in subsection 5.3.

5.2. *Completely bounded maps*

In this subsection, we shall give an introduction to the category of operator spaces from a point of view advantageous to the goals of this paper. The immediate purpose is to give context to completely bounded maps for use when we return to cohomology in the next subsection. Besides the standard monoraphs which appeared after the beginning of the 21st century[130], [131], [132], [133], [134], there is quite a good bit of the basic theory in the lecture notes of Runde[46]. Proofs of all the results in this subsection can be found in one or more of these references.

5.2.1. *Banach spaces*

Why are normed spaces important? One answer is that $\mathbf{R}^n$ is a vector space with a norm, so you can take derivatives and integrals. Thus, normed spaces are important because you can do calculus.

Why is completeness important? Three basic principles of functional analysis are based in some way on completeness:

- Hahn-Banach theorem (which depends on Zorn's lemma)
- Open mapping theorem (which depends on Baire category)
- Uniform boundedness theorem (which depends on Baire category)

Some examples which will be of interest to us are:

- $C[0,1]$, L^p, ℓ^p, c_0, $C(\Omega)$, $L^p(\Omega,\mu)$, $C_0(\Omega)$
- $B(X,Y)$, Banach algebras, Operator algebras $\mathcal{A} \subset B(X)$
- Operator spaces; C*-algebras, Ternary rings of operators

The Hahn-Banach theorem states that every bounded linear functional on a subspace of a Banach space has a bounded extension to the larger space with the same norm.

Definition 5.1. A Banach space is **injective** if every bounded linear operator from a subspace of a Banach space into it has a bounded extension to the larger space with the same norm.

Theorem 5.5 (Kelley 1952[135]). *A Banach space is injective if and only if it is isometric to $C(\Omega)$, where Ω is extremely disconnected.*

Injectivity in a category of Banach spaces is intimately related to contractive projections. Our first example of this is the following easily verified fact.

Remark 5.1. A Banach space is injective if and only if there is an injective Banach space containing it and a contractive projection of that space onto it.

5.2.2. *Operator spaces*

Every Banach space X is commutative, by which we mean that by the Hahn-Banach theorem, $X \subset C(X_1^*)$ (where X_1^* denotes the unit ball of X^*), that is, X is a linear subspace of a **commutative** C*-algebra. Hence: $C(\Omega)$ is the "mother of all Banach spaces."

Definition 5.2. An **operator space** (or noncommutative Banach space, or quantized Banach space) is a linear subspace of a C*-algebra (or $B(H)$).

Hence: $B(H)$ is the "mother of all operator spaces."

An operator space may be viewed as a C*-algebra for which the multiplication and involution have been ignored. You must replace these by some other structure. What should this structure be and what can you do with it? The answer lies in the morphisms. In the category of Banach spaces, the morphisms are the bounded linear maps. In the category of operator spaces, the morphisms are the **completely bounded maps**.

Operator spaces are intermediate between Banach spaces and C*-algebras. The advantage of operator spaces over C*-algebras is that they allow the use of finite dimensional tools and isomorphic invariants ("local theory"). C*-algebras are too rigid: morphisms are contractive, norms are unique. Operator space theory opens the door to a massive transfer of technology coming from Banach space theory. This quantization process has benefitted operator algebra theory mainly (as opposed to Banach space theory).

Definition 5.3. An operator space is **injective** if every completely contractive linear operator from a subspace of an operator space into it has a completely contractive extension to the larger space. (See the beginning of section 8 for the definition of completely bounded map)

Theorem 5.6 (Arveson-Wittstock-Paulsen Hahn-Banach). $B(H)$ *is injective in the category of operator spaces and completely contractive maps.*

Our second example of the connection of contractive projections to injectivity is the following early verified fact.

Remark 5.2. $X \subset B(H)$ is injective if and only if there is a completely contractive projection $P : B(H) \to B(H)$ with $P(B(H)) = X$.

5.2.3. *Injective and mixed injective operator spaces*

Theorem 5.7. *An operator space is injective if and only if it is completely isometric to a corner of a C*-algebra. In particular, it is completely isometric to a ternary ring of operators (TRO).*

Theorem 5.8. *Every operator space has an injective envelope.*

There are also the related notions of ternary envelope and Shilov boundary, which are useful but are not needed here, so will not be defined. The following concept will appear later in part 3.

Definition 5.4. An operator space is a **mixed injective** if every completely contractive linear operator from a subspace of an operator space into it has a bounded contractive extension to the larger space.

Remark 5.3. $X \subset B(H)$ is a mixed injective if and only if there is a contractive projection $P : B(H) \to B(H)$ with $P(B(H)) = X$.

5.2.4. *Applications of operator space theory*

The following theorems highlight the profound applications that operator space theory has had to operator algebra theory.

- **Similarity problems**. The Halmos problem is solved in the following theorem. For progress on and a description of the Kadison problem, see[136].

  Theorem 5.9 (Pisier 1997[137]). *There exists a polynomially bounded operator which is not similar to a contraction.*

- **Tensor products**

  Theorem 5.10 (Junge-Pisier 1995[138]). *$B(H)$ is not a nuclear pair. That is, $B(H) \otimes B(H)$ does not have a unique C^*-norm.*

  To quote Pisier[139]: "This is a good case study. The proof is based on the solution to a problem which would be studied for its own sake: whether the set of finite dimensional operator spaces is separable (it is not)."

- **Operator amenable groups**. A well-known theorem of Johnson, which is false for the Fourier algebra, states that a group is amenable if and only if its group algebra is an amenable Banach algebra under convolution.

  Theorem 5.11 (Ruan 1995[140]). *A group is amenable if and only if its Fourier algebra is amenable in the operator space formulation.*

- Operator local reflexivity. A well known theorem of Lindenstrauss-Rosenthal with many applications states that every Banach space is locally reflexive (Finite dimensional subspaces of the second dual can be approximated by finite dimensional subspaces of the space itself). One application occurs in the proof that the second dual of a JB*-triple is a JB*-triple, with the consequence being the Gelfand-Naimark for JB*-triples (see subsection 7.1.2). Not all operator spaces are locally reflexive (in the appropriate operator space sense). However, there is the following theorem.

  Theorem 5.12 (Effros-Junge-Ruan 2000[141]). *The dual of a C^*-algebra is a locally reflexive operator space.*

5.3. *Completely bounded cohomology*

After this brief excursion into the world of operator spaces, we return to cohomology.

Let A be a C*-algebra and let S be a von Neumann algebra, both acting on the same Hilbert space H with $A \subset S$. We can view S as a dual normal A-module with A acting on S by left and right multiplication. We now have

- completely bounded n-cochains $L^n_{cb}(A, S)$
- coboundary operators Δ_n,
- completely bounded n-coboundaries B^n_{cb},
- completely bounded n-cocycles Z^n_{bc}
- completely bounded cohomology groups $H^n_{cb}(A, S)$.

For a C*-algebra A and a von Neumann algebra S with $A \subset S \subset B(H)$ we thus have two new cohomology theories:

- the completely bounded theory H^n_{cb}
- the completely bounded normal theory H^n_{cbw}

By straightforward analogues of Theorems 5.1 and 5.2, the four cohomology groups

$$H^n_{cb}(A, S)\ ,\ H^n_{cbw}(A, S)\ ,\ H^n_{cb}(R, S)\ ,\ H^n_{cbw}(R, S)$$

are mutually isomorphic, where R is the ultraweak closure of A.

Theorem 5.13 (Christensen-Effros-Sinclair 1987[142]).

$H^n_{cb}(R, B(H)) = 0\ \forall n \geq 1$ ($R =$ *any von Neumann algebra acting on* H).

Theorem 5.14 (Christensen-Sinclair 1987[143]).

$$H^n_{cb}(R, R) = 0\ \forall n \geq 1\ (R = \textit{any von Neumann algebra}).$$

5.3.1. *Advances using complete boundedness*

A noteworthy quote from[127] is the following: "Cohomology and complete boundedness have enjoyed a symbiotic relationship where advances in one have triggered progress in the other."

Theorems 5.15 and 5.16 are also due to Christensen-Effros-Sinclair[142].

Theorem 5.15. $H^n_c(R, R) = 0\ \forall n \geq 1$ *(*$R =$ *von Neumann algebra of type* I*,* II_∞*,* III*, or of type* II_1 *and stable under tensoring with the hyperfinite factor).*

Theorem 5.16. *$H_c^n(R, B(H)) = 0$ $\forall n \geq 1$ (R = von Neumann algebra of type I, II_∞, III, or of type II_1 and stable under tensoring with the hyperfinite factor, acting on a Hilbert space H).*

The following theorem is the culmination of earlier work in[144] and[145].

Theorem 5.17 (Sinclair-Smith[146]). *$H_c^n(R, R) = 0$ $\forall n \geq 1$ (R = von Neumann algebra of type II_1 with a Cartan subalgebra and a separable[31] predual).*

The following theorem was proved in[148] and[149] and was new only for $n \geq 3$. For $n = 1$, the two cases are in[29],[28] and[145], while for $n = 2$ one can refer to[143] and[150].

Theorem 5.18 (Christensen-Pop-Sinclair-Smith $n \geq 3$ 2003). *$H_c^n(R, R) = H_c^n(R, B(H)) = 0$ $\forall n \geq 1$ (R = von Neumann algebra <u>factor</u> of type II_1 with property Γ, acting on a Hilbert space H).*

A more recent result is the main result in[151].

Theorem 5.19. *Let M and N be von Neumann algebras of type II_1. Then $H_c^2(M\overline{\otimes}N, M\overline{\otimes}N) = 0$.*

We can now add a third problem to our previous two (Problems 5.1,5.2). A candidate for a counterexample is the factor arising from the free group on 2 generators.

Problem 5.3. *$H_c^n(R, R)) = 0$ $\forall n \geq 2$? (R is a von Neumann algebra of type II_1).*

5.3.2. *Another approach (Paulsen)*

A different approach, which is popular among algebraists is taken in the paper[152]. We quote from the review of this paper in Mathematical Reviews[153].

"Let A be a subalgebra in the algebra B(H) of bounded operators in a Hilbert space H, and X be a linear subspace in B(H) such that AXA⊂X. Then X can be considered as an A-bimodule, and the Hochschild cohomology groups for the pair (A,X) can be constructed in the usual way, but with the cocycles being n-linear maps from A to X satisfying certain extra 'complete boundedness' conditions."

---

[31] The separability assumption was removed in 2009[147]

"Continuing the study of A. Ya. Khelemskii[154], himself and others, Paulsen presents two new alternative presentations of these completely bounded Hochschild cohomologies, one of them as a relative Yoneda cohomology, i.e. as equivalence classes of relatively split resolutions, and the second as a derived functor, making it similar to EXT. These presentations make clear the importance of the notions of relative injectivity, projectivity and amenability, which are introduced and studied."

"Using the relative injectivity of von Neumann algebras, the author proves the triviality of completely bounded Hochschild cohomologies for all von Neumann algebras. The Yoneda representation makes the proofs of a number of classical results more transparent."

5.4. *Perturbation of Banach algebras*

One of the most striking applications of the cohomology theory of Banach algebras lies in the area of perturbation theory.

5.4.1. *Theorem of Johnson and Raeburn-Taylor*

Definition 5.5. A Banach algebra is **stable** if any other product making it into a Banach algebra, which is sufficiently close to the original product gives rise to a Banach algebra which is topologically algebraically isomorphic to the original Banach algebra..

Let's make this more precise. If m is a Banach algebra multiplication on A, then its norm is defined by $\|m(x,y)\| \leq \|m\|\|x\|\|y\|$. The following theorem was proved simultaneously and independently by Johnson[155] and the team of Raeburn and Taylor[156].

Theorem 5.20. *For any Banach algebra A with multiplication m, if $H^2(A,A) = H^3(A,A) = 0$, then there exists $\varepsilon > 0$ such that if m_1 is another Banach algebra multiplication on A such that $\|m_1 - m\| < \varepsilon$ then (A, m_1) and (A, m) are topologically algebraically isomorphic.*

For a stimulating discussion of this topic and many others, we recommend[154]. For a similar problem on operator algebras one can look into[157].

5.4.2. *Perturbation for nonassociative Banach algebras*

The origin of perturbation theory for general non associative algebras is deformation theory. Let c_{ij}^k be the structure constants of a finite dimensional Lie algebra L. Let $c_{ij}^k(\varepsilon) \to c_{ij}^k$. Stability in this context means $(L, c_{ij}^k(\varepsilon))$ is isomorphic to (L, c_{ij}^k) if ε is sufficiently small.

Theorem 5.21 (Gerstenhaber 1964[158]). *Finite dimensional semisimple Lie algebras are stable.*

Perturbation results similar to Theorem 5.20 have been proved recently for some non associative Banach algebras in[159].

Theorem 5.22 (Dosi 2009). *If L is a Banach Lie algebra and $H^2(L, L) = H^3(L, L) = 0$, then L is a stable Banach Lie algebra.*

An entirely similar result for Banach Jordan algebras is also proved in[159]. Of necessity, the definition of cohomology for Jordan algebras in[159] is made only in dimensions 3 or less.

6. Cohomology of triple systems

In this section we shall describe the results which the author learned about in preparing this mini course. Since the author of this paper is not (yet) an expert in the purely algebraic side of cohomology theory, he is going to rely on information obtained from reviews of these papers in Mathematical Reviews.

6.1. *Cohomology of finite dimensional triple systems*

6.1.1. *Cohomology of Lie triple systems*

The earliest work on cohomology of triple systems seems to be[160], of which the following description is taken from its review in Mathematical Reviews[161]. Four decades later, the second paper on the subject appeared[162].

A Lie triple system T is a subspace of a Lie algebra L closed under the ternary operation $[xyz] = [x, [y, z]]$ or, equivalently, it is the subspace of L consisting of those elements x such that $\sigma(x) = -x$, where σ is an involution of L. A T-module M is a vector space such that the vector-space direct sum $T \oplus M$ is itself a Lie triple system in such a way that

(1) T is a subsystem

(2) $[xyz] \in M$ if any of x, y, z is in M
(3) $[xyz] = 0$ if two of x, y, z are in M.

A universal Lie algebra $L_u(T)$ and an $L_u(T)$-module M_s can be constructed in such a way that both are operated on by an involution σ and so that T and M consist of those elements of $L_u(T)$ and M_s which are mapped into their negatives by σ.

Now suppose L is a Lie algebra with involution σ and N is an L-σ module. Then σ operates on $H^n(L, N)$ so that

$$H^n(L, N) = H^n_+(L, N) \oplus H^n_-(L, N)$$

with both summands invariant under σ. The cohomology of the Lie triple system is defined by $H^n(T, M) = H^n_+(L_u(T), M_s)$. Harris investigates these groups for $n = 0, 1, 2$.

(1) $H^0(T, M) = 0$ for all T and M
(2) $H^1(T, M) =$ derivations of T into M modulo inner derivations
(3) $H^2(T, M) =$ factor sets of T into M modulo trivial factor sets.

Turning to the case of finite-dimensional simple T and ground field of characteristic 0, one has the Whitehead lemmas $H^1(T, M) = 0 = H^2(T, M)$ and Weyl's theorem: Every finite-dimensional module is semi-simple. Also, if in addition, the ground field Φ is algebraically closed, then $H^3(T, \Phi)$ is 0 or not 0, according as $L_u(T)$ is simple or not.

The following is the verbatim review[163] of[162].

"This is a study of representations of Lie triple systems, both ordinary and restricted. The theory is based on the connection between Lie algebras and Lie triple systems. In addition, the authors begin the study of the cohomology theory for Lie triple systems and their restricted versions. They also sketch some future applications and developments of the theory."

6.1.2. *Cohomology of associative triple systems*

The following is the verbatim review[164] of[35].

"A cohomology for associative triple systems is defined, with the main purpose to get quickly the cohomological triviality of finite-dimensional separable objects over fields of characteristic $\neq 2$, i.e., in particular the Whitehead lemmas and the Wedderburn principal theorem.

This is achieved by embedding an associative triple system A in an associative algebra $U(A)$ and associating with every trimodule M for A a bimodule M_u for $U(A)$ such that the cohomology groups $H^n(A, M)$ are subgroups of the classical cohomology groups $H^n(U(A), M_u)$.

Since $U(A)$ is chosen sufficiently close to A, in order to inherit separability, the cohomological triviality of separable A is an immediate consequence of the associative algebra theory.

The paper does not deal with functorialities, nor with the existence of a long exact cohomology sequence."

6.1.3. *Wedderburn decomposition*

The classical Wedderburn decomposition is the following theorem, which we state in its Banach algebra form from[165]. It has been generalized, not only to non associative algebras, see for example[25] and[42], but to some classes of Banach algebras, see for example[166].

Theorem 6.1. *Let A be a Banach algebra with radical R. If the dimension of A is finite, there is a subalgebra B of A such that $A = B + R$ and $B \cap R = 0$. Moreover, if A is commutative, then B is necessarily unique.*

The following definition is the model to use in extending this concept to infinite dimensional triples.

Definition 6.1. If A is a Banach algebra with radical R, then $A = B \oplus R$ is a strong Wedderburn decomposition of A if B is a closed subalgebra of A.

The author knows of only two papers dealing with this concept in the context of triple systems, namely,[167] and[85]. The following, including Theorem 6.2 is the verbatim review[168] of[167].

The Wedderburn principal theorem, known for Lie triple systems, is proved for alternative triple systems and pairs. If i is an involution of an alternative algebra B, then $\langle xyz\rangle := (x \cdot i(y)) \cdot z$ is an alternative triple $(x, y, z \in B)$.

A polarisation of an alternative triple A is a direct sum of two submodules $A^1 \oplus A^{-1}$ with $\langle A^1A^{-1}A^1\rangle \subset A^1$, $\langle A^{-1}A^1A^{-1}\rangle \subset A^{-1}$ and

$$\begin{aligned}\langle A^1A^1A^1\rangle &= \langle A^{-1}A^{-1}A^{-1}\rangle = \langle A^1A^1A^{-1}\rangle = \langle A^{-1}A^{-1}A^1\rangle \\ &= \langle A^{-1}A^1A^1\rangle = \langle A^1A^{-1}A^{-1}\rangle = \{0\}.\end{aligned}$$

An alternative pair is an alternative triple with a polarisation.

Theorem 6.2. *If A is a finite-dimensional alternative triple system (or an alternative pair) over a field K, R the radical and A/R separable, then $A = B \oplus R$, where B is a semisimple subtriple (subpair) of A with $B = A/R$.*

The following description of[85] is taken from the abstract and the review[169]. This paper summarizes some properties of Jordan pairs, states some results about some groups defined by Jordan pairs, and constructs a Lie algebra to a Jordan pair. This construction is a generalization of the well-known Koecher-Tits-construction. The radical of this Lie algebra is calculated in terms of the given Jordan pair and a Wedderburn decomposition theorem for Jordan pairs (and triples) in the characteristic zero case is proved.

More precisely, an observation of Koecher that the theorem of Levi for Lie algebras of characteristic 0 implies the Wedderburn principal theorem for Jordan algebras is extended to Jordan pairs (and Jordan triples) V over a field of characteristic 0. In addition, the authors show that any two Wedderburn splittings of V are conjugate under a certain normal subgroup of the automorphism group of V.

6.1.4. *Cohomology of algebras and triple systems*

The reader may be wondering about the cohomology of Jordan triple systems. As noted in 4.3, there is a hint of this in the paper of McCrimmon[126]. Building on the 1 and 2 dimensional cohomology for quadratic Jordan algebras,[126] broaches the analogous construction for Jordan triple systems.

There is also a more general approach in the paper of Seibt[170]. The following is from the introduction to[170].

"The classical cohomologies of unital associative algebras and of Lie algebras have both a double algebraic character: They are embedded in all of the machinery of derived functors, and they allow full extension theoretic interpretations (**Yoneda**-interpretation) of the higher cohomology groups—which seems natural since the coefficient category for cohomology is actually definable in terms of singular extension theory.

If one wants to define a uniform cohomology theory for (linear) nonassociative algebras and triple systems which "generalizes" these two classical cohomologies one may proceed either via derived functors[171] or via singular extension theory[119, 120]

The purpose of this paper which adopts the first point of view is to discuss compatibility questions with the second one."

6.2. *Cohomology of Banach triple systems—Prospectus*

In this subsection we propose, albeit quite vaguely and briefly, some places to look for extending some of the preceding material to infinite dimensional Banach triples of various kinds as well as infinite dimensional non associative Banach algebras[32]. We begin with the following quotation from[172].

"A veritable army of researchers took the theory of derivations of operator algebras to dizzying heights—producing a theory of cohomology of operator algebras as well as much information about automorphisms of operator algebras." —Dick Kadison (*Which Singer is that?* 2000)

In addition to associative algebras, cohomology groups are defined for Lie algebras and, to some extent, for Jordan algebras. Since the structures of Jordan derivations and Lie derivations on von Neumann algebras are well understood, isn't it time to study the higher dimensional non associative cohomology of a von Neumann algebra? This subsection includes an introduction to the second Jordan cohomology groups of a von Neumann algebra.

Also included, with or without comment, are some papers related to topics discussed in this survey which the author has downloaded and thinks might be worth exploring. In some cases possible connections between papers are suggested.

6.2.1. *Jordan 2-cocycles on von Neumann algebras*

Let M be a von Neumann algebra. A **Hochschild 2-cocycle** is a bilinear map $f : M \times M \to M$ satisfying

$$af(b,c) - f(ab,c) + f(a,bc) - f(a,b)c = 0 \tag{1}$$

An example is a **Hochschild 2-coboundary**:

$$f(a,b) = a\mu(b) - \mu(ab) + \mu(a)b \quad \text{for some} \quad \mu : M \to M \text{ linear} .$$

A **Jordan 2-cocycle** is a bilinear map $f : M \times M \to M$ satisfying

$$f(a,b) = f(b,a) \text{ (symmetric)}$$

$$f(a^2, a \circ b) + f(a,b) \circ a^2 + f(a,a) \circ (a \circ b) \tag{2}$$

[32]See the recent monograph Miguel Cabrera Garcia, Angel Rodriguez Palacios, Non-associative normed algebras. Vol. 1. The Vidav-Palmer and Gelfand-Naimark theorems. Encyclopedia of Mathematics and its Applications, 154. *Cambridge University Press, Cambridge,* 2014. xxii+712 pp.

$$-f(a^2 \circ b, a) - f(a^2, b) \circ a - (f(a, a) \circ b) \circ a = 0$$

An example is a **Jordan 2-coboundary**:$f(a, b) = a \circ \mu(b) - \mu(a \circ b) + \mu(a) \circ b$ for some $\mu : M \to M$ linear. We have the following cohomology groups and their interpretations:

$$H^1(M, M) = \frac{\text{1-cocycles}}{\text{1-coboundaries}} = \frac{\text{derivations}}{\text{inner derivations}}$$

$$H^1_J(M, M) = \frac{\text{Jordan 1-cocycles}}{\text{Jordan 1-coboundaries}} = \frac{\text{Jordan derivations}}{\text{inner Jordan derivations}}$$

$$H^2(M, M) = \frac{\text{2-cocycles}}{\text{2-coboundaries}} = \frac{\text{null extensions}}{\text{split null extensions}}$$

$$H^2_J(M, M) = \frac{\text{Jordan 2-cocycles}}{\text{Jordan 2-coboundaries}} = \frac{\text{Jordan null extensions}}{\text{Jordan split null extensions}}$$

For "almost all" von Neumann algebras, $H^2(M, M) = 0$. For finite dimensional M, the Albert-Penico-Taft theorem (step 7 in subsection 4.3.1) shows that $H^2_J(M, M) = 0$? This is still unknown for infinite dimensional M.

There are at least two elegant approaches to this problem: Jordan classification (subsection 4.3.1) and Lie algebras (not discussed in this survey). We present now a pedestrian approach based on solving linear equations and motivated by the proof of reduction IV at the end of subsection 4.3.1. We first illustrate with two known associative cases.

Test case 1

Let h be a Hochschild 1-cocycle, that is, a linear map

$$h : M_n(\mathbf{C}) \to M_n(\mathbf{C})$$

satisfying $h(ab) - ah(b) - h(a)b = 0$. To show that there is an element $x \in M_n(\mathbf{C})$ such that $h(a) = xa - ax$, it is enough to prove this with $a \in \{e_{ij}\}$. With

$$x = \sum_{p,q} x_{pq} e_{pq} \tag{3}$$

and γ_{ijpq} defined by

$$h(e_{ij}) = \sum_{p,q} \gamma_{ijpq} e_{pq}, \tag{4}$$

we arrive at the system of linear vector equations

$$\sum_{p,q} \gamma_{ijpq} e_{pq} = \sum_{p,q} \delta_{qi} x_{pq} e_{pj} - \sum_{p,q} \delta_{jp} x_{pq} e_{iq} \tag{5}$$

with n^2 unknowns x_{ij}. Then any solution of (5) proves the result.

Test case 2

Let h be a Hochschild 2-cocycle, that is, a bilinear map

$$h : M_n(\mathbf{C}) \times M_n(\mathbf{C}) \to M_n(\mathbf{C})$$

satisfying $ah(b,c) - h(ab,c) + h(a,bc) - h(a,b)c = 0$. To show that there is a linear transformation $\mu : M_n(\mathbf{C}) \to M_n(\mathbf{C})$ such that

$$h(a,b) = \mu(ab) - a\mu(b) - \mu(a)b,$$

it is enough to prove that this holds with $a, b \in \{e_{ij}\}$, that is

$$h(e_{ij}, e_{kl}) = \delta_{jk}\mu(e_{il}) - e_{ij}\mu(e_{kl}) - \mu(e_{ij})e_{kl}. \tag{6}$$

With $\mu(e_{ij}) = \sum_{k,l} \mu_{ijkl} e_{kl}$ and γ_{ijklpq} defined by

$$h(e_{ij}, e_{kl}) = \sum_{p,q} \gamma_{ijklpq} e_{pq},$$

we arrive at the system of n^6 linear equations

$$\sum_{p,q} \gamma_{ijklpq} e_{pq} = \sum_{p,q} \delta_{jk}\mu_{ilpq} - \sum_{p,q} \delta_{jp}\mu_{klpq} e_{iq} - \sum_{p,q} \delta_{qk}\mu_{ijpq} e_{pl}, \tag{7}$$

with n^4 unknowns μ_{ijkl}. Then any solution of (7) proves (6).

We propose to use this method to prove that $H^2_J(M, M) = 0$ if M is a finite von Neumann algebra of type I. (A separate proof would be needed for commutative M.) Let $M = M_2(L^\infty(\Omega))$ be a finite von Neumann algebra of type I_n with $n = 2$. Let f be a Jordan 2-cocycle, that is, a symmetric bilinear map $f : M \times M \to M$ with

$$\begin{aligned} f(a^2, ab) + f(a,b)a^2 + f(a,a)(ab) \\ - f(a^2 b, a) - f(a^2, b)a - (f(a,a)b)a = 0. \end{aligned} \tag{8}$$

(To save space, ab temporarily denotes the Jordan product in the associative algebra M)

To show that there is a linear transformation $\mu : M \to M$ such that

$$f(a,b) = \mu(ab) - a\mu(b) - \mu(a)b, \tag{9}$$

it is enough to prove, for $a, b \in Z(M)$,

$$f(ae_{ij}, be_{kl}) = \delta_{jk}\mu(abe_{il}) - ae_{ij}\mu(be_{kl}) - \mu(ae_{ij})be_{kl}. \tag{10}$$

With $\mu(ae_{ij}) = \sum_{k,l} \mu_{ijkl}(a)e_{kl}$ and $\gamma_{ijklpq}(a,b) \in Z(M)$ defined by

$$f(ae_{ij}, be_{kl}) = \sum_{p,q} \gamma_{ijklpq}(a,b)e_{pq}, \tag{11}$$

we arrive at the system of n^6 linear vector equations with $3n^4$ unknowns $\mu_{ijkl}(ab), \mu_{ijkl}(a), \mu_{ijkl}(b)$

$$\begin{aligned} 2\sum_{p,q} \gamma_{ijklpq}(a,b)e_{pq} &= \delta_{jk} \sum_{p,q} \mu_{ilpq}(ab)e_{pq} \\ &\quad - \sum_{p,q} a(\delta_{jp}\mu_{klpq}(b)e_{iq} + \delta_{iq}\mu_{klpq}(b)e_{pj}) \\ &\quad - \sum_{p,q} b(\delta_{qk}\mu_{ijpq}(a)e_{pl} + \delta_{lp}\mu_{ijpq}(a)e_{kq}). \end{aligned} \tag{12}$$

Then any solution of (12) proves (10) and hence (9). (Warning: it is not clear that this approach will be successful!)

We shall now describe some properties of Jordan 2-cocycles. Just as every derivation is a Jordan derivation, it is also true that every symmetric Hochschild 2-cocycle is a Jordan 2-cocyle. Recall that every Jordan derivation on a semisimple Banach algebra is a derivation (Sinclair). If every Jordan 2-cocycle was a Hochschild 2-cocycle, we would have $H^2(M,M) = H^2_J(M,M)$. Fortunately, we have the following examples.

Proposition 6.1. *Let M be a von Neumann algebra.*

(a) *Let $f : M \times M \to M$ be defined by $f(a,b) = a \circ b$. Then f is a Jordan 2-cocycle with values in M, which is not a Hochschild 2-cocycle unless M is commutative.*

(b) *If M is finite with a faithful normal finite trace tr, then $f : M \times M \to M_*$ defined by $f(a,b)(x) = tr((a \circ b)x)$ is a Jordan 2-cocycle with values in M_* which is not a Hochschild 2-cocycle unless M is commutative.*

It can also be shown that if f is a Jordan 2-cocycle on the von Neumann algebra M, then $f(1,x) = xf(1,1)$ for every $x \in M$ and $f(1,1)$ belongs to the center of M.

Let us show the proof of (a) of Proposition 6.1. Recall the definition of Jordan 2-cocycle

$$\begin{aligned} f(a^2, a \circ b) + f(a,b) \circ a^2 + f(a,a) \circ (a \circ b) \\ - f(a^2 \circ b, a) - f(a^2, b) \circ a - (f(a,a) \circ b) \circ a = 0 \end{aligned} \tag{13}$$

Let $f(a,b) = a \circ b$. The equation (13) reduces to

$$a^2 \circ (a \circ b) + (a \circ b) \circ a^2 + a^2 \circ (a \circ b) - (a^2 \circ b) \circ a - (a^2 \circ b) \circ a - (a^2 \circ b) \circ a,$$

which is zero by the Jordan axiom, so f is a Jordan 2-cocycle. If this f were a Hochschild 2-cocycle, we would have

$$c(a \circ b) - (ca) \circ b + c \circ (ab) - (c \circ a)b = 0,$$

which reduces to $[[c,b],a] = 0$ and therefore $[M,M] \subset Z(M)$ (the center of M). Since $M = Z(M) + [M,M]$, M is commutative. This proves (a).

While we are at it, let us prove that every symmetric Hochschild 2-cocycle is a Jordan 2-cocycle. The Jordan 2-cocycle identity reduces to

$$\begin{aligned} f(a^2, ab) &+ f(a^2, ba) + f(a,b)a^2 + a^2 f(a,b) \\ &+ f(a,a)ab/2 + baf(a,a)/2 - f(a^2b, a) - f(ba^2, a) \\ &- f(a^2, b)a - af(a^2, b) - bf(a,a)a/2 - af(a,a)b/2 = 0 \quad (14) \end{aligned}$$

In (1), replace a by a^2 and c by a to obtain

$$a^2 f(b,a) - f(a^2b, a) + f(a^2, ba) - f(a^2, b)a = 0.$$

Thus, taking into account that f is symmetric, four of the twelve terms in (14) sum to zero. In (1), replace c by a^2 to obtain

$$af(b, a^2) - f(ab, a^2) + f(a, ba^2) - f(a,b)a^2 = 0.$$

Thus, four more of the twelve terms in (14) sum to zero. It remains to show that

$$f(a,a)ab + baf(a,a) - bf(a,a)a - af(a,a)b = 0. \quad (15)$$

In (1), replace b and c by a to obtain $af(a,a) = f(a,a)a$ from which (15) follows.

6.2.2. *Ternary weak amenability for group algebras and Fourier algebras*

The purpose of the summary in this subsection is to suggest that the concepts of triple derivation and ternary weak amenability be explored in the contexts of group algebras and Fourier algebras.

Group algebras $L^1(G)$ and Fourier algebras $A(G)$ of a locally compact group G are closely related and share some techniques with operator algebras, as do the measure algebra $M(G)$ and the Fourier-Stieltjes algebra $B(G)$. The Fourier algebra and the Fourier-Stieltjes algebras were introduced in the thesis of Pierre Eymard[173] and have provided fertile ground for the extension of many classical harmonic analysis results to the noncommutative setting.

Johnson showed very early that G is an amenable group if and only if $L^1(G)$ is an amenable Banach algebra[32]. Two decades later he showed that $L^1(G)$ is weakly amenable[174]. A decade after that it was shown that $M(G)$ is weakly amenable if and only if G is discrete (and therefore $M(G) = \ell^1(G)$) with the consequence that $M(G)$ is amenable if and only if G is discrete[175].

The Fourier algebra $A(G)$ is amenable if and only if G has an abelian group of finite index. The "if" was proved in[176] and the "only if" in[177]. The following was also proved in the latter reference: $B(G)$ is amenable if and only if G has a compact abelian subgroup of finite index.

A characterization for weak amenability for the Fourier algebras is not yet known, but there are several partial results which can be found in the survey[178], which contains much other information, some of which has been summarized here. We mention only one such result: If G is a compact group, then $A(G)$ is weakly amenable if and only if the connected component of the identity in G is abelian. The "only if" part was proved in[179] and the "if" part holds for all locally compact groups, as shown in[177].

6.2.3. *Cohomology of commutative JB*-triples and TROs*

The first three cohomology groups of a commutative C^*-algebra with values in a Banach module were studied in the pioneering paper[30]. The cohomology groups of a finite dimensional associative triple system with values in the relevant module were studied in the[35]. By some combination of the techniques in these two papers, it should be possible to develop the cohomology of a commutative JB^*-triple with values in a module.

More generally, the cohomology of an arbitrary TRO (ternary ring of operators) should be pursued. Once this has been worked out for TROs, the author feels that the enveloping TRO of a JC^*-triple, discussed in section 11 may be helpful in working out the theory for JC^*-triples, as it seems to be for working out the K-theory of JB^*-triples, studied in[180].

6.2.4. *Some other avenues to pursue*

- Wedderburn decompositions for JB*-triples (Kühn-Rosendahl[85])
- Low dimensional cohomology for JBW*-triples and algebras-perturbation (Dosi[159], McCrimmon[126])

- Structure group of JB*-triple (McCrimmon[126], Meyberg[24])
- Alternative Banach triples (Carlsson[167], Braun[181])
- Completely bounded triple cohomology (Bunce-Feely-Timoney[9], Bunce-Timoney[10], Christensen-Effros-Sinclair[142])
- Lie algebraic techniques. Applications of the Koecher-Kantor-Tits construction[39]

As an illustration of the last bullet, we shall now give a preview of the forthcoming paper[40], which develops a cohomology theory for Jordan triples based on the corresponding theory for Lie algebras (cf. Theorem 2.9).

Let V be a non-degenerate Jordan triple and let $L(V)$ be its TKK Lie algebra as defined in[39] p.45, where

$$L(V) = V \oplus V_0 \oplus V$$

and the Lie product is given by

$$[(x,h,y),(u,k,v)] = (hu - kx,\ [h,k] + x\Box\, v - u\Box\, y,\ k^\natural y - h^\natural v).$$

We note that $V_0 = \text{Span}\{V\Box\, V\}$ is a Lie subalgebra of $L(V)$ and for $h = \sum_i a_i \Box\, b_i \in V_0$, the map $h^\natural : V \to V$ is defined by

$$h^\natural = \sum_i b_i \Box\, a_i.$$

Given a Lie algebra L and an L-module X, we denote the action of L on X by

$$(\ell, x) \in L \times X \mapsto \ell.x \in X$$

so that

$$[\ell, \ell'].x = \ell.(\ell'.x) - \ell'.(\ell.x).$$

Let V be a Jordan triple and M be a triple V-module as defined in subsection 3.6.2, but for convenience, we omit the numeric subscripts of the triple actions on M. Given $a, b \in V$, the box operator $a\Box\, b : V \to V$ can also be considered as a mapping from M to M. Similarly, for $u \in V$ and $m \in M$, the "*box operators*"

$$u\Box\, m,\ m\Box\, u : V \longrightarrow M$$

are defined in a natural way as $v \mapsto \{u, m, v\}$ and $v \mapsto \{m, u, v\}$ respectively. In view of the main identity for Jordan triple modules, for $a\Box\, b \in V_0$, we have (as mentioned above)

$$[a\Box\, b, u\Box\, m] = \{a, b, u\}\Box\, m - u\Box\, \{m, a, b\}$$

and

$$[a\square b, m\square u] = \{a,b,m\}\square u - m\square \{u,a,b\}$$

for $u \in V$ and $m \in M$. We also have

$$[u\square m, a\square b] = \{u,m,a\}\square b - m\square \{u,a,b\}$$

and similar identity for $[m\square u, a\square b]$. Let M_0 be the linear span of

$$\{u\square m, n\square v : u,v \in V, m,n \in M\}$$

in the vector space $L(V,M)$ of linear maps from V to M. Extending the above product by linearity, we can define an action of V_0 on M_0 by

$$(h,\varphi) \in V_0 \times M_0 \mapsto [h,\varphi] \in M_0.$$

Lemma 6.1. *M_0 is a V_0-module of the Lie algebra V_0.*

Proof. We are required to show that

$$[[h,k],\varphi] = [h,[k,\varphi]] - [k,[h,\varphi]]. \tag{16}$$

We can assume that $h = a\square b$, $k = c\square d$ and $\varphi = w\square m$ or $m\square w$. We assume $\varphi = w\square m$, the other case being similar. For the left side of (16), we have

$$\begin{aligned}[[a\square b, u\square v], w\square m] &= [\{abu\}\square v - u\square \{vab\}, w\square m]\\ &= \{\{abu\}vw\}\square m - w\square \{m\{abu\}v\}\\ &- \{u\{vab\}w\}\square m + w\square \{mu\{vab\}\}\\ &= (\{\{abu\}vw\} - \{u\{vab\}w\})\square m\\ &- w\square (\{m\{abu\}v\} - \{mu\{vab\}\}.\end{aligned}$$

For the right side of (16), we have

$$\begin{aligned}&[a\square b, [u\square v, w\square m]] - [u\square v, [a\square b, w\square m]]\\ &= [a\square b, \{uvw\}\square m - w\square \{muv\}] - [u\square v, \{abw\}\square m - w\square \{mab\}]\\ &= \{ab\{uvw\}\}\square w - \{uvw\}\square \{mab\} - \{abw\}\square \{muv\} + w\square \{\{muv\}ab\}\\ &- \{uv\{abw\}\}\square m + \{abw\}\square \{muv\} + \{uvw\}\square \{mab\} - w\square \{\{mab\}uv\}\\ &= (\{ab\{uvw\}\} - \{uv\{abw\}\})\square m - w\square (\{\{mab\}uv\} - \{\{muv\}ab\}).\end{aligned}$$

(16) now follows from the main identity for Jordan triple modules.

Let V be Jordan triple and $L(V)$ its TKK Lie algebra. Given a triple V-module M, we now construct a corresponding Lie module $\mathcal{L}(M)$ of the

Lie algebra $L(V)$ as follows. Define $\mathcal{L}(M) = M \oplus M_0 \oplus M$ and define the action

$$((a,h,b),(m,\varphi,n)) \in L(V) \times \mathcal{L}(M) \mapsto (a,h,b).(m,\varphi,n) \in \mathcal{L}(M)$$

by $(a,h,b).(m,\varphi,n) = (hm - \varphi a,\ [h,\varphi] + a\square n - m\square b,\ \varphi^\natural b - h^\natural(n))$ where, for $h = \sum_i a_i \square b_i$ and $\varphi = \sum_i u_i \square m_i + \sum_j n_j \square v_j$, we have the following natural definitions

$$\begin{aligned} hm &= \sum_i \{a_i, b_i, m\}, \\ \varphi a &= \sum_i \{u_i, m_i, a\} + \sum_j \{n_j, v_j, a\}, \\ \varphi^\natural &= \sum_i m_i \square u_i + \sum_j v_j \square n_j. \end{aligned}$$

The following theorem is the starting point of the paper[40].

Theorem 6.3. *Let V be a Jordan triple and let $L(V)$ be its TKK Lie algebra. Let M be a triple V-module. Then $\mathcal{L}(M)$ is a Lie $L(V)$-module.*

Part III
Quantum functional analysis

7. Contractive projections

The purpose of this section is to describe the role which contractive projections have played in the theory of JB^*-triples, which is the primary algebraic structure of interest in this survey. Contractive projections on Banach spaces have been the focus of much study. For a survey see[182]. We are concerned here with the interplay between algebraic structure and contractive projections.

7.1. *Projective stability*

A well-known and useful result in the structure theory of operator triple systems is the "contractive projection principle," that is, the fact that the range of a contractive projection on a JB^*-triple is linearly isometric in a natural way to another JB^*-triple. The genesis of the role of contractive projections in Banach triple systems lies in the following theorem. This was preceded by a similar principle in a geometric context[183], which was unknown to the authors of[184].

Theorem 7.1 (Friedman and Russo 1985[184]). *The range of a contractive projection on a C^*-algebra is linearly isometric to a JC^*-triple, that is, a linear subspace of $B(H)$ which is closed under the symmetric triple product $xy^*z + zy^*x$.*

Thus, as shown later and described below, the category of JB^*-triples and contractions is stable under contractive projections. To put this result in proper prospective, let $\mathcal{B}$ be the category of Banach spaces and contractions. We say that a sub-category $\mathcal{S}$ of $\mathcal{B}$ is **projectively stable** if it has the property that whenever A is an object of $\mathcal{S}$ and X is the range of an idempotent morphism of $\mathcal{S}$ on A, then X is isometric (that is, isomorphic in $\mathcal{S}$) to an object in $\mathcal{S}$.

Examples of projectively stable categories are the following.

- L_1, contractions (Grothendieck 1955[185])
- L^p, $1 \leq p < \infty$, contractions (Douglas 1965[186], Ando 1966[187], Bernau-Lacey 1974[188], Tzafriri 1969[189])
- C^*-algebras, completely positive unital maps (Choi-Effros 1977[190])
- ℓ_p, $1 \leq p < \infty$, contractions (Lindenstrauss-Tzafriri 1977[191])
- JC^*-algebras, positive unital maps (Effros-Stormer 1979[192])
- $TROs$ (ternary rings of operators), complete contractions (Youngson 1983[193])
- JB^*-triples, contractions (Kaup 1984[194], Stacho 1982[183], Friedman-Russo 1985[184])
- ℓ^p-direct sums of $C_p(H)$, $1 \leq p < \infty$, H Hilbert space, contractions (Arazy-Friedman, 1978[195], 2000[196])
- ℓ^p-direct sums of $L^p(\Omega, H)$, $1 \leq p < \infty$, H Hilbert space, contractions (Raynaud 2004[197])
- ℓ^p-direct sums of $C_p(H)$, $1 \leq p \neq 2 < \infty$, H Hilbert space, complete contractions (LeMerdy-Ricard-Roydor 2009[198])

It follows immediately that if $\mathcal{S}$ is projectively stable, then so is the category $\mathcal{S}_*$ of spaces whose dual spaces belong to $\mathcal{S}$. It should be noted that $TROs$, C^*-algebras and JC^*-algebras are not stable under contractive projections and JB^*-triples are not stable under bounded projections.

7.1.1. *More about JB*-triples*

JB*-triples are generalizations of JB*-algebras and C*-algebras. The axioms can be said to come from geometry in view of Kaup's Riemann mapping theorem[115]. Kaup showed in 1983 that JB*-triples are exactly those Banach spaces whose open unit ball is a bounded symmetric domain. Kaup's holomorphic characterization of JB*-triples directly led to the proof of the projective stability of JB*-triples mentioned above.

Many authors have studied the interplay between JB*-triples and infinite dimensional holomorphy. Contractive projections have proved to be a valuable tool for the study of problems on JB*-triples (Gelfand-Naimark theorem[199], structure of inner ideals[200], operator space characterization of TROs[201], to name a few). They are justified both as a natural generalization of operator algebras as well as because of their connections with complex geometry.

Preduals of JBW*-triples have been called pre-symmetric spaces[202] and have been proposed as mathematical models of physical systems[203]. In this model the operations on the physical system are represented by contractive projections on the pre-symmetric space.

JB*-triples first arose in Koecher's proof[204],[205] of the classification of bounded symmetric domains in $\mathbf{C}^n$. The original proof of this fact, done in the 1930's by Cartan, used Lie algebras and Lie groups, techniques which do not extend to infinite dimensions. On the other hand, to a large extent, the Jordan algebra techniques do so extend, as shown by Kaup and Upmeier.[33]

7.1.2. *Application: Gelfand Naimark theorem for JB*-triples*

Theorem 7.2 (Friedman and Russo 1986[199]). *Every JB*-triple is isometrically isomorphic to a subtriple of a direct sum of Cartan factors.*

The theorem was not unexpected. However, the proof required new techniques because of the lack of an order structure on a JB*-triple.

- **Step 1: February 1983 Friedman-Russo[184]**
  Let $P : A \to A$ be a linear projection of norm 1 on a JC*-triple A. Then $P(A)$ is a JB*-triple under $\{xyz\}_{P(A)} = P(\{xyz\})$ for $x, y, z \in P(A)$.
- **Step 2: April 1983 Friedman-Russo[206]**

[33]The opposite is true concerning cohomology. Lie algebra cohomology is well developed, Jordan algebra cohomology is not

Same hypotheses. Then P is a conditional expectation in the sense that

$$P\{PaPbPc\} = P\{Pa, b, Pc\}$$

and

$$P\{PaPbPc\} = P\{aPbPc\}.$$

- **Step 3: May 1983 Kaup**[194]
  Let $P : U \to U$ be a linear projection of norm 1 on a JB*-triple U. Then $P(U)$ is a JB*-triple under $\{xyz\}_{P(U)} = P(\{xyz\}_U)$ for $x, y, z \in P(U)$. Also, $P\{PaPbPc\} = P\{Pa, b, Pc\}$ for $a, b, c \in U$, which extends one of the formulas in the previous step.
- **Step 4: February 1984 Friedman-Russo**[207]
  Every JBW*-triple splits into atomic and purely non-atomic ideals.
- **Step 5: August 1984 Dineen**[208]
  The bidual of a JB*-triple is a JB*-triple.
- **Step 6: October 1984 Barton-Timoney**[209]
  The bidual of a JB*-triple is a JBW*-triple, that is, the triple product is separately weak*-continuous.
- **Step 7: December 1984 Horn**[210]
  Every JBW*-triple factor of type I is isomorphic to a Cartan factor. More generally, every JBW*-triple of type I is isomorphic to a direct sum of L^∞ spaces with values in a Cartan factor.
- **Step 8: March 1985 Friedman-Russo**[199]
  Putting it all together

  $$\pi : U \to U^{**} = A \oplus N = (\oplus_\alpha C_\alpha) \oplus N = \sigma(U^{**}) \oplus N$$

  implies that $\sigma \circ \pi : U \to A = \oplus_\alpha C_\alpha$ is an isometric isomorphism.

Consequences of the Gelfand-Naimark theorem

- Every JB*-triple is isomorphic to a subtriple of a JB*-algebra.
- In every JB*-triple, $\|\{xyz\}\| \leq \|x\|\|y\|\|z\|$
- Every JB*-triple contains a unique norm-closed ideal J such that U/J is isomorphic to a JC*-triple and J is purely exceptional, that is, every homomorphism of J into a C*-algebra is zero.

7.1.3. *Preservation of type*

The following results are due to Chu-Neal-Russo[211], and simultaneously, to Bunce-Peralta[212]. The corresponding (classical) result for von Neumann algebras, due to Tomiyama[213], required, of necessity, that the range of the projection be a subalgebra, since the category of von Neumann algebras and positive normal contractions is not projectively stable.

Theorem 7.3. *Let P be a normal contractive projection on a JBW^*-triple Z of type I. Then $P(Z)$ is of type I.*

Theorem 7.4. *Let P be a normal contractive projection on a semifinite JBW^*-triple Z. Then $P(Z)$ is a semifinite JW^*-triple.*

7.2. *Projective rigidity*

By considering the converse of projective stability, one is lead to the following definition.

A sub-category $\mathcal{S}$ of $\mathcal{B}$ is **projectively rigid** if it has the property that whenever A is an object of $\mathcal{S}$ and X is a subspace of A which is isometric to an object in $\mathcal{S}$, then X is the range of an idempotent morphism of $\mathcal{S}$ on A.

Examples of projectively rigid categories are the following.

- ℓ_p, $1 < p < \infty$, contractions (Pelczynski 1960[214])
- L^p, $1 \leq p < \infty$, contractions (Douglas 1965[186], Ando 1966[187], Bernau-Lacey 1974[188])
- $C_p, 1 \leq p < \infty$, contractions (Arazy-Friedman 1977[215])
- Preduals of von Neumann algebras, contractions (Kirchberg 1993[216])
- Preduals of $TROs$, complete contractions (Ng-Ozawa 2002[217])
- Preduals of JBW*-triples, contractions[34] (Neal-Russo 2008[218])
- ℓ^p-direct sums of $C_p(H)$, $1 \leq p \neq 2 < \infty$, H Hilbert space, complete contractions (LeMerdy-Ricard-Roydor 2009[198])

Theorem 7.5 (Neal-Russo). *The category of preduals of JBW^*-triples with no summands of the form $L^1(\Omega, H)$ where H is a Hilbert space of dimension at least two, is projectively rigid.*

---

[34]with one exception, see Theorem 7.5

7.3. *Structural projections*

7.3.1. *Structure of inner ideals*

In a series of papers, mainly by Edwards and Rüttimann (cf.[219]), the ideal structure of C*-algebras and JB*-algebras has been thoroughly studied, and a surprising geometric characterization of the closed inner ideals among the closed subtriples of a JB*-triple has been established: a norm-closed subtriple B of a JB*-triple A is an inner ideal if and only if every bounded linear functional on B has a unique norm-preserving linear extension to A.

Loos and Neher ([220], [221]) have introduced the notions of complementation and structural projection in the purely algebraic setting of Jordan pairs and Jordan *-triples. Let A be an anisotropic Jordan *-triple; then for every element a in A the quadratic mapping Q(a) is defined by Q(a)x:={axa}. The kernel Ker(B) of a subset B in A is the subspace of A consisting of all elements annihilated by the mappings Q(b) as b ranges over B. A subtriple B is said to be complemented if $A = B \oplus \mathrm{Ker}(B)$. A linear projection P on A is said to be structural if it satisfies Q(Pa) = PQ(a)P for all a in A, which occurs if and only if its range B = P(A) is a complemented subtriple of A.

In[200], the study of structural projections on JBW*-triples is continued. It is shown that a structural projection on a JBW*-triple is necessarily contractive and weak* continuous, and that every weak* closed inner ideal in a JBW*-triple A is a complemented subtriple of A and therefore the range of a unique structural projection.

7.3.2. *Geometric characterization*

Since structural projections are contractive and weak* continuous, every structural projection is the adjoint of a contractive linear projection on the predual of A. The contractive projections P on A_* that arise in this way also have the property that they are neutral, in that, if x is an element of for which $\|Px\|$ and $\|x\|$ coincide then Px and x also coincide. It follows from the results of[222] that the mapping $P \mapsto P^*(A)$ is a bijection from the family of neutral projections on A_* for which P^*A is a subtriple of A onto the complete lattice of weak*-closed inner ideals in A.

The L-orthogonal complement $G^\diamond$ of a subset G of a complex Banach space E is the set of elements x in E such that, for all elements y in G, $\|x \pm y\| = \|x\| + \|y\|$: A contractive projection P on E is said to be a GL-projection if the L-orthogonal complement $PE^\diamond$ of the range PE of P is contained in the kernel of P. It is shown in[223] that, for a JBW^*-triple A with pre dual A_*, a linear projection R on A is structural if and only if it is the adjoint of a neutral GL-projection on A, thereby giving a purely geometric characterization of structural projections.

7.3.3. *Physical interpretation*

The predual of a JBW*-triple A, also called a pre-symmetric space, has been proposed as a model of the state space of a statistical physical system, in which contractive linear mappings on A_* represent operations or filters on the system. For a discussion of this model we refer to[223] and for an application to decoherent states we refer to[224].

In the theory of Banach spaces and in the study of state spaces of physical systems, much effort has been devoted to the investigation of when particular closed subspaces of a Banach space are the ranges of contractive projections. Since the range of a contractive projection on a pre-symmetric space is itself a pre-symmetric space, this problem is particularly relevant in the case of pre-symmetric spaces. The main results of[224] are used to determine under what circumstances the closed subspace of A_* generated by a family $\{x_j : j \in \Lambda\}$ of elements of A_* of norm one which are pairwise decoherent (see[224] for the definition) is the range of a contractive projection. (Theorem 7.5 might be relevant here.)

8. Hilbertian operator spaces

8.1. *Operator spaces revisited*

We are going to quote some definitions and theorems from[132]. Operator space theory may also be considered as noncommutative Banach space theory. Although the subject had its genesis in the papers[225] and[226] (announcement in[227]), the first impetus appeared in the thesis of Ruan[228]. An operator space is a Banach space together with an isometric embedding in $B(H)$. While the objects of this category are Banach spaces, it is the morphisms, namely, the completely bounded maps, which are more important.

An operator space is a subspace X of $B(H)$. Its *operator space structure* is given by the sequence of norms on the set of matrices $M_n(X)$ with entries from X, determined by the identification

$$M_n(X) \subset M_n(B(H)) = B(H \oplus H \oplus \cdots \oplus H).$$

A linear mapping $\varphi : X \to Y$ between two operator spaces is *completely bounded* if the induced mappings $\varphi_n : M_n(X) \to M_n(Y)$ defined by $\varphi_n([x_{ij}]) = [\varphi(x_{ij})]$ satisfy $\|\varphi\|_{\text{cb}} := \sup_n \|\varphi_n\| < \infty$.

Some simple examples of completely bounded maps are restrictions of *-homomorphisms and multiplication operators.

Theorem 8.1. *Every completely bounded map is the product of the above two mentioned maps.*

Some special types of completely bounded maps which we will consider are complete contractions, complete isometries, and complete isomorphisms. Additionally, a complete semi-isometry is defined to be a complete contraction which is isometric.

Remark 8.1. The space of completely bounded maps with the completely bounded norm $CB(X,Y)$, $\|\cdot\|_{\text{cb}}$ is a Banach space.

Definition 8.1. The completely bounded Banach-Mazur distance is defined by

$$\mathrm{d}_{\text{cb}}(E,F) = \inf\{\|u\|_{\text{cb}} \cdot \|u^{-1}\|_{\text{cb}} : u : E \to F \text{ complete isomorphism }\}.$$

Besides the pioneering work of Stinespring and Arveson mentioned above, many tools were developed in the 1970s and 1980s by a number of operator algebraists. As noted above, an abstract framework was developed in 1988 in the thesis of Ruan. Besides[132] we have already mentioned four other monographs on the subject ([130],[131],[133],[134]).

8.2. *Classical operator spaces*

"Neoclassical" and "modern" operator spaces will appear later.

Two important examples of Hilbertian operator spaces (:= operator spaces isometric to Hilbert space) are the row and column spaces R, C, and their finite-dimensional versions R_n, C_n. In $B(\ell_2)$, *column Hilbert space* $C := \overline{\text{sp}}\{e_{i1} : i \geq 1\}$ and *row Hilbert space* $R := \overline{\text{sp}}\{e_{1j} : j \geq 1\}$. R and C are Banach isometric, but not completely isomorphic: $\mathrm{d}_{\text{cb}}(R,C) = \infty$. R_n and C_n are completely isomorphic, but not completely isometric: $\mathrm{d}_{\text{cb}}(R_n, C_n) = n$.

R, C, R_n, C_n are examples of *homogeneous* operator spaces, that is, operator spaces E for which $\forall u : E \to E$, $\|u\|_{\text{cb}} = \|u\|$. Another important example of an Hilbertian homogeneous operator space is $\Phi(I)$. $\Phi(I) = \overline{\text{sp}}\{V_i : i \in I\}$, where the V_i are bounded operators on a Hilbert space satisfying the canonical anti-commutation relations. In some special cases, the notations $\Phi_n := \Phi(\{1, 2, \ldots, n\})$, and $\Phi = \Phi(\{1, 2, \ldots\})$ are used.

If $E_0 \subset B(H_0)$ and $E_1 \subset B(H_1)$ are operator spaces whose underlying Banach spaces form a compatible pair in the sense of interpolation theory, then the Banach space $E_0 \cap E_1$, with the norm

$$\|x\|_{E_0 \cap E_1} = \max(\|x\|_{E_0}, \|x\|_{E_1})$$

equipped with the operator space structure given by the embedding $E_0 \cap E_1 \ni x \mapsto (x, x) \in E_0 \oplus E_1 \subset B(H_0 \oplus H_1)$ is called the *intersection* of E_0 and E_1 and is denoted by $E_0 \cap E_1$. Examples are $R \cap C$, $\Phi = \cap_0^\infty H_\infty^{m,L}$, $\Phi_n = \cap_1^n H_n^k$. ($H_\infty^{m,L}$ and H_n^k are defined below.) The definition of intersection extends easily to arbitrary families of compatible operator spaces.

We already noted that R, C, R_n, C_n are homogenous operator spaces. So are $\min(E)$ and $\max(E)$ and $\Phi(I)$, where $\min(E)$ is defined by $E \subset C(T) \subset B(H)$, $\|(a_{ij})\|_{M_n(\min(E))} = \sup_{\xi \in B_{E^*}} \|(\xi(a_{ij}))\|_{M_n}$ and $\max(E)$ is defined by

$$\|(a_{ij})\|_{M_n(\max(E))} = \sup\{\|(u(a_{ij}))\|_{M_n(B(H_u))} : u : E \to B(H_u),\ \|u\| \leq 1\}.$$

These two spaces have a stronger property:

$$F \xrightarrow{u} \min(E) \Rightarrow\rightarrow \|u\|_{\text{cb}} = \|u\| \text{ and } \max(E) \xrightarrow{v} G \Rightarrow \|v\|_{\text{cb}} = \|v\|,$$

and the identity map $\min(E) \to \max(E)$ is completely contractive.

The following is an application of the Russo-Dye theorem[229].

Proposition 8.1. *Let E be a Hilbertian operator space. Then E is homogeneous if and only if $\|U\|_{cb} = 1\ \forall$ unitary $U : E \to E$.*

The following are considered to be among the "classical" Banach spaces ℓ_p, c_0, L_p, $C(K)$. A "second generation" of such spaces would contain Orlicz, Sobolev, and Hardy spaces, the disk algebra, and the Schatten p-classes. By analogy, some classical operator spaces are R, C, $\min(\ell_2)$, $\max(\ell_2)$, OH, Φ and their finite dimensional versions R_n, C_n, $\min(\ell_2^n)$, $\max(\ell_2^n)$, OH_n, Φ_n. Note that the classical operator spaces are all Hilbertian.

Proposition 8.2. *The classical operator spaces are mutually completely non-isomorphic. If E_n, F_n are n-dimensional versions, then*

$$d_{cb}(E_n, F_n) \to \infty.$$

8.3. *Neoclassical operator spaces*

The "neoclassical" operator spaces H_n^k appeared in[8] (announcement in[230]), from which the following three theorems were taken.

Theorem 8.2. *There is a family of 1-mixed injective Hilbertian operator spaces H_n^k, $1 \leq k \leq n$, of finite dimension n, with the following properties:*

(a) *H_n^k is a subtriple of the Cartan factor of type 1 consisting of all $\binom{n}{k}$ by $\binom{n}{n-k+1}$ complex matrices.*

(b) *Let Y be a JW^*-triple of rank 1 (necessarily atomic).*

  (i) *If Y is of finite dimension n then it is isometrically completely contractive to some H_n^k.*

  (ii) *If Y is infinite dimensional then it is isometrically completely contractive to $B(H, \mathbf{C})$ or $B(\mathbf{C}, K)$.*

(c) *H_n^n (resp. H_n^1) coincides with R_n (resp. C_n).*

(d) *For $1 < k < n$, H_n^k is not completely semi-isometric to R_n or C_n.*

Example 1: H_3^2 is the subtriple of $B(\mathbf{C}^3)$ consisting of all matrices of the form

$$\begin{bmatrix} 0 & a & -b \\ -a & 0 & c \\ b & -c & 0 \end{bmatrix}$$

and is completely semi-isometric to the Cartan factor $A(\mathbf{C}^3)$ of 3 by 3 anti-symmetric complex matrices.

Example 2: H_4^3 is the subtriple of $B(\mathbf{C}^6, \mathbf{C}^4)$ consisting of all matrices of the form

$$\begin{bmatrix} 0 & 0 & 0 & -d & c & -b \\ 0 & d & -c & 0 & 0 & a \\ -d & 0 & b & 0 & -a & 0 \\ c & -b & 0 & a & 0 & 0 \end{bmatrix}.$$

Theorem 8.3. *Let X be a 1-mixed injective operator space which is atomic. Then X is completely semi-isometric to a direct sum of Cartan factors of types 1 to 4 and the spaces H_n^k.*

Theorem 8.4. *Let Y be an atomic w^*-closed JW^*-subtriple of a W^*-algebra.*

(a) *If Y is irreducible and of rank at least 2, then it is completely isometric to a Cartan factor of type 1–4 or the space $Diag(B(H, K), B(K, H))$.*

(b) *If Y is of finite dimension n and of rank 1, then it is completely isometric to Diag$(H_n^{k_1}, \ldots, H_n^{k_m})$, for appropriately chosen bases, and where $k_1 > k_2 > \cdots > k_m$.*

(c) *Y is completely semi-isometric to a direct sum of the spaces in (a) and (b). If Y has no infinite dimensional rank 1 summand, then it is completely isometric to a direct sum of the spaces in (a) and (b).*

We now describe, in more detail, the spaces H_n^k, including the next two theorems,[231].

A frequently mentioned result of Friedman and Russo states that if a subspace X of a C*-algebra A is the range of a contractive projection on A, then X is isometric to a JC*-triple, that is, a norm closed subspace of $B(H, K)$ stable under the triple product $ab^*c + cb^*a$. If X is atomic (in particular, finite-dimensional), then it is isometric to a direct sum of Cartan factors of types 1 to 4. This latter result fails, as it stands, in the category of operator spaces.

Nevertheless, there exists a family of n-dimensional Hilbertian operator spaces H_n^k, $1 \leq k \leq n$, generalizing the row and column Hilbert spaces R_n and C_n such that, in the above result, if X is atomic, the word "isometric" can be replaced by "completely semi-isometric," provided the spaces H_n^k are allowed as summands along with the Cartan factors. The space H_n^k is contractively complemented in some $B(K)$, and for $1 < k < n$, is not completely (semi-)isometric to either of the Cartan factors $B(\mathbf{C}, \mathbf{C}^n) = H_n^1$ or $B(\mathbf{C}^n, \mathbf{C}) = H_n^n$. These spaces appeared in a slightly different form and context in a memoir of Arazy and Friedman[195].

The construction of H_n^k is as follows. Let I denote a subset of $\{1, 2, \ldots, n\}$ of cardinality $|I| = k - 1$. The number of such I is $q := \binom{n}{k-1}$. Let J denote a subset of $\{1, 2, \ldots, n\}$ of cardinality $|J| = n - k$. The number of such J is $p := \binom{n}{n-k}$. The space H_n^k is the linear span of matrices $b_i^{n,k}$, $1 \leq i \leq n$, given by

$$b_i^{n,k} = \sum_{I \cap J = \emptyset, (I \cup J)^c = \{i\}} \varepsilon(I, i, J) e_{J,I},$$

where $e_{J,I} = e_J \otimes e_I = e_J e_I^t \in M_{p,q}(\mathbf{C}) = B(\mathbf{C}^q, \mathbf{C}^p)$, and $\varepsilon(I, i, J)$ is the signature of the permutation taking $(i_1, \ldots, i_{k-1}, i, j_1, \ldots, j_{n-k})$ to $(1, \ldots, n)$. Since the $b_i^{n,k}$ are the image under a triple isomorphism (actually ternary isomorphism) of a rectangular grid in a JC*-triple of rank one, they form an orthonormal basis for H_n^k.

Theorem 8.5. *H_n^k is a homogeneous operator space.*

Theorem 8.6. $d_{cb}(H_n^k, H_n^1) = \sqrt{\frac{kn}{n-k+1}}$, *for* $1 \leq k \leq n$.

The latter theorem is proved using the relation with creation operators. Let $C_h^{n,k}$ denote the wedge (or creation) operator from $\wedge^{k-1}\mathbf{C}^n$ to $\wedge^k\mathbf{C}^n$ given by

$$C_h^{n,k}(h_1 \wedge \cdots \wedge h_{k-1}) = h \wedge h_1 \wedge \cdots \wedge h_{k-1}.$$

Letting $\mathcal{C}^{n,k}$ denote the space $\mathrm{sp}\{C_{e_i}^{n,k}\}$, we have the following.

Proposition 8.3. H_n^k *is completely isometric to* $\mathcal{C}^{n,k}$.

8.4. *Modern operator spaces*

This subsection is a description of the results of[232], which considers the following two problems, leading to the notion of "modern operator space."

- Classify all infinite dimensional rank 1 JC*-triples up to complete isometry (**algebraic**)
  ANSWER: Φ, $H_\infty^{m,R}$, $H_\infty^{m,L}$, $H_\infty^{m,R} \cap H_\infty^{m,L}$
- Give a suitable "classification" of all Hilbertian operator spaces which are contractively complemented in a C*-algebra or normally contractively complemented in a W*-algebra (**analytic**)
  ANSWER: Φ, $\mathbf{C}$, $\mathbf{R}$, $\mathbf{C} \cap \mathbf{R}$

It is known that $\mathbf{R}$ and $\mathbf{C}$ are the only **completely** contractively complemented Hilbertian operator spaces[233].

8.4.1. *Rank one JC*-triples*

In order to attack the operator space structure theory of rank 1 JC*-triples, let us first review some of the basic concepts about them, as well as those concerning ternary rings of operators (TROs).

A *JC*-triple* is a norm closed complex linear subspace of $B(H,K)$ (equivalently, of a C^*-algebra) which is closed under the operation $a \mapsto aa^*a$. JC^*-triples were defined and studied (using the name J^*-algebra) as a generalization of C^*-algebras by Harris[82] in connection with function theory on infinite dimensional bounded symmetric domains. By a polarization identity (involving $\sqrt{-1}$), any JC^*-triple is closed under the triple product

$$(a, b, c) \mapsto \{a\}bc := \frac{1}{2}(ab^*c + cb^*a), \qquad (1)$$

under which it becomes a Jordan triple system. A linear map which preserves the triple product (1) will be called a *triple homomorphism.* Cartan factors are examples of JC^*-triples, as are C^*-algebras, and Jordan C^*-algebras. We shall only make use of Cartan factors of type 1, that is, spaces of the form $B(H,K)$ where H and K are complex Hilbert spaces.

A special case of a JC^*-triple is a *ternary algebra*, that is, a subspace of $B(H,K)$ closed under the *ternary product* $(a,b,c)\mapsto ab^*c$. A *ternary homomorphism* is a linear map ϕ satisfying $\phi(ab^*c)=\phi(a)\phi(b)^*\phi(c)$. These spaces are also called ternary rings of operators and abbreviated TRO. TROs have come to play a key role in operator space theory, serving as the algebraic model in the category. (The algebraic models for the categories of order-unit spaces, operator systems, and Banach spaces, are respectively Jordan C^*-algebras, C^*-algebras, and JB^*-triples.) For TROs, a ternary isomorphism is the same as a complete isometry.

Every JW^*-triple of rank one is isometric to a Hilbert space and every maximal collinear family of partial isometries corresponds to an orthonormal basis. Conversely, every Hilbert space with the abstract triple product $\{xyz\}:=((x|y)z+(z|y)x)/2$ can be realized as a JC^*-triple of rank one in which every orthonormal basis forms a maximal family of mutually collinear minimal partial isometries. Collinearity of v and w means:

$$vv^*w+wv^*v=w \text{ and } ww^*v+vw^*w=v.$$

8.4.2. *Operator space structure of Hilbertian JC*-triples*

We shall now outline the proof from[232] of the classification of infinite dimensional Hilbertian operator spaces up to complete isometry.

- The general setting: Y is a JC^*-subtriple of $B(H)$ which is Hilbertian in the operator space structure arising from $B(H)$, and $\{u_i : i\in\Omega\}$ is an orthonormal basis consisting of a maximal family of mutually collinear partial isometries of Y.
- We let T and A denote the TRO and the C^*-algebra respectively generated by Y. For any subset $G\subset\Omega$, $(uu^*)_G := \prod_{i\in G}u_iu_i^*$ and $(u^*u)_G := \prod_{i\in G}u_i^*u_i$. The elements $(uu^*)_G$ and $(u^*u)_G$ lie in the weak closure of A and more generally in the left and right linking von Neumann algebras of T.

  Fix $m\geq 0$. To construct $H_\infty^{m,R}$ we make a temporary assumption on the ternary envelope of Y.
- **Assume** $(u^*u)_G\neq 0$ for $|G|\leq m+1$ and $(u^*u)_G=0$ for $|G|\geq m+2$.

- Define elements which are indexed by an arbitrary pair of subsets I, J of Ω satisfying

$$|\Omega - I| = m + 1, \ |J| = m, \tag{2}$$

  as follows:
  $u_{IJ} = (uu^*)_{I-J} u_{c_1} u_{d_1}^* u_{c_2} u_{d_2}^* \cdots u_{c_s} u_{d_s}^* u_{c_{s+1}} (u^* u)_{J-I}$, where $I \cap J = \{d_1, \ldots, d_s\}$ and $(I \cup J)^c = \{c_1, \ldots, c_{s+1}\}$.
- After an appropriate choice $\varepsilon(IJ)$ of signs, the map $\varepsilon(IJ)u_{IJ} \to E_{JI}$ is a ternary isomorphism (and hence complete isometry) from the norm closure of $\mathrm{sp}_C\, u_{IJ}$ to the norm closure of $\mathrm{sp}_C\, \{E_{JI}\}$, where E_{JI} denotes an elementary matrix, whose rows and columns are indexed by the sets J and I, with a 1 in the (J, I)-position.
- This map can be extended to a ternary isomorphism from the w*-closure of $\mathrm{sp}_C\, u_{IJ}$ onto the Cartan factor of type I consisting of all $\aleph_0$ by $\aleph_0$ complex matrices which act as bounded operators on ℓ_2. By restriction, Y is completely isometric to a subtriple $\tilde{Y}$, of this Cartan factor of type 1.

- **Definition 8.2.** We shall denote the space $\tilde{Y}$ above by $H_\infty^{m,R}$.

- An entirely symmetric argument (with J infinite and I finite) under an entirely symmetric assumption on Y defines the space $H_\infty^{m,L}$.
- Having constructed the spaces $H_\infty^{m,R}$ and $H_\infty^{m,L}$, we can now state the following theorem.

Theorem 8.7. *Let Y be a JC^*-subtriple of $B(H)$ which is a separable infinite dimensional Hilbertian operator space. Then Y is completely isometric to one of the following spaces:*

$$\Phi, \quad H_\infty^{m,R}, \quad H_\infty^{m,L}, \quad H_\infty^{m,R} \cap H_\infty^{n,L}.$$

8.4.3. *Further properties of $H_\infty^{m,R}$ and $H_\infty^{m,L}$*

We now describe, from[232] the relation of these spaces with Fock space and compute some completely bounded Banach-Mazur distances.

If H is a separable Hilbert space, $l_m(h) : H^{\wedge m} \to H^{\wedge m+1}$ is the creation operator $l_m(h)x = h \wedge x$. Creation operators form a linear space $\mathcal{C}^m = \overline{\mathrm{sp}}\{l_m(e_i)\}$, where $\{e_i\}$ is an orthonormal basis, and inherit an operator space structure from $B(H^{\wedge m}, H^{\wedge m+1})$. Annihilation operators $\mathcal{A}^m$ consists of the adjoints of the creation operators on $H^{\wedge m-1}$.

Lemma 8.1. *$H_\infty^{m,R}$ is completely isometric to $\mathcal{A}^{m+1}$ and $H_\infty^{m,L}$ is completely isometric to $\mathcal{C}^m$.*

Remark 8.2. Every finite or infinite dimensional separable Hilbertian JC*-subtriple Y is completely isometric to a finite or infinite intersection of spaces of creation and annihilation operators, as follows

(a) If Y is infinite dimensional, then it is completely isometric to one of $\mathcal{A}^m$, $\mathcal{C}^m$, $\mathcal{A}^m \cap \mathcal{C}^k$, $\cap_{k=1}^\infty \mathcal{C}^k$
(b) If Y is of dimension n, then Y is completely isometric to $\cap_{j=1}^m \mathcal{C}^{k_j}$, where $n \geq k_1 > \cdots > k_m \geq 1$

Theorem 8.8. *For m, $k \geq 1$,*

(a) $d_{cb}(H_\infty^{m,R}, H_\infty^{k,R}) = d_{cb}(H_\infty^{m,L}, H_\infty^{k,L}) = \sqrt{\frac{m+1}{k+1}}$ *when* $m \geq k$
(b) $d_{cb}(H_\infty^{m,R}, H_\infty^{k,L}) = \infty$
(c) $d_{cb}(H_\infty^{m,R}, \Phi) = d_{cb}(H_\infty^{m,L}, \Phi) = \infty$

8.4.4. *Contractively complemented Hilbertian operator spaces*

Theorem 8.9. *Suppose Y is a separable infinite dimensional Hilbertian operator space which is contractively complemented (resp. normally contractively complemented) in a C*-algebra A (resp. W*-algebra A) by a projection P. Then, replacing Y with its support (defined in[232]), Y is completely isometric to either R, C, $R \cap C$, or Φ.*

Theorem 8.10 (Converse). *The operator spaces $R, C, R \cap C$, and Φ are each essentially (defined in[232]) normally contractively complemented in a von Neumann algebra.*

9. Quantum operator algebras

9.1. *Operator space characterization of TROs*

A natural object to characterize in the context of operator spaces are the so called *ternary rings of operators* (TRO's). These are subspaces of $B(H)$ which are closed under the ternary product xy^*z. TRO's, like C*-algebras, carry a natural operator space structure. In fact, every TRO is (completely) isometric to a corner $pA(1-p)$ of a C*-algebra A. TRO's are important because, as shown by Ruan[234], the injectives in the category of operator spaces are TRO's (corners of injective C*-algebras). (For the dual version of this result see[235].) Injective envelopes of operator systems and of operator spaces ([236],[234]) have proven to be important tools, see for example[237]. The characterization of TRO's among operator spaces is the subject of this subsection.

Closely related to TRO's are the so called JC*-triples, norm closed subspaces of $B(H)$ which are closed under the triple product $(xy^*z + zy^*x)/2$. These generalize the class of TRO's and have the property, as shown by Harris in[82], that isometries coincide with algebraic isomorphisms. It is not hard to see this implies that the algebraic isomorphisms in the class of TRO's are complete isometries, since for each TRO A, $M_n(A)$ is a JC*-triple (For the converse of this, see[238] Proposition 2.1). As a consequence, if an operator space X is completely isometric to a TRO, then the induced ternary product on X is unique, *i.e.*, independent of the TRO.

Relevant to this subsection is another property shared by all JC*-triples (and hence all TRO's). For any Banach space X, we denote by X_0 its open unit ball: $\{x \in X : \|x\| < 1\}$. The open unit ball of every JC*-triple is a *bounded symmetric domain.* This is equivalent to saying that it has a transitive group of biholomorphic automorphisms. It was shown by Koecher in finite dimensions (see[205]) and Kaup[115] in the general case that this is a defining property for the slightly larger class of JB*-triples. The only illustrative basic examples of JB*-triples which are not JC*-triples are the space $H_3(\mathcal{O})$ of 3 x 3 hermitian matrices over the octonians and a certain subtriple of $H_3(\mathcal{O})$. These are called *exceptional* triples, and they cannot be represented as a JC*-triple. This holomorphic characterization has been useful because, as noted earlier, it gives an elegant proof, due to Kaup[194], that the range of a contractive projection on a JB*-triple is isometric to another JB*-triple.

Motivated by this characterization for JB*-triples, a holomorphic characterization of TRO's up to complete isometry is given in[201] and stated below. As a consequence, a holomorphic operator space characterization of C*-algebras is obtained as well. It should be mentioned that Upmeier (for the category of Banach spaces) in[239] and El Amin-Campoy-Palacios (for the category of Banach algebras) in[240], gave different but still holomorphic characterizations of C*-algebras up to isometry.

Theorem 9.1. *Let $A \subset B(H)$ be an operator space and suppose that $M_n(A)_0$ is a bounded symmetric domain for all $n \geq 2$. Then A is ternary isomorphic and completely isometric to a TRO.*

Theorem 9.2. *Let $A \subset B(H)$ be an operator space and suppose that*

$M_n(A)_0$ is a bounded symmetric domain for all $n \geq 2$ and A_0 is of tube type. Then A is completely isometric to a C^*-algebra.

9.2. *Holomorphic characterization of operator algebras*

In the category of operator spaces, that is, subspaces of the bounded linear operators $B(H)$ on a complex Hilbert space H together with the induced matricial operator norm structure, objects are equivalent if they are completely isometric, *i.e.* if there is a linear isomorphism between the spaces which preserves this matricial norm structure. Since operator algebras, that is, subalgebras of $B(H)$, are motivating examples for much of operator space theory, it is natural to ask if one can characterize which operator spaces are operator algebras. One satisfying answer was given by Blecher, Ruan and Sinclair in[241], where it was shown that among operator spaces A with a (unital but not necessarily associative) Banach algebra product, those which are completely isometric to operator algebras are precisely the ones whose multiplication is completely contractive with respect to the Haagerup norm on $A \otimes A$ (For a completely bounded version of this result, see[242]).

The main result of[12] drops the algebra assumption on A in the Blecher-Ruan-Sinclair theorem in favor of a holomorphic assumption. Using only natural conditions on holomorphic vector fields on Banach spaces, an algebra product is constructed on A which is completely contractive and unital, so that the result of[241] can be applied.

We state this result as Theorem 9.3. In this theorem, for any element v in the symmetric part (recalled below) of a Banach space X, h_v denotes the corresponding complete holomorphic vector field on the open unit ball of X.

Theorem 9.3. *An operator space A is completely isometric to a unital operator algebra if and only there exists an element v in the completely symmetric part (defined below) of A such that:*

(1) $h_v(x+v) - h_v(x) - h_v(v) + v = -2x$ for all $x \in A$

(2) For all $X \in M_n(A)$, $\|V - h_V(X)\| \leq \|X\|^2$.

This result is thus an instance where the consideration of a ternary product, called the partial triple product, which arises from the holomorphic structure via the symmetric part of the Banach space, leads to results for binary products. Examples of this phenomenon occurred in[243],[244] where this technique is used to describe the algebraic properties of isometries of certain operator algebras. The technique was also used in[245] to show that Banach spaces with holomorphically equivalent unit balls are linearly isometric (see[246] for an exposition of[245]).

Suppose that X is a TRO (i.e., a closed subspace of $B(H)$ closed under the ternary product ab^*c) which contains an element v satisfying $xv^*v = vv^*x = x$ for all $x \in X$. Then it is trivial that X becomes a unital C*-algebra for the product xv^*y, involution vx^*v, and unit v. By comparison, the above result starts only with an operator space X containing a distinguished element v in the completely symmetric part of X (defined below) having a unit-like property. This is to be expected since the result of[241] fails in the absence of a unit element.

The main technique in the proof of this result is to use a variety of elementary isometries on n by n matrices over A (most of the time, $n = 2$) and to exploit the fact that isometries of arbitrary Banach spaces preserve the partial triple product. The first occurrence of this technique appears in the construction, for each n, of a contractive projection P_n on $K\overline{\otimes}A$ ($K=$ compact operators on separable infinite dimensional Hilbert space) with range $M_n(A)$ as a convex combination of isometries. The completely symmetric part of A is defined to be the intersection of A (embedded in $K\overline{\otimes}A$) and the symmetric part of $K\overline{\otimes}A$. It is the image under P_1 of the symmetric part of $K\overline{\otimes}A$. It follows from[201] that the completely symmetric part of A is a TRO, which is a crucial tool in this work. The binary product $x \cdot y$ on A is constructed using properties of isometries on 2 by 2 matrices over A and it is shown that the symmetrized product can be expressed in terms of the partial Jordan triple product on A. Specifically,

$$\begin{bmatrix} y \cdot x & 0 \\ 0 & 0 \end{bmatrix} = 2\left\{ \begin{bmatrix} x & 0 \\ 0 & 0 \end{bmatrix} \begin{bmatrix} 0 & v \\ 0 & 0 \end{bmatrix} \begin{bmatrix} 0 & y \\ 0 & 0 \end{bmatrix} \right\}$$

and

$$\{xvy\} = \frac{1}{2}(y \cdot x + x \cdot y).$$

According to[247], "The one-sided multipliers of an operator space X are a key to the 'latent operator algebraic structure' in X." The unified approach through multiplier operator algebras developed in[247] leads to simplifications of known results and applications to quantum M-ideal theory. They also state "With the extra structure consisting of the additional matrix norms on an operator algebra, one might expect to not have to rely as heavily on other structure, such as the product." Theorem 9.3 is certainly in the spirit of this statement.

Another approach to operator algebras is[248], in which the set of operator algebra products on an operator space is shown to be in bijective correspondence with the space of norm one quasi-multipliers on the operator space.

9.2.1. *Symmetric part of a Banach space*

We review the construction and properties of the partial Jordan triple product in an arbitrary Banach space. Let X be a complex Banach space with open unit ball X_0. Every holomorphic function $h : X_0 \to X$, also called a holomorphic vector field, is locally integrable, that is, the initial value problem

$$\frac{\partial}{\partial t}\varphi(t,z) = h(\varphi(t,z)) \ , \ \varphi(0,z) = z,$$

has a unique solution for every $z \in X_0$ for t in a maximal open interval J_z containing 0. A **complete holomorphic vector field** is one for which $J_z = (-\infty, \infty)$ for every $z \in X_0$.

It is a fact that every complete holomorphic vector field is the sum of the restriction of a skew-Hermitian bounded linear operator A on X and a function h_a of the form $h_a(z) = a - Q_a(z)$, where Q_a is a quadratic homogeneous polynomial on X.

The **symmetric part** of X is the orbit of 0 under the set of complete holomorphic vector fields, and is denoted by $S(X)$. It is a closed subspace of X and is equal to X precisely when X has the structure of a JB^*-triple (by[115]).

Examples of the symmetric parts of some classical Banach spaces are given in the seminal paper R. Braun, W. Kaup, H. Upmeier, On the automorphisms of circular and Reinhardt domains in complex Banach spaces, Manuscripta Math. 25 (1978) no. 2, 97-133. Other examples are given in references 183, 246, 244 and in A. M. Peralta, L.L. Stach, Von Neumann algebra preduals satisfy the linear biholomorphic property, to appear in *Math. Scand.*

If $a \in S(X)$, we can obtain a symmetric bilinear form on X, also denoted by Q_a via the polarization formula

$$Q_a(x,y) = \frac{1}{2}\left(Q_a(x+y) - Q_a(x) - Q_a(y)\right)$$

and then the partial Jordan triple product $\{\cdot,\cdot,\cdot\} : X \times S(X) \times X \to X$ is defined by $\{x, a, z\} = Q_a(x, z)$. The space $S(X)$ becomes a JB^*-triple in this triple product.

It is also true that the "main identity" (1) (see subsection 3.6.2) holds whenever $a, y, b \in S(X)$ and $x, z \in X$. The partial triple product behaves well under the actions of isometries and contractive projections (see[246] 5.2,5.3).

10. Universal enveloping TROs and C*-algebras

10.1. *Operator space structure of JC^*-triples and TROs*

The paper[9], together with its sequel[10], initiates a systematic investigation of the operator space structure of JC^*-triples via a study of the TROs they generate. The approach is through the introduction and development of a variety of universal objects (TROs and C^*-algebras). Explicit descriptions of operator space structures of Cartan factors of arbitrary dimension are obtained as a consequence. It should be noted that Theorem 10.1 and its corollary, as well as the computation of the universal enveloping TROs of all Cartan factors of finite dimension, were also obtained roughly around the same time in[11]. Much of this subsection is taken literally from the introduction to[9].

It is necessary to widen the definition of JC^*-triple to include any complex Banach space with a surjective isometry onto a subspace of a TRO which is closed under the (Jordan) triple product $\{a, b, c\} = (ab^*c+cb^*a)/2$. A contractively complemented subspace of $B(H)$ is an example of such a JC^*-triple, not necessarily a JC^*-subtriple of $B(H)$. In the reflexive case, operator space structure of JC^*-triples of this kind has been investigated in[8], [231], and[232].

The aim of[9] is to explore the operator space structures of an arbitrary JC^*-triple, and, quoting the authors of[9], "to place the ground-breaking results of[8], [231], [232] in the general setting of a full theory." This new device (universal TRO of a JC^*-triple) facilitates a general inquiry into the operator space structure of JC^*-triples.

Theorem 10.1 (Theorem 3.1 of[9]). *Let E be a JC^*-triple. Then there is a unique pair $(C^*(E), \alpha_E)$ where $C^*(E)$ is a C^*-algebra and $\alpha_E : E \to C^*(E)$ is an injective triple homomorphism, with the following properties:*

(a) $\alpha_E(E)$ generates $C^*(E)$ as a C^*-algebra;

(b) for each triple homomorphism $\pi : E \to A$, where A is a C^*-algebra, there is a unique $*$-homomorphism $\tilde{\pi} : C^*(E) \to A$ with $\tilde{\pi} \circ \alpha_E = \pi$.

Corollary 10.1 (Corollary 3.2 of[9]). *Let E be a JC^*-triple. Then there is a unique pair $(T^*(E), \alpha_E)$ where $T^*(E)$ is a TRO and $\alpha_E : E \to T^*(E)$ is an injective triple homomorphism, with the following properties:*

(a) $\alpha_E(E)$ generates $T^*(E)$ as a TRO;

(b) for each triple homomorphism $\pi : E \to T$, where T is aTRO, there is a unique TRO-homomorphism $\tilde{\pi} : T^*(E) \to T$ with $\tilde{\pi} \circ \alpha_E = \pi$.

In a technical tour de force, $T^*(E)$ is explicitly calculated for the Cartan factors of any dimension: Hilbert space[9] Theorem 5.1, spin factor[9] Lemma 5.2, rectangular of rank 2 or more[9] Theorem 5.4, symmetric and antisymmetric[9] Theorem 5.5. The corresponding references in[11] to these results, but only for the finite dimensional cases as noted above, are: Hilbert space[11] Theorem 3.15, spin factor[11] Theorem 3.5, rectangular of rank 2 or more[11] Theorem 3.13, symmetric and antisymmetric[11] Theorem 3.6).

As a technical aid of possibly independent interest, the notion of a reversible JC^*-triple is also introduced in[9]. and the universal reversibility (or not) of all Cartan factors, is also determined. Building on the definition of reversibility for Jordan algebras leads to the following definition.

Definition 10.1 (Definition 4.1 of[9]). *A JC^*-subtriple E of a TRO T is said to be reversible in T if*

$$a_1 a_2^* a_3 \cdots a_{2n}^* a_{2n+1} + a_{2n+1} a_{2n}^* \cdots a_2^* a_1 \in E$$

whenever $a_1, \ldots a_{2n+1} \in E$. A JC^*-triple E is universally reversible if $\alpha_E(E)$ is reversible in $T^*(E)$.

Theorem 10.2 (Theorem 5.6 of[9]). *Spin factors of dimension greater than 4 and Hilbert spaces of dimension greater than 2 are not universally reversible. All other Cartan factors are universally reversible.*

In the next theorem, operator space structures of JC^*-triples arising from concrete triple embeddings in C^*-algebras are considered, making use of universal TROs and establishing links with injective envelopes and triple envelopes, and concentrating on Cartan factors. It is necessary first to refine the definition of operator space structure.

Definition 10.2 (Definition 6.1 of[9]). *A JC-operator space structure on a JC^*-triple E is an operator space structure determined by a linear isometry from E onto a JC^*-subtriple of $B(H)$.*

Another important definition is that of an operator space ideal. This is a norm closed ideal $\mathcal{I}$ of $T^*(E)$ for which $\mathcal{I} \cap \alpha_E(E) = \{0\}$. For each operator space ideal $\mathcal{I}$, we have the JC-operator space structure $E_{\mathcal{I}}$ on E determined by the isometric embedding $E \mapsto T^*(E)/\mathcal{I}$ given by $x \mapsto \alpha_E(x) + \mathcal{I}$.

Theorem 10.3 (Theorem 6.8 of[9]). *Let E be a Cartan factor which is not a Hilbert space. Then $\mathcal{I} \leftrightarrow E_{\mathcal{I}}$ is a bijective correspondence between the operator space ideals of $T^*(E)$ and the JC-operator space structures of E.*

The case of a Hilbert space, excluded from Theorem 10.3 is covered in[10] Theorem 2.7. It follows from this that the triple envelope, in the sense of[238], of a JC^*-triple E which is isometric to a Cartan factor is identified with the TRO generated by E. The following theorem shows that distinct JC-operator space structures on a Hilbert space cannot be completely isometric.

Theorem 10.4 (Theorem 2.6 of[10]). *For a Hilbert space E and operator space ideals $\mathcal{I}$ and $\mathcal{J}$ of $T(E)$, the following are equivalent:*

(a) $E_{\mathcal{I}} = E_{\mathcal{J}}$
(b) $\mathcal{I} = \mathcal{J}$
(c) $E_{\mathcal{I}}$ and $E_{\mathcal{J}}$ are completely isometric.

It is also shown in[10] Theorem 3.4 how infinite-dimensional Hilbertian JC-operator spaces are determined explicitly by their finite-dimensional subspaces, and that, in turn, they impose very rigid constraints upon the operator space structure of their finite-dimensional subspaces. Finally, the operator space ideals of the universal TRO of a Hilbert space are identified[10] Theorem 3.7, as well as the corresponding injective envelopes[10] Theorem 4.4.

Building on[9,10] the paper[249] establishes more tools for the study of JC*-triples via the TRO and C*-algebra they generate. The paper focusses on two basic technical tools introduced in the earlier work of the authors, namely, the universal TRO and the property of universal reversibility. The properties developed involve on the one hand ideals and on the other hand tensor products.

In turn, the results of[9,10,249] are used in[250] to extend Theorem 10.2 to general JC^*-triples ([250] Theorem 4.7), provide the structure of universally reversible JW*-triples and W*-TROs ([250] Theorem 4.14), and show under mild conditions that a triple homomorphism on a TRO decomposes into a sum of a TRO homomorphism and a TRO antihomorphism ([250] Proposition 5.14).

Finally we make four remarks about the paper[11]. First, the methods for computing the universal enveloping TROs of the Cartan factors (in finite dimensions only) are different from those of[9] in that they use grids and depend strongly on the results on grids in[8]. This approach is potentially very useful in future applications (for example, cohomology).

Second, because the paper[11] works in the slightly more general context of JB^*-triples, they are able to obtain a new proof of a version of the Gelfand-Naimark theorem for JB^*-triples ([199] Theorem 2).

Proposition 10.1 (Corollary 2.6 of[11]). *Any JB^*-triple Z contains a unique purely exceptional ideal J, such that Z/J is JB^*-triple isomorphic to a JC^*-triple.*

Third, they define a radical for universally reversible JC^*-triples and use it to give the structure of the universal enveloping TRO of a universally reversible TRO in a C^*-algebra with a TRO antiautomorphism of order 2, generalizing[9] Corollary 4.5. More generally, radicals have been defined and studied for Jordan triple systems in, for example,[24,251,252].

Finally, the results of[11] are used in[180] to show that the classification of the symmetric spaces, derived initially from Lie theory and then from Jordan theory, can be achieved by K-theoretic methods. This in spite of the fact that JB^*-triples do not, in general, behave well under the formation of tensor products.

References

1. T. Ho, J. Martinez-Moreno, A. M. Peralta and B. Russo, Derivations on real and complex JB *-triples, *J. London Math. Soc.* **65**, 85 (2002).
2. T. Ho, A. M. Peralta and B. Russo, Ternary weakly amenable C*-algebras and JB*-triples, *Q. J. Math.* **64**, 1109 (2013).
3. A. M. Peralta and B. Russo, Automatic continuity of derivations on C*-algebras and JB*-triples, *J. Algebra* **399**, 960 (2014).
4. H. Cartan and S. Eilenberg, *Homological algebra* (Princeton University Press, Princeton, N. J., 1956).
5. D. B. Fuks, *Cohomology of infinite-dimensional Lie algebras* (Contemporary Soviet Mathematics, Consultants Bureau, New York, 1986).
6. A. W. Knapp, *Lie groups, Lie algebras, and cohomology*, Mathematical Notes, Vol. 34 (Princeton University Press, Princeton, NJ, 1988).
7. G. Hochschild, *Review of[4] (MR0077480)*.

8. M. Neal and B. Russo, Contractive projections and operator spaces, *Trans. Amer. Math. Soc.* **355**, 2223 (2003).
9. L. J. Bunce, B. Feely and R. M. Timoney, Operator space structure of JC^*-triples and TROs, I, *Math. Z.* **270**, 961 (2012).
10. L. J. Bunce and R. M. Timoney, On the operator space structure of Hilbert spaces, *Bull. Lond. Math. Soc.* **43**, 1205 (2011).
11. D. Bohle and W. Werner, The universal enveloping ternary ring of operators of a JB*-triple system, *Proc. Edinb. Math. Soc.* **57**, 347 (2014).
12. M. Neal and B. Russo, A holomorphic characterization of operator algebras, *Math. Scand.* **115** (2), 229 (2014).
13. I. M. Singer and J. Wermer, Derivations on commutative normed algebras, *Math. Ann.* **129**, 260 (1955).
14. S. Sakai, On a conjecture of Kaplansky, *Tôhoku Math. J.* **12**, 31 (1960).
15. B. E. Johnson, Continuity of derivations on commutative algebras, *Amer. J. Math.* **91**, 1 (1969).
16. M. P. Thomas, The image of a derivation is contained in the radical, *Ann. of Math.* **128**, 435 (1988).
17. M. Mathieu and G. J. Murphy, Derivations mapping into the radical, *Arch. Math. (Basel)* **57**, 469 (1991).
18. M. Mathieu and V. Runde, Derivations mapping into the radical. II, *Bull. London Math. Soc.* **24**, 485 (1992).
19. M. Brešar and M. Mathieu, Derivations mapping into the radical. III, *J. Funct. Anal.* **133**, 21 (1995).
20. G. Hochschild, Semi-simple algebras and generalized derivations, *Amer. J. Math.* **64**, 677 (1942).
21. W. F. Santos and E. Moskowitz, Martin, Gerhard hochschild (1915–2010), *Notices Amer. Math. Soc.* **58**, 1078 (2011), Walter Ferrer Santos and Martin Moskowitz, coordinating editors.
22. W. F. Santos, *Gerhard Hochschild (1915/2010): A mathematician of the XXth Century (arXiv:1104.0335v1 [math.HO])*, (2 Apr 2011).
23. N. Jacobson, Abstract derivation and Lie algebras, *Trans. Amer. Math. Soc.* **42**, 206 (1937).
24. K. Meyberg, *Lectures on algebras and triple systems* (The University of Virginia, Charlottesville, Va., 1972), Notes on a course of lectures given during the academic year 1971–1972.
25. N. Jacobson, *Lie algebras* (Interscience Tracts in Pure and Applied Mathematics, No. 10, Interscience Publishers (a division of John

Wiley & Sons), New York-London, 1962).
26. N. Jacobson, Derivation algebras and multiplication algebras of semisimple Jordan algebras, *Ann. of Math.* **50**, 866 (1949).
27. A. M. Sinclair, Jordan homomorphisms and derivations on semisimple Banach algebras., *Proc. Amer. Math. Soc.* **24**, 209 (1970).
28. S. Sakai, Derivations of W^*-algebras, *Ann. of Math.* **83**, 273 (1966).
29. R. V. Kadison, Derivations of operator algebras, *Ann. of Math.* **83**, 280 (1966).
30. H. Kamowitz, Cohomology groups of commutative Banach algebras, *Trans. Amer. Math. Soc.* **102**, 352 (1962).
31. E. Christensen, Extensions of derivations. II, *Math. Scand.* **50**, 111 (1982).
32. B. E. Johnson, *Cohomology in Banach algebras* (American Mathematical Society, Providence, R.I., 1972), Memoirs of the American Mathematical Society, No. 127.
33. A. Sinclair, Barry Edward Johnson 1937–2002, *Bull. London Math. Soc.* **36**, 559 (2004).
34. W. G. Lister, Ternary rings, *Trans. Amer. Math. Soc.* **154**, 37 (1971).
35. R. Carlsson, Cohomology of associative triple systems, *Proc. Amer. Math. Soc.* **60**, 1 (1976).
36. B. Zalar, On the structure of automorphism and derivation pairs of B^*-triple systems, in *Topics in operator theory, operator algebras and applications (Timişoara, 1994)*, eds. A. Gheondea, . Strătilă and D. Timotin (Rom. Acad., Bucharest, 1995) pp. 265–271.
37. M. V. Velasco and A. R. Villena, Derivations on Banach pairs, *Rocky Mountain J. Math.* **28**, 1153 (1998).
38. W. G. Lister, A structure theory of Lie triple systems, *Trans. Amer. Math. Soc.* **72**, 217 (1952).
39. C.-H. Chu, *Jordan structures in geometry and analysis*, Cambridge Tracts in Mathematics, Vol. 190 (Cambridge University Press, Cambridge, 2012).
40. C.-H. Chu and B. Russo, *Cohomology of Jordan triples and Lie algebras.* To appear in a forthcoming volume of Contemporary Mathematics.
41. H. Upmeier, *Symmetric Banach manifolds and Jordan C^*-algebras*, North-Holland Mathematics Studies, Vol. 104 (North-Holland Publishing Co., Amsterdam, 1985), Notas de Matemática [Mathematical Notes], 96.
42. N. Jacobson, *Structure and representations of Jordan algebras*

(American Mathematical Society Colloquium Publications, Vol. XXXIX, American Mathematical Society, Providence, R.I., 1968).

43. H. Braun and M. Koecher, *Jordan-Algebren* (Die Grundlehren der mathematischen Wissenschaften in Einzeldarstellungen mit besonderer Berücksichtigung der Anwendungsgebiete, Band 128, Springer-Verlag, Berlin-New York, 1966).
44. O. Bratteli and D. W. Robinson, *Operator algebras and quantum statistical mechanics. Vol. 1* (Springer-Verlag, New York-Heidelberg, 1979), C^*- and W^*-algebras, algebras, symmetry groups, decomposition of states, Texts and Monographs in Physics.
45. S. Sakai, *Operator algebras in dynamical systems*, Encyclopedia of Mathematics and its Applications, Vol. 41 (Cambridge University Press, Cambridge, 1991), The theory of unbounded derivations in C^*-algebras.
46. V. Runde, *Lectures on amenability*, Lecture Notes in Mathematics, Vol. 1774 (Springer-Verlag, Berlin, 2002).
47. J. R. Ringrose, Automatic continuity of derivations of operator algebras, *J. London Math. Soc.* **5**, 432 (1972).
48. A. Connes, Classification of injective factors. Cases II_1, II_∞, III_λ, $\lambda \neq 1$, *Ann. of Math.* **104**, 73 (1976).
49. U. Haagerup, All nuclear C^*-algebras are amenable, *Invent. Math.* **74**, 305 (1983).
50. F. J. Murray and J. Von Neumann, On rings of operators, *Ann. of Math.* **37**, 116 (1936).
51. R. V. Kadison and Z. Liu, A note on derivations of Murray–von Neumann algebras, *Proc. Natl. Acad. Sci. USA* **111**, 2087 (2014).
52. R. V. Kadison and Z. Liu, Derivations of Murray–von Neumann algebras, *Math. Scand.* **115**, 206 (2014).
53. Z. Liu, On some mathematical aspects of the Heisenberg relation, *Sci. China Math.* **54**, 2427 (2011).
54. R. V. Kadison, Z. Liu and A. Thom, *A note on commutators in algebras of unbounded operators,* submitted.
55. U. Haagerup and H. Schultz, Brown measures of unbounded operators affiliated with a finite von Neumann algebra, *Math. Scand.* **100**, 209 (2007).
56. I. E. Segal, A non-commutative extension of abstract integration, *Ann. of Math.* **57**, 401 (1953).
57. A. F. Ber, V. I. Chilin and F. A. Sukochev, Non-trivial derivations on commutative regular algebras, *Extracta Math.* **21**, 107 (2006).

58. S. Albeverio, S. A. Ayupov and K. K. Kudaybergenov, Non-commutative Arens algebras and their derivations, *J. Funct. Anal.* **253**, 287 (2007).
59. S. Albeverio, S. A. Ayupov and K. K. Kudaybergenov, Derivations on the algebra of measurable operators affiliated with a type I von Neumann algebra, *Siberian Adv. Math.* **18**, 86 (2008).
60. S. Albeverio, S. A. Ayupov and K. K. Kudaybergenov, Structure of derivations on various algebras of measurable operators for type I von Neumann algebras, *J. Funct. Anal.* **256**, 2917 (2009).
61. S. A. Ayupov and K. K. Kudaybergenov, Derivations on algebras of measurable operators, *Infin. Dimens. Anal. Quantum Probab. Relat. Top.* **13**, 305 (2010).
62. A. F. Ber, B. de Pagter and F. A. Sukochev, Derivations in algebras of operator-valued functions, *J. Operator Theory* **66**, 261 (2011).
63. A. F. Ber, V. I. Chilin and F. A. Sukochev, Continuity of derivations of algebras of locally measurable operators, *Integral Equations Operator Theory* **75**, 527 (2013).
64. A. F. Ber, V. I. Chilin and F. A. Sukochev, Continuous derivations on algebras of locally measurable operators are inner, *Proc. Lond. Math. Soc.* **109**, 65 (2014).
65. B. E. Johnson, Symmetric amenability and the nonexistence of lie and jordan derivations, *Math. Proc. Cambridge Philos. Soc.* **120**, 455 (1996).
66. J. Alaminos, M. Brešar and A. R. Villena, The strong degree of von Neumann algebras and the structure of Lie and Jordan derivations, *Math. Proc. Cambridge Philos. Soc.* **137**, 441 (2004).
67. J. Cuntz, On the continuity of semi-norms on operator algebras, *Math. Ann.* **220**, 171 (1976).
68. S. Hejazian and A. Niknam, Modules, annihilators and module derivations of JB*-algebras, *Indian J. Pure Appl. Math.* **27**, 129 (1996).
69. H. Upmeier, Derivations of Jordan C^*-algebras, *Math. Scand.* **46**, 251 (1980).
70. C. R. Miers, Lie derivations of von Neumann algebras, *Duke Math. J.* **40**, 403 (1973).
71. C. R. Miers, Lie triple derivations of von Neumann algebras, *Proc. Amer. Math. Soc.* **71**, 57 (1978).
72. M. Mathieu and A. R. Villena, The structure of Lie derivations on C^*-algebras, *J. Funct. Anal.* **202**, 504 (2003).

73. J. Alaminos, M. Mathieu and A. R. Villena, Symmetric amenability and lie derivations, *Math. Proc. Cambridge Philos. Soc.* **137**, 433 (2004).
74. H. G. Dales, *Review of*[73] *(MR2092069).*
75. C. J. Read, The decomposition of Lie derivations, *Irish Math. Soc. Bull.* **51**, 11 (2003).
76. J. R. Ringrose, The cohomology of operator algebras: a survey, *Bull. London Math. Soc.* **28**, 225 (1996).
77. C. Hou, Cohomology of a class of Kadison-Singer algebras, *Sci. China Math.* **53**, 1827 (2010).
78. C. Hou and C. Wei, Completely bounded cohomology of non-selfadjoint operator algebras, *Acta Math. Sci. Ser. B Engl. Ed.* **27**, 25 (2007).
79. P. Ji and W. Qi, Characterizations of Lie derivations of triangular algebras, *Linear Algebra Appl.* **435**, 1137 (2011).
80. I. M. Zhuraev, The structure of Lie derivations of unbounded operator algebras of type I, *Uzbek. Mat. Zh.* **1**, 71 (2011).
81. I. Z. Abdullaev, n-Lie derivations on von Neumann algebras, *Uzbek. Mat. Zh.* **5-6**, 3 (1992).
82. L. A. Harris, Bounded symmetric homogeneous domains in infinite dimensional spaces, in *Proceedings on Infinite Dimensional Holomorphy (Internat. Conf., Univ. Kentucky, Lexington, Ky., 1973)*, (Springer, Berlin, 1974).
83. L. A. Harris, A generalization of C^*-algebras, *Proc. London Math. Soc.* **42**, 331 (1981).
84. T. J. Barton and Y. Friedman, Bounded derivations of JB*-triples, *Quart. J. Math. Oxford Ser. (2)* **41**, 255 (1990).
85. O. Kühn and A. Rosendahl, Wedderburnzerlegung für Jordan-Paare, *Manuscripta Math.* **24**, 403 (1978).
86. C. Pearcy and D. Topping, On commutators in ideals of compact operators, *Michigan Math. J.* **18**, 247 (1971).
87. J. W. Bunce and W. L. Paschke, Derivations on a C^*-algebra and its double dual, *J. Funct. Anal.* **37**, 235 (1980).
88. R. Pluta and B. Russo, *Triple derivations on von Neumann algebras*, Studia Math., to appear.
89. K. J. Dykema and N. J. Kalton, Sums of commutators in ideals and modules of type II factors, *Ann. Inst. Fourier (Grenoble)* **55**, 931 (2005).
90. R. V. Kadison, Local derivations, *J. Algebra* **130**, 494 (1990).

91. B. E. Johnson, Local derivations on C^*-algebras are derivations, *Trans. Amer. Math. Soc.* **353**, 313 (2001).
92. S. A. Ayupov, K. K. Kudaĭbergenov and B. O. Nurzhanov, Local derivations of the algebra of τ-measurable operators, *Uzbek. Mat. Zh.* **2**, 20 (2009).
93. S. A. Ayupov, K. K. Kudaĭbergenov and B. O. Nurzhanov, Local derivations of the algebra of measurable operators affiliated with von Neumann algebras of type I, *Uzbek. Mat. Zh.* **3**, 9 (2010).
94. S. Albeverio, S. A. Ayupov, K. K. Kudaybergenov and B. O. Nurjanov, Local derivations on algebras of measurable operators, *Commun. Contemp. Math.* **13**, 643 (2011).
95. C. J. K. Batty and L. Molnár, On topological reflexivity of the groups of $*$-automorphisms and surjective isometries of $B(H)$, *Arch. Math.* **67**, 415 (1996).
96. M. Mackey, Local derivations on Jordan triples, *Bull. Lond. Math. Soc.* **45**, 811 (2013).
97. M. Burgos, F. J. Fernández-Polo, J. J. Garcés and A. M. Peralta, Local triple derivations on C^*-algebras, *Comm. Algebra* **42**, 1276 (2014).
98. M. Burgos, F. J. Fernández-Polo and A. M. Peralta, Local triple derivations on C*-algebras and JB*-triples, *Bull. Lond. Math. Soc.* **46**, 709 (2014).
99. F. J. Fernández-Polo, A. M. Salas and A. M. Peralta, *Local triple derivations on real C*-algebras and JB*-triples. To appear in Bull. Malays. Math. Sci. Soc.*
100. M. Brešar and P. Šemrl, On local automorphisms and mappings that preserve idempotents, *Studia Math.* **113**, 101 (1995).
101. D. R. Larson and A. R. Sourour, Local derivations and local automorphisms of $B(X)$, in *Operator theory: operator algebras and applications, Part 2 (Durham, NH, 1988)*, eds. W. B. Arveson and R. G. Douglas, Proc. Sympos. Pure Math., Vol. 51 (Amer. Math. Soc., Providence, RI, 1990) pp. 187–194.
102. P. Semrl, Local automorphisms and derivations on $B(H)$, *Proc. Amer. Math. Soc.* **125**, 2677 (1997).
103. S. Ayupov and K. Kudaybergenov, 2-local derivations and automorphisms on $B(H)$, *J. Math. Anal. Appl.* **395**, 15 (2012).
104. S. Ayupov and K. K. Kudaybergenov, *2-local derivations on von Neumann algebras*, preprint, arXiv:1402.3368 (2014).
105. M. Burgos, F. J. Fernández-Polo, J. J. Garcés and A. M. Peralta,

*A Kowalski-Slodkowski theorem for 2-local *-homomorphisms on von Neumann algebras*, preprint, arXiv:1404.7597 (2014).

106. M. J. Burgos, F. J. Fernández-Polo, J. J. Garcés and A. M. Peralta, 2-local triple homomorphisms on von neumann algebras and JBW*-triples, *J. Math. Anal. Appl.* **426**, 43 (2015).
107. S. O. Kim and J. S. Kim, Local automorphisms and derivations on certain C^*-algebras, *Proc. Amer. Math. Soc.* **133**, 3303 (2005).
108. J.-H. Liu and N.-C. Wong, 2-local automorphisms of operator algebras, *J. Math. Anal. Appl.* **321**, 741 (2006).
109. L. Molnár, 2-local isometries of some operator algebras, *Proc. Edinb. Math. Soc.* **45**, 349 (2002).
110. L. Molnár, Local automorphisms of operator algebras on Banach spaces, *Proc. Amer. Math. Soc.* **131**, 1867 (2003).
111. F. Pop, On local representations of von Neumann algebras, *Proc. Amer. Math. Soc.* **132**, 3569 (2004).
112. S. O. Kim and J. S. Kim, Local automorphisms and derivations on $\mathbb{M}_n$, *Proc. Amer. Math. Soc.* **132**, 1389 (2004).
113. A. M. Peralta, A note on 2-local representations of C*-algebras, *Operators and Matrices*, **9** (2), 343 (2015).
114. K. Kudaybergenov, T. Oikhberg, A. Peralta and B. Russo, *2-local triple derivations on von Neumann algebras*, preprint, arXiv:1407.3878 (2014).
115. W. Kaup, A Riemann mapping theorem for bounded symmetric domains in complex Banach spaces, *Math. Z.* **183**, 503 (1983).
116. G. Hochschild, On the cohomology groups of an associative algebra, *Ann. of Math.* **46**, 58 (1945).
117. C. A. Weibel, *An introduction to homological algebra*, Cambridge Studies in Advanced Mathematics, Vol. 38 (Cambridge University Press, Cambridge, 1994).
118. C. A. Weibel, History of homological algebra, in *History of topology*, ed. I. M. James (North-Holland, Amsterdam, 1999) pp. 797–836.
119. M. Gerstenhaber, A uniform cohomology theory for algebras, *Proc. Nat. Acad. Sci. U.S.A.* **51**, 626 (1964).
120. N. D. Glassman, Cohomology of nonassociative algebras, *Pacific J. Math.* **33**, 617 (1970).
121. N. D. Glassman, Cohomology of Jordan algebras, *J. Algebra* **15**, 167 (1970).
122. N. Jacobson, Jordan algebras, in *Report of a conference on linear algebras, June, 1956*, ed. A. A. Albert (National Academy of Sciences

– National Research Council, Washington, Public. 502, 1957) pp. 12–19.
123. N. Jacobson, General representation theory of Jordan algebras, *Trans. Amer. Math. Soc.* **70**, 509 (1951).
124. A. J. Penico, The Wedderburn principal theorem for Jordan algebras, *Trans. Amer. Math. Soc.* **70**, 404 (1951).
125. K. McCrimmon, Representations of quadratic Jordan algebras, *Trans. Amer. Math. Soc.* **153**, 279 (1971).
126. K. McCrimmon, Compatible Peirce decompositions of Jordan triple systems, *Pacific J. Math.* **103**, 57 (1982).
127. A. M. Sinclair and R. R. Smith, A survey of Hochschild cohomology for von Neumann algebras, in *Operator algebras, quantization, and noncommutative geometry*, eds. R. S. Doran and R. V. Kadison, Contemp. Math., Vol. 365 (Amer. Math. Soc., Providence, RI, 2004) pp. 383–400.
128. B. E. Johnson, R. V. Kadison and J. R. Ringrose, Cohomology of operator algebras. III. Reduction to normal cohomology, *Bull. Soc. Math. France* **100**, 73 (1972).
129. A. Connes, On the cohomology of operator algebras, *J. Functional Analysis* **28**, 248 (1978).
130. E. G. Effros and Z.-J. Ruan, *Operator spaces*, London Mathematical Society Monographs. New Series, Vol. 23 (The Clarendon Press, Oxford University Press, New York, 2000).
131. V. Paulsen, *Completely bounded maps and operator algebras*, Cambridge Studies in Advanced Mathematics, Vol. 78 (Cambridge University Press, Cambridge, 2002).
132. G. Pisier, *Introduction to operator space theory*, London Mathematical Society Lecture Note Series, Vol. 294 (Cambridge University Press, Cambridge, 2003).
133. D. P. Blecher and C. Le Merdy, *Operator algebras and their modules—an operator space approach*, London Mathematical Society Monographs. New Series, Vol. 30 (The Clarendon Press, Oxford University Press, Oxford, 2004), Oxford Science Publications.
134. A. Y. Helemskii, *Quantum functional analysis*, University Lecture Series, Vol. 56 (American Mathematical Society, Providence, RI, 2010), Non-coordinate approach.
135. J. L. Kelley, Banach spaces with the extension property, *Trans. Amer. Math. Soc.* **72**, 323 (1952).
136. G. Pisier, *Similarity problems and completely bounded maps*, Lecture

Notes in Mathematics, Vol. 1618, expanded edn. (Springer-Verlag, Berlin, 2001), Includes the solution to "The Halmos problem".

137. G. Pisier, A polynomially bounded operator on Hilbert space which is not similar to a contraction, *J. Amer. Math. Soc.* **10**, 351 (1997).
138. M. Junge and G. Pisier, Bilinear forms on exact operator spaces and $B(H) \otimes B(H)$, *Geom. Funct. Anal.* **5**, 329 (1995).
139. G. Pisier, Operator spaces and similarity problems, in *Proceedings of the International Congress of Mathematicians, Vol. I*, eds. A. Louis, U. Rehmann and P. Schneider (Berlin, 1998) pp. 429–452.
140. Z.-J. Ruan, The operator amenability of $A(G)$, *Amer. J. Math.* **117**, 1449 (1995).
141. E. G. Effros, M. Junge and Z.-J. Ruan, Integral mappings and the principle of local reflexivity for noncommutative L^1-spaces, *Ann. of Math.* **151**, 59 (2000).
142. E. Christensen, E. G. Effros and A. Sinclair, Completely bounded multilinear maps and C^*-algebraic cohomology, *Invent. Math.* **90**, 279 (1987).
143. E. Christensen and A. Sinclair, On the Hochschild cohomology for von Neumann algebras, unpublished manuscript 1987.
144. F. Pop and R. R. Smith, Cohomology for certain finite factors, *Bull. London Math. Soc.* **26**, 303 (1994).
145. E. Christensen, F. Pop, A. M. Sinclair and R. R. Smith, On the cohomology groups of certain finite von Neumann algebras, *Math. Ann.* **307**, 71 (1997).
146. A. M. Sinclair and R. R. Smith, Hochschild cohomology for von Neumann algebras with Cartan subalgebras, *Amer. J. Math.* **120**, 1043 (1998).
147. J. M. Cameron, Hochschild cohomology of II_1 factors with Cartan maximal abelian subalgebras, *Proc. Edinb. Math. Soc.* **52**, 287 (2009).
148. E. Christensen, F. Pop, A. M. Sinclair and R. R. Smith, Hochschild cohomology of factors with property Γ, *Ann. of Math.* **158**, 635 (2003).
149. E. Christensen, F. Pop, A. M. Sinclair and R. R. Smith, Property Γ factors and the Hochschild cohomology problem, *Proc. Natl. Acad. Sci. USA* **100**, 3865 (2003).
150. E. Christensen, Finite von Neumann algebra factors with property Γ, *J. Funct. Anal.* **186**, 366 (2001).
151. F. Pop and R. R. Smith, Vanishing of second cohomology for tensor products of type II_1 von Neumann algebras, *J. Funct. Anal.* **258**, 2695 (2010).

152. V. I. Paulsen, Relative Yoneda cohomology for operator spaces, *J. Funct. Anal.* **157**, 358 (1998).
153. G. V. Rozenblum, *Review of*[152] *(MR1638320).*
154. A. Y. Helemskii, *The homology of Banach and topological algebras*, Mathematics and its Applications (Soviet Series), Vol. 41 (Kluwer Academic Publishers Group, Dordrecht, 1989), Translated from the Russian by Alan West.
155. B. E. Johnson, Perturbations of Banach algebras, *Proc. London Math. Soc.* **34**, 439 (1977).
156. I. Raeburn and J. L. Taylor, Hochschild cohomology and perturbations of Banach algebras, *J. Functional Analysis* **25**, 258 (1977).
157. R. V. Kadison and D. Kastler, Perturbations of von Neumann algebras. I. Stability of type, *Amer. J. Math.* **94**, 38 (1972).
158. M. Gerstenhaber, On the deformation of rings and algebras, *Ann. of Math.* **79**, 59 (1964).
159. A. Dosi, Perturbations of nonassociative Banach algebras, *Rocky Mountain J. Math.* **39**, 509 (2009).
160. B. Harris, Cohomology of Lie triple systems and Lie algebras with involution, *Trans. Amer. Math. Soc.* **98**, 148 (1961).
161. G. Leger, *Review of*[160] *(MR01203013).*
162. T. L. Hodge and B. J. Parshall, On the representation theory of Lie triple systems, *Trans. Amer. Math. Soc.* **354**, 4359 (2002).
163. P. Koshlukov, *Review of*[162] *(MR19268806).*
164. P. Seibt, *Review of*[35] *(MR0430026).*
165. W. G. Bade and P. C. Curtis, Jr., The Wedderburn decomposition of commutative Banach algebras, *Amer. J. Math.* **82**, 851 (1960).
166. H. G. Dales, *Banach algebras and automatic continuity*, London Mathematical Society Monographs. New Series, Vol. 24 (The Clarendon Press, Oxford University Press, New York, 2000), Oxford Science Publications.
167. R. Carlsson, Der Wedderburnsche Hauptsatz für alternative Tripelsysteme und Paare, *Math. Ann.* **228**, 233 (1977).
168. A. H. Boers, *Review of*[167] *(MR0476821).*
169. A. Thedy, *Review of*[85] *(MR0496757).*
170. P. Seibt, Cohomology of algebras and triple systems, *Comm. Algebra* **3**, 1097 (1975).
171. M. Barr and G. S. Rinehart, Cohomology as the derived functor of derivations, *Trans. Amer. Math. Soc.* **122**, 416 (1966).
172. R. V. Kadison, Which singer is that?, in *Surveys in differential*

geometry, ed. S.-T. Yau, Surv. Differ. Geom., Vol. 7 (Int. Press, Somerville, MA, 2000) pp. 347–373.

173. P. Eymard, L'algèbre de Fourier d'un groupe localement compact, *Bull. Soc. Math. France* **92**, 181 (1964).
174. B. E. Johnson, Weak amenability of group algebras, *Bull. London Math. Soc.* **23**, 281 (1991).
175. H. G. Dales, F. Ghahramani and A. Y. Helemskii, The amenability of measure algebras, *J. London Math. Soc.* **66**, 213 (2002).
176. A. T.-M. Lau, R. J. Loy and G. A. Willis, Amenability of Banach and C^*-algebras on locally compact groups, *Studia Math.* **119**, 161 (1996).
177. B. E. Forrest and V. Runde, Amenability and weak amenability of the Fourier algebra, *Math. Z.* **250**, 731 (2005).
178. N. Spronk, Amenability properties of Fourier algebras and Fourier-Stieltjes algebras: a survey, in *Banach algebras 2009*, eds. R. J. Loy, V. Runde and A. Soltysiak, Banach Center Publ., Vol. 91 (Polish Acad. Sci. Inst. Math., Warsaw, 2010) pp. 365–383.
179. B. E. Forrest, E. Samei and N. Spronk, Weak amenability of Fourier algebras on compact groups, *Indiana Univ. Math. J.* **58**, 1379 (2009).
180. D. Bohle and W. Werner, A K-theoretic approach to the classification of symmetric spaces, to appear in J. of Pure and Applied Algebra, http://www.sciencedirect.com/science/article/pii/S0022404915000201.
181. R. B. Braun, A Gelfand-Neumark theorem for C^*-alternative algebras, *Math. Z.* **185**, 225 (1984).
182. B. Randrianantoanina, Norm-one projections in Banach spaces, *Taiwanese J. Math.* **5**, 35 (2001), International Conference on Mathematical Analysis and its Applications (Kaohsiung, 2000).
183. L. L. Stachó, A projection principle concerning biholomorphic automorphisms, *Acta Sci. Math. (Szeged)* **44**, 99 (1982).
184. Y. Friedman and B. Russo, Solution of the contractive projection problem, *J. Funct. Anal.* **60**, 56 (1985).
185. A. Grothendieck, Une caractérisation vectorielle-métrique des espaces L^1, *Canad. J. Math.* **7**, 552 (1955).
186. R. G. Douglas, Contractive projections on an L_1-space, *Pacific J. Math.* **15**, 443 (1965).
187. T. Ando, Contractive projections in L_p-spaces, *Pacific J. Math.* **17**, 391 (1966).
188. S. J. Bernau and H. E. Lacey, The range of a contractive projection on an L_p-space, *Pacific J. Math.* **53**, 21 (1974).

189. L. Tzafriri, Remarks on contractive projections in L_p-spaces, *Israel J. Math.* **7**, 9 (1969).
190. M. D. Choi and E. G. Effros, Injectivity and operator spaces, *J. Functional Analysis* **24**, 156 (1977).
191. J. Lindenstrauss and L. Tzafriri, *Classical Banach spaces. I* (Springer-Verlag, Berlin-New York, 1977), Sequence spaces, Ergebnisse der Mathematik und ihrer Grenzgebiete, Vol. 92.
192. E. G. Effros and E. Stormer, Positive projections and Jordan structure in operator algebras, *Math. Scand.* **45**, 127 (1979).
193. M. A. Youngson, Completely contractive projections on C^*-algebras, *Quart. J. Math. Oxford Ser.* **34**, 507 (1983).
194. W. Kaup, Contractive projections on Jordan C^*-algebras and generalizations, *Math. Scand.* **54**, 95 (1984).
195. J. Arazy and Y. Friedman, Contractive projections in C_1 and C_∞, *Mem. Amer. Math. Soc.* **13**, iv+165 (1978).
196. J. Arazy and Y. Friedman, Contractive projections in C_p, *Mem. Amer. Math. Soc.* **95**, vi+109 (1992).
197. Y. Raynaud, The range of a contractive projection in $L_p(H)$, *Rev. Mat. Complut.* **17**, 485 (2004).
198. C. Le Merdy, É. Ricard and J. Roydor, Completely 1-complemented subspaces of Schatten spaces, *Trans. Amer. Math. Soc.* **361**, 849 (2009).
199. Y. Friedman and B. Russo, The Gelfand-Naĭmark theorem for JB*-triples, *Duke Math. J.* **53**, 139 (1986).
200. C. M. Edwards, K. McCrimmon and G. T. Rüttimann, The range of a structural projection, *J. Funct. Anal.* **139**, 196 (1996).
201. M. Neal and B. Russo, Operator space characterizations of C^*-algebras and ternary rings, *Pacific J. Math.* **209**, 339 (2003).
202. C. M. Edwards, Rigidly collinear pairs of structural projections on a JBW*-triple, *Acta Sci. Math. (Szeged)* **72**, 205 (2006).
203. Y. Friedman, *Physical applications of homogeneous balls*, Progress in Mathematical Physics, Vol. 40 (Birkhäuser Boston, Inc., Boston, MA, 2005), With the assistance of Tzvi Scarr.
204. M. Koecher, *An elementary approach to bounded symmetric domains* (Rice University, Houston, Tex., 1969).
205. O. Loos, *Bounded symmetric domains and Jordan pairs* (Univ. of California, Irvine, Ca, 1977).
206. Y. Friedman and B. Russo, Conditional expectation without order, *Pacific J. Math.* **115**, 351 (1984).

207. Y. Friedman and B. Russo, Structure of the predual of a JBW^*-triple, *J. Reine Angew. Math.* **356**, 67 (1985).
208. S. Dineen, The second dual of a JB* triple system, in *Complex analysis, functional analysis and approximation theory (Campinas, 1984)*, ed. J. Mujica, North-Holland Math. Stud., Vol. 125 (North-Holland, Amsterdam, 1986) pp. 67–69.
209. T. Barton and R. M. Timoney, Weak*-continuity of Jordan triple products and its applications, *Math. Scand.* **59**, 177 (1986).
210. G. Horn, Classification of JBW*-triples of type I, *Math. Z.* **196**, 271 (1987).
211. C.-H. Chu, M. Neal and B. Russo, Normal contractive projections preserve type, *J. Operator Theory* **51**, 281 (2004).
212. L. J. Bunce and A. M. Peralta, Images of contractive projections on operator algebras, *J. Math. Anal. Appl.* **272**, 55 (2002).
213. J. Tomiyama, On the projection of norm one in W^*-algebras. III, *Tôhoku Math. J.* **11**, 125 (1959).
214. A. Pelczyński, Projections in certain Banach spaces, *Studia Math.* **19**, 209 (1960).
215. J. Arazy and Y. Friedman, The isometries of $C_p^{n,m}$ into C, *Israel J. Math.* **26**, 151 (1977).
216. E. Kirchberg, On nonsemisplit extensions, tensor products and exactness of group C^*-algebras, *Invent. Math.* **112**, 449 (1993).
217. P. W. Ng and N. Ozawa, A characterization of completely 1-complemented subspaces of noncommutative L_1-spaces, *Pacific J. Math.* **205**, 171 (2002).
218. M. Neal and B. Russo, Existence of contractive projections on preduals of JBW*-triples, *Israel J. Math.* **182**, 293 (2011).
219. C. M. Edwards and G. T. Rüttimann, Structural projections on JBW*-triples, *J. London Math. Soc.* **53**, 354 (1996).
220. O. Loos, On the socle of a Jordan pair, *Collect. Math.* **40**, 109 (1989).
221. O. Loos and E. Neher, Complementation of inner ideals in Jordan pairs, *J. Algebra* **166**, 255 (1994).
222. C. M. Edwards and G. T. Rüttimann, A characterization of inner ideals in JB*-triples, *Proc. Amer. Math. Soc.* **116**, 1049 (1992).
223. C. M. Edwards, R. V. Hügli and G. T. Rüttimann, A geometric characterization of structural projections on a JBW*-triple, *J. Funct. Anal.* **202**, 174 (2003).
224. C. M. Edwards and R. V. Hügli, Decoherence in pre-symmetric spaces, *Rev. Mat. Complut.* **21**, 219 (2008).

225. W. F. Stinespring, Positive functions on C^*-algebras, *Proc. Amer. Math. Soc.* **6**, 211 (1955).
226. W. B. Arveson, Subalgebras of C^*-algebras, *Acta Math.* **123**, 141 (1969).
227. W. B. Arveson, On subalgebras of C^*-algebras, *Bull. Amer. Math. Soc.* **75**, 790 (1969).
228. Z.-J. Ruan, Subspaces of C^*-algebras, *J. Funct. Anal.* **76**, 217 (1988).
229. B. Russo and H. A. Dye, A note on unitary operators in C^*-algebras, *Duke Math. J.* **33**, 413 (1966).
230. M. Neal and B. Russo, Contractive projections and operator spaces, *C. R. Acad. Sci. Paris Sér. I Math.* **331**, 873 (2000).
231. M. Neal and B. Russo, Representation of contractively complemented Hilbertian operator spaces on the Fock space, *Proc. Amer. Math. Soc.* **134**, 475 (2006).
232. M. Neal, É. Ricard and B. Russo, Classification of contractively complemented Hilbertian operator spaces, *J. Funct. Anal.* **237**, 589 (2006).
233. A. G. Robertson, Injective matricial Hilbert spaces, *Math. Proc. Cambridge Philos. Soc.* **110**, 183 (1991).
234. Z.-J. Ruan, Injectivity of operator spaces, *Trans. Amer. Math. Soc.* **315**, 89 (1989).
235. E. G. Effros, N. Ozawa and Z.-J. Ruan, On injectivity and nuclearity for operator spaces, *Duke Math. J.* **110**, 489 (2001).
236. M. Hamana, Injective envelopes of operator systems, *Publ. Res. Inst. Math. Sci.* **15**, 773 (1979).
237. D. P. Blecher and V. I. Paulsen, Multipliers of operator spaces, and the injective envelope, *Pacific J. Math.* **200**, 1 (2001).
238. M. Hamana, Triple envelopes and Šilov boundaries of operator spaces, *Math. J. Toyama Univ.* **22**, 77 (1999).
239. H. Upmeier, A holomorphic characterization of C^*-algebras, in *Functional analysis, holomorphy and approximation theory II*, ed. G. I. Zapata, North-Holland Math. Stud., Vol. 86 (North-Holland, Amsterdam, 1984) pp. 427–467.
240. K. El Amin, A. Morales Campoy and A. Rodríguez Palacios, A holomorphic characterization of C^* and JB^*-algebras, *Manuscripta Math.* **104**, 467 (2001).
241. D. P. Blecher, Z.-J. Ruan and A. M. Sinclair, A characterization of operator algebras, *J. Funct. Anal.* **89**, 188 (1990).
242. D. P. Blecher, A completely bounded characterization of operator

algebras, *Math. Ann.* **303**, 227 (1995).

243. J. Arazy, Isometries of Banach algebras satisfying the von Neumann inequality, *Math. Scand.* **74**, 137 (1994).
244. J. Arazy and B. Solel, Isometries of nonselfadjoint operator algebras, *J. Funct. Anal.* **90**, 284 (1990).
245. W. Kaup and H. Upmeier, Banach spaces with biholomorphically equivalent unit balls are isomorphic, *Proc. Amer. Math. Soc.* **58**, 129 (1976).
246. J. Arazy, An application of infinite-dimensional holomorphy to the geometry of Banach spaces, in *Geometrical aspects of functional analysis (1985/86)*, eds. J. Lindenstrauss and V. D. Milman, Lecture Notes in Math., Vol. 1267 (Springer, Berlin, 1987) pp. 122–150.
247. D. P. Blecher and V. Zarikian, Multiplier operator algebras and applications, *Proc. Natl. Acad. Sci. USA* **101**, 727 (2004).
248. M. Kaneda, Quasi-multipliers and algebrizations of an operator space, *J. Funct. Anal.* **251**, 346 (2007).
249. L. J. Bunce and R. M. Timoney, On the universal TRO of a JC^*-triple, ideals and tensor products, *Q. J. Math.* **64**, 327 (2013).
250. L. J. Bunce and R. M. Timoney, Universally reversible JC^*-triples and operator spaces, *J. Lond. Math. Soc.* **88**, 271 (2013).
251. O. Loos, Representations of Jordan triples, *Trans. Amer. Math. Soc.* **185**, 199 (1973).
252. E. Neher, On the radical of J^*-triples, *Math. Z.* **211**, 323 (1992).

Weighted inequalities and extrapolation*

Javier Duoandikoetxea

Departamento de Matemáticas, Universidad del País Vasco-Euskal Herriko Unibertsitatea (UPV/EHU), Apartado 644, 48080 Bilbao, Spain
E-mail: javier.duoandikoetxea@ehu.eus

The extrapolation theorem of J. L. Rubio de Francia is one of the main results in the theory of weighted inequalities. Its proof requires a reverse factorization property and a construction of A_1 weights. We present a simple proof written in the setting of Muckenhoupt bases and discuss some extensions. In the case of the classical A_p weights the proof leads to the sharp form of the extrapolation theorem in terms of the A_p-constant of the weights.

Keywords: Weighted inequalities; extrapolation.

1. Introduction

The Hardy-Littlewood maximal operator was introduced by G. H. Hardy and J. E. Littlewood in 1930 in the one-dimensional case (actually in $(0, +\infty)$) and extended to higher dimensions by N. Wiener in 1939 (see[1] and[2], respectively). Although there are several variants, we shall use the following definition for the operator: for f locally integrable in $\mathbf{R}^n$,

$$Mf(x) = \sup_{x \in B} \frac{1}{|B|} \int_B |f|,$$

where the supremum is taken over the balls containing x and can be infinity. Here and in the sequel, $|A|$ denotes the Lebesgue measure of the set A.

In both cases the authors proved the basic results: M is bounded on L^p for $1 < p \leq \infty$ and is of weak-type $(1, 1)$. These results are well known and can be found in many textbooks (for instance, in[3] and[4] of the references at the end of this paper).

It was in 1972 that B. Muckenhoupt obtained the full characterization of the weighted inequalities satisfied by the Hardy-Littlewood maximal

*Supported in part by the grants MTM2011-24054 of MINECO (Spain) and FEDER, and IT-641-13 of the Basque Gouvernment.

**2000 Mathematics Subject Classification:* 42B25.

operator by means of the so-called A_p condition ([5]). The significance of the A_p classes became apparent soon after Muckenhoupt's work when it was shown that the weighted inequalities obtained for the Hardy-Littlewood operator were shared by some of the classical operators of Fourier Analysis like the Hilbert transform and more generally singular integrals and Calderón-Zygmund operators, Littlewood-Paley operators and other square functions, Bochner-Riesz operators at the critical index, and others.

A remarkable property of the A_p classes was found by J. L. Rubio de Francia ([6],[7]), who named it *extrapolation theorem.* An extrapolation theorem for weights basically says that the L^p-weighted inequalities with some appropriate class of weights for just one value of p imply weighted inequalities for (all the) other values of p. Extrapolation will be the main topic adressed in this paper.

Rubio de Francia authored with J. García-Cuerva the first book having weighted and vector-valued inequalities as its main theme ([4]), published in 1985. From then on many textbooks incorporate the basic theory of weighted inequalities and the reader will easily find references. The book[8] contains an advanced treatment of weighted inequalities from the point of view of extrapolation.

Muckenhoupt's theory and subsequent developments were extended by B. Jawerth and formulated in the framework of bases of open sets in $\mathbf{R}^n$ (see[9]). It can be further extended and $\mathbf{R}^n$ can be replaced by a measure space with σ-finite measure.

There are several proofs of the extrapolation theorem. Nevertheless, they share the essential ingredients: some kind of factorization of weights and a construction of A_1 weights. The author proposed a proof in[10] with a modified form of factorization. Apart from being shorter than other proofs, it has the added value of being sharp in the sense of the size of the constants of the weights. Such proof can be carried out directly to the context of Muckenhoupt bases and this will be the setting adopted in this paper.

First we recall the definitions and some immediate properties of the A_p weights for Muckenhoupt bases (Section 2), next we introduce the factorization we need (Section 3) and the construction of A_1 weights (Section 4). In Section 5 we present the proof of the extrapolation theorem and in Section 6 we show how to obtain sharp dependence of the constants. After some comments on the extrapolation to $L^1(w)$ in Section 7, we discuss the extension of the proposed factorization beyond the A_p classes and its consequences to the extrapolation theorem in Section 8. Finally, we adapt the given proofs to some two-weighted extrapolation results in Section 9.

We denote by $L^p(w)$ the L^p space with respect to the measure $w(x)dx$ and by $\|f\|_{p,w}$ the norm of f in such space. We write $w(A)$ for the integral of w on the measurable set A.

2. Muckenhoupt bases and the A_p condition

A basis $\mathcal{B}$ in $\mathbf{R}^n$ is a collection of open sets B contained in $\mathbf{R}^n$. Given a basis $\mathcal{B}$, the maximal operator associated with $\mathcal{B}$ is defined by

$$M_{\mathcal{B}}f(x) = \sup_{B\in\mathcal{B}:x\in B} \frac{1}{|B|}\int_B |f(y)|dy$$

if $x \in \cup_{B\in\mathcal{B}} B$ and $M_{\mathcal{B}}f(x) = 0$ otherwise.

Definition 2.1. We say that a nonnegative locally integrable function w is in $A_{p,\mathcal{B}}$ for $1 < p < \infty$ if

$$[w]_{A_{p,\mathcal{B}}} := \sup_{B\in\mathcal{B}} \left(\frac{w(B)}{|B|}\right)\left(\frac{w^{1-p'}(B)}{|B|}\right)^{p-1} < +\infty,$$

We say that w is in $A_{1,\mathcal{B}}$ if

$$M_{\mathcal{B}}w(x) \le Cw(x) \quad \text{a.e.}$$

and we denote by $[w]_{A_{1,\mathcal{B}}}$ the smallest value of C in the right-hand side.

The value $[w]_{A_{p,\mathcal{B}}}$ is the A_p-constant of the weight w.

The usual Hardy-Littlewood maximal operator corresponds therefore to the basis of all open Euclidean balls in $\mathbf{R}^n$. We shall drop the subscript $\mathcal{B}$ when we refer to such operator and to the weights associated to it.

Definition 2.2. We say that $\mathcal{B}$ is a Muckenhoupt basis if for all $1 < p < \infty$, $M_{\mathcal{B}}$ is bounded on $L^p(w)$ for $w \in A_{p,\mathcal{B}}$.

With this definition, B. Muckenhoupt's theorem in[5] can be restated saying that the basis formed by the open Euclidean balls in $\mathbf{R}^n$ is a Muckenhoupt basis.

Several properties of the A_p weights are immediately deduced from the definition. The following proposition shows some of them to be mentioned later in this paper with indication of the size of the constants of the involved weights.

Prop 2.1. Let $1 < p < \infty$.

(i) If $q < p$, then $A_{q,\mathcal{B}} \subset A_{p,\mathcal{B}}$ and $[w]_{A_{p,\mathcal{B}}} \leq [w]_{A_{q,\mathcal{B}}}$.

(ii) If $w \in A_{p,\mathcal{B}}$, then $w^{1-p'} \in A_{p',\mathcal{B}}$ and $[w^{1-p'}]_{A_{p',\mathcal{B}}} = [w]_{A_{p,\mathcal{B}}}^{p'-1}$.

(iii) If $w_0, w_1 \in A_{1,\mathcal{B}}$ then $w_0 w_1^{1-p} \in A_{p,\mathcal{B}}$ and

$$[w]_{A_{p,\mathcal{B}}} \leq [w_0]_{A_{1,\mathcal{B}}} [w_1]_{A_{1,\mathcal{B}}}^{p-1}. \tag{1}$$

To prove the third property we use the following consequence of the definition: for $u \in A_{1,\mathcal{B}}$,

$$u(x)^{-1} \leq [u]_{A_{1,\mathcal{B}}} \left(\frac{u(B)}{|B|} \right)^{-1} \quad \text{a.e. } x \in B. \tag{2}$$

Inserting this bound into the $A_{p,\mathcal{B}}$ condition we get the property.

3. Factorization

In[8] the third property in Proposition 2.1 is called *reverse factorization.* Its converse (*factorization*) is also true but it is not immediate. Muckenhoupt conjectured it (for M) and P. Jones found the first proof ([11]). Rubio de Francia's approach in the above mentioned papers simplified the proof.

Only reverse factorization is needed to prove the extrapolation theorem. But in our proof we shall use other forms of reverse factorization that are described in the following proposition.

Prop 3.1. *(a)* Let $1 \leq p < p_0 < \infty$. If $w \in A_{p,\mathcal{B}}$ and $u \in A_{1,\mathcal{B}}$, then wu^{p-p_0} is in $A_{p_0,\mathcal{B}}$ and

$$[wu^{p-p_0}]_{A_{p_0,\mathcal{B}}} \leq [w]_{A_{p,\mathcal{B}}} [u]_{A_{1,\mathcal{B}}}^{p_0-p}.$$

(b) Let $1 < p_0 < p < \infty$. If $w \in A_{p,\mathcal{B}}$ and $u \in A_{1,\mathcal{B}}$, then $(w^{p_0-1}u^{p-p_0})^{\frac{1}{p-1}}$ is in $A_{p_0,\mathcal{B}}$ and

$$[(w^{p_0-1}u^{p-p_0})^{1/(p-1)}]_{A_{p_0,\mathcal{B}}} \leq [w]_{A_{p,\mathcal{B}}}^{\frac{p_0-1}{p-1}} [u]_{A_{1,\mathcal{B}}}^{\frac{p-p_0}{p-1}}.$$

These are also immediate consequences of the definition of $A_{p,\mathcal{B}}$, property (2) of $A_{1,\mathcal{B}}$ weights and Hölder's inequality, and we leave the proofs to the reader. Actually, they are dual of each other and the second one can be obtained from the first one and the second property of Proposition 2.1.

4. Building A_1 weights

The factorization theorem suggests that it should be interesting to have a way to produce $A_{1,\mathcal{B}}$ weights. The first characterization of the A_1 weights (for M) is due to R. Coifman and R. Rochberg.

Theorem 4.1 ([12]). (a) *Let μ a finite measure such that $M\mu(x) < +\infty$ a.e. and let $0 \leq \delta < 1$. Then $(M\mu(x))^\delta \in A_1$ and $[(M\mu(x))^\delta]_{A_1}$ depends on δ, not on μ.*

(b) *If $w \in A_1$, then $w(t) = Mg(t)^\delta k(t)$, where g is a locally integrable function, $0 \leq \delta < 1$ and k and k^{-1} are bounded.*

The interesting part is the construction of the A_1 weights and it is crucial that the A_1-constant of the weight depends only on δ.

Unfortunately, the construction of Coifman and Rochberg cannot be extended in general to all the Muckenhoupt bases. For instance, if we consider the operator associated to the basis of rectangles in $\mathbf{R}^n$ with sides parallel to the coordinate axes (strong maximal function), the corresponding statement is false (see[13]). On the other hand, (a) holds for the basis of balls centered at the origin considered in[14], but (b) fails. Actually, in the latter case the pointwise inequality $f(x) \leq (M_\mathcal{B} f^{1/\delta})^\delta(x)$ need not hold and this makes the result unusable for the proof of the extrapolation theorem in such a case.

Despite the disparate behaviour of the Muckenhoupt bases with respect to the Coifman-Rochberg characterization, the A_1 weights needed in the proof of the extrapolation theorem can be built using a method invented by J. L. Rubio de Francia which can be carried out in very general situations. This construction has been called *Rubio de Francia algorithm.*

Theorem 4.2. *Let S be a nonnegative sublinear operator bounded on $L^p(\mu)$ with norm $\|S\|_{p,\mu}$, where $p \geq 1$ and μ is a positive measure on some measurable space. Given $f \in L^p(\mu)$, define*

$$Rf(x) = \sum_{k=0}^{\infty} \frac{S^k f(x)}{(2\|S\|_{p,\mu})^k},$$

where S^0 is the identity and $S^k f = S(S^{k-1} f)$. Then

(i) *$f(x) \leq Rf(x)$ a.e.;*

(ii) *$\|Rf\|_{p,\mu} \leq 2\|f\|_{p,\mu}$;*

(iii) *$S(Rf)(x) \leq 2\|S\|_{p,\mu} Rf(x)$ for almost every x.*

The series defining Rf clearly converges in norm and the properties are easy to check. If we choose S to be $M_\mathcal{B}$, $1 < p < \infty$, and $w \in A_{p,\mathcal{B}}$, for each $f \in L^p(w)$ we built an $A_{1,\mathcal{B}}$ weight.

5. Extrapolation

The original statement of the extrapolation theorem was given for sublinear operators. But it turns out that it can be stated for pairs of functions, free of operators. It seems that this observation appears in[15]*Remark 1.11 for the first time. Working with pairs of functions offers several advantages as can be seen, for instance, in Sections 3.2 and 3.3 of[8] and does not add any difficulty to the proof, so that we adopt such a point of view in our statements.

Theorem 5.1. *Assume that for some* $1 \le p_0 < \infty$ *and all* $w \in A_{p_0,\mathcal{B}}$ *it holds that*

$$\|g\|_{p_0,w} \le C\|f\|_{p_0,w}, \tag{1}$$

for some pairs of nonnegative functions (f,g) *and with a constant* C *depending on the* $A_{p_0,\mathcal{B}}$*-constant of* w *but not otherwise on* w *itself. Then for all* $w \in A_{p,\mathcal{B}}, 1 < p < \infty$, *the following inequality holds:*

$$\|g\|_{p,w} \le C\|f\|_{p,w},$$

whenever the left-hand side is finite.

Proof. *Case* $p < p_0$. Let $f \in L^p(w)$ for $w \in A_{p,\mathcal{B}}$. Built Rf using the Rubio de Francia algorithm of Theorem 4.2. From the first reverse factorization in Proposition 3.1 we deduce that $w(Rf)^{p-p_0} \in A_{p_0,\mathcal{B}}$. Then

$$\begin{aligned}\|g\|_{p,w} &= \left(\int_{\mathbf{R}^n} g^p\, w\, (Rf)^{\frac{p}{p_0}(p-p_0)}(Rf)^{\frac{p}{p_0}(p_0-p)}\right)^{1/p}\\ &\le \left(\int_{\mathbf{R}^n} g^{p_0} w(Rf)^{p-p_0}\right)^{\frac{1}{p_0}} \|Rf\|_{p,w}^{1-\frac{p}{p_0}}\\ &\le C_1 \left(\int_{\mathbf{R}^n} f^{p_0} w(Rf)^{p-p_0}\right)^{\frac{1}{p_0}} \|f\|_{p,w}^{1-\frac{p}{p_0}}\\ &\le C_1\|f\|_{p,w},\end{aligned}$$

where we applied respectively Hölder's inequality, the assumption (1) with the weight $w(Rf)^{p-p_0}$, and properties (ii) and (i) of the Rubio de Francia algorithm.

Case $p > p_0$. Using duality we have

$$\left(\int_{\mathbf{R}^n} g^p w\right)^{\frac{p_0}{p}} = \sup\{\int_{\mathbf{R}^n} g^{p_0} hw : h \ge 0 \text{ and } \int_{\mathbf{R}^n} h^{\frac{p}{p-p_0}} w = 1\}.$$

For such an h let us define H by $H^{p'}w^{1-p'} = h^{\frac{p}{p-p_0}}w$. Then $H \in L^{p'}(w^{1-p'})$ with norm 1. Built RH using the Rubio de Francia algorithm. From the second reverse factorization in Proposition 3.1 it follows that $w^{\frac{p_0-1}{p-1}}(RH)^{\frac{p-p_0}{p-1}} \in A_{p_0,\mathcal{B}}$. Then

$$\begin{aligned}\int_{\mathbf{R}^n} g^{p_0}hw &= \int_{\mathbf{R}^n} g^{p_0}w^{\frac{p_0-1}{p-1}}H^{\frac{p-p_0}{p-1}} \le \int_{\mathbf{R}^n} g^{p_0}w^{\frac{p_0-1}{p-1}}(RH)^{\frac{p-p_0}{p-1}}\\ &\le C_1\int_{\mathbf{R}^n} f^{p_0}w^{\frac{p_0-1}{p-1}}(RH)^{\frac{p-p_0}{p-1}}\\ &\le C_1\|f\|_{p,w}^{p_0}\|RH\|_{p',w^{1-p'}}^{\frac{p-p_0}{p-1}} \le C_2\|f\|_{p,w}^{p_0},\end{aligned}$$

where we used properties (i) and (ii) of the Rubio de Francia algorithm respectively in the first and last inequality, the hypothesis (1) with the weight $w^{\frac{p_0-1}{p-1}}(RH)^{\frac{p-p_0}{p-1}}$ in the second and Hölder's inequality in the third. □

Remark 5.1. For the A_p classes the given proof can be written with some modifications using the A_1 weights given by the Coifman-Rochberg construction in Theorem 4.1. But we need the strong property that every $w \in A_p$ is in A_q for some $q < p$ (openness property). For instance, in the case $p < p_0$ instead of $w(Rf)^{p-p_0}$ we would use $w(Mf)^{p-p_0}$. This can be written as $w[(Mf)^{(p_0-p)/(p_0-q)}]^{q-p_0}$, which is in A_{p_0} since $w \in A_q$ and $(Mf)^{(p_0-p)/(p_0-q)} \in A_1$. It is worth noting that the openness property of the weights does not extend in general to Muckenhoupt bases, as shown by the counterexample given in [14].

The proof provided in [3] follows a similar idea but limits the first part of the proof ($p < p_0$) to the A_1 weights. The order of the exposition is crucial in that approach because the result for all values of p is obtained from the A_1 weights. It is customary to split the proof into two parts but the reader will find in [8] a unified proof avoiding such splitting.

Apart from the necessity of the openness property of the A_p classes, using the A_1 weights of Theorem 4.1 has the disadvantage of being specific to the Hardy-Littlewood maximal operator since the theorem is not generalizable to other settings.

An apparently more general form of extrapolation is obtained by assuming that the index of the class of weights and the exponent of the Lebesgue space are different. If fact, it is a simple corollary to the previous theorem.

Corollary 5.2. *Assume that for some $1 \leq s_0 < \infty$, some $p_0 > 0$ and all $w \in A_{s_0}$ we have*

$$\|g\|_{p_0,w} \leq C\|f\|_{p_0,w},$$

for a family of pairs of nonnegative functions (f, g). Then for $p > p_0/s_0$ and $s = ps_0/p_0$,

$$\|g\|_{p,w} \leq C\|f\|_{p,w},$$

for all $w \in A_s$.

Write the assumption as

$$\|g^{p_0/s_0}\|_{s_0,w} \leq C\|f^{p_0/s_0}\|_{s_0,w},$$

and apply Theorem 5.1 to the pairs $(f^{p_0/s_0}, g^{p_0/s_0})$ to prove this corollary.

A definitely more general version of extrapolation is obtained if the exponents on both sides of the inequalities are permitted to be different (off-diagonal extrapolation). Thus we have the following theorem.

Theorem 5.3. *Let $1 \leq p_0 < \infty$ and $0 < q_0, r_0 < \infty$. Let $\gamma = \dfrac{1}{r_0} + \dfrac{1}{p_0'}$. Assume that for $w^{r_0} \in A_{r_0\gamma}$ it holds that*

$$\|g\|_{q_0,w^{q_0}} \leq C\|f\|_{p_0,w^{p_0}}.$$

Then for all $1 < p < \infty$ and $0 < q, r < \infty$, such that

$$\frac{1}{q} - \frac{1}{q_0} = \frac{1}{r} - \frac{1}{r_0} = \frac{1}{p} - \frac{1}{p_0},$$

and all $w^r \in A_{r\gamma}$ we have

$$\|g\|_{q,w^q} \leq C\|f\|_{p,w^p}.$$

The proof of the theorem follows the same ideas as that of Theorem 5.1 and can be found in[10].

The case $q_0 = r_0 > p_0$ is due to E. Harboure, R. Macías and C. Segovia for the $A_{p,q}$ classes of Muckenhoupt and Wheeden, related to the boundedness of the fractional integral operator I_α from $L^p(w^p)$ to $L^q(w^q)$ when $1/q - 1/p = \alpha/n$ (see[16]).

The case $q_0 < r_0 = p_0$ serves to obtain by iteration a short proof of a multilinear extrapolation theorem of L. Grafakos and J. M. Martell in[17].

6. Sharp bounds in terms of the A_p constant

In addition to the boundedness result, years later it became interesting to know which was the growth of the operator norm of M in $L^p(w)$ in terms of the A_p constant of w. The following theorem is due to S. Buckley.

Theorem 6.1 ([18]). *For $1 < p < \infty$ the following inequality holds:*

$$\|Mf\|_{p,w} \leq C_p[w]_{A_p}^{\frac{1}{p-1}} \|f\|_{p,w},$$

with C_p independent of w. The exponent is sharp.

Andrei Lerner gave a remarkably simple proof of this result, valid even in a more general setting (see [19]). Apart from giving the sharp estimate in terms of the size of the exponent of the A_p constant, Lerner's approach provides a new proof of Muckenhoupt's theorem.

Remark 6.1. The sharp weak-type inequality is

$$\lambda w(\{x : Mf(x) > \lambda\})^{1/p} \leq C_p[w]_{A_p}^{\frac{1}{p}} \|f\|_{p,w}.$$

This is already contained in Muckenhoupt's original proof in [5]. Interpolating with the trivial L^∞ estimate we obtain the following consequence: for $1 \leq q < p < \infty$ and $w \in A_q$,

$$\|Mf\|_{p,w} \leq C_{p,q}[w]_{A_q}^{\frac{1}{p}} \|f\|_{p,w}.$$

The sharp dependence of the norms in $L^p(w)$ of several classical operators in terms of the size of the exponent of the A_p constant is currently known and has been an active area of research in recent years. For instance, the answer for Calderón-Zygmund operators was obtained by T. Hytönen only in 2010. The paper [20] contains the proof of his result and a historical account of previous work with many references.

A common feature of the techniques used to obtain the sharp behaviour of the norm is that it is enough to obtain it for a critical value of p (usually $p = 2$, but not always) and extrapolate. To be useful in this context, the extrapolation theorem must include the dependence on the A_p constant, a version obtained for the first time by O. Dragičević et al. in [21]. The proof of Theorem 5.1 gives also this sharp version just keeping track of the dependence of the involved constants. We state here the theorem for pairs of functions as we did for the theorems in the previous section and for Muckenhoupt bases.

Theorem 6.2. *Let Φ be an increasing function in $(0,+\infty)$. Assume that for some $1 \le p_0 < \infty$ and all $w \in A_{p_0,\mathcal{B}}$ we have*

$$\|g\|_{p_0,w} \le \Phi([w]_{A_{p_0,\mathcal{B}}})\|f\|_{p_0,w}, \tag{1}$$

for a family of pairs of nonnegative functions (f,g). Then for all $w \in A_{p,,\mathcal{B}}, 1 < p < \infty$, it holds that

$$\|g\|_{p,w} \le \begin{cases} 2^{\frac{p_0-p}{p_0}}\Phi([w]_{A_p,\mathcal{B}}(2\|M_{\mathcal{B}}\|_{p,w})^{p_0-p})\|f\|_{p,w}, & \text{if } p < p_0; \\ 2^{\frac{p-p_0}{p_0(p-1)}}\Phi([w]_{A_p,\mathcal{B}}^{\frac{p_0-1}{p-1}}(2\|M_{\mathcal{B}}\|_{p',w^{1-p'}})^{\frac{p-p_0}{p-1}})\|f\|_{p,w}, & \text{if } p > p_0. \end{cases}$$

Proof. We specify the constants appearing in the proof of Theorem 5.1 using (1).

Case $p < p_0$. The constant C_1 appearing in that proof may be taken as

$$C_1 = \|R\|_{p,w}^{1-\frac{p}{p_0}}\Phi([w(Rf)^{p-p_0}]_{A_{p_0,\mathcal{B}}}).$$

Taking into account that R has been built with $M_{\mathcal{B}}$ acting on $f \in L^p(w)$, from Proposition 3.1 and Theorem 4.2 we deduce the result.

Case $p > p_0$. The constant C_1 appearing in that proof may be taken as

$$C_1 = \Phi([w^{\frac{p_0-1}{p-1}}(RH)^{\frac{p-p_0}{p-1}}]_{A_{p_0},\mathcal{B}})^{p_0},$$

and then $C_2 = C_1\|R\|_{p',w^{1-p'}}^{\frac{p-p_0}{p-1}}$. In this case R has been built with $M_{\mathcal{B}}$ acting on $H \in L^{p'}(w^{1-p'})$. Now we need C_2^{1/p_0} and we use again Proposition 3.1 and Theorem 4.2. □

When we consider the Hardy-Littlewood maximal operator, we can insert Buckley's bound from Theorem 6.1 into the obtained bounds to get

$$\|g\|_{p,w} \le K_1\Phi(K_2[w]_{A_p}^{\max(1,\frac{p_0-1}{p-1})})\|f\|_{p,w},$$

for constants K_1 and K_2 depending only on p, p_0 and the dimension n, but not on w. This is the result in[21].

7. Extrapolation to L^1

In the statement of Theorem 5.1 the value $p_0 = 1$ is accepted in the assumption, but $p = 1$ is excluded from the conclusion if $p_0 > 1$. It is clear that is has to be this way, because the usual Hardy-Littlewood maximal operator itself is bounded on $L^1(w)$ only if w is identically zero.

A look at the proof of the theorem shows that actually the construction of the $A_{1,\mathcal{B}}$ weights in the Rubio de Francia algorithm needs the maximal operator to be bounded on the corresponding $L^p(w)$. There are Muckenhoupt bases for which the corresponding maximal operator $M_{\mathcal{B}}$ is bounded on $L^1(w)$ for nontrivial w. Apart from easy examples consisting for instance of a single set, a less trivial example appears in[14]. It is provided by the basis consisting of the balls centered at the origin. In that paper it is proved that the corresponding maximal operator is bounded on $L^1(w)$ for a class of weights strictly contained in $A_{1,\mathcal{B}}$, but not reduced to the trivial case in which w is identically zero.

We can then say that the result in the extrapolation theorem is valid for $p = 1$ if we restrict the conclusion to the class of weights for which $M_{\mathcal{B}}$ is bounded on $L^1(w)$. This result is a particular case of the theorem appearing in the next section.

8. Beyond the Muckenhoupt bases

In the exposition of the previous sections we considered weights defined through a Muckenhoupt basis. With small changes we can extend the results to a more general setting. We already gave a quite general statement for the Rubio de Francia algorithm so that we should consider the possibility of suitably extending the factorization theorems.

Definition 8.1. Let S be a nonnegative sublinear operator. Define for $1 \le p < \infty$ the classes

$$W_p(S) = \{w : S \text{ is bounded on } L^p(w)\}.$$

Let the constant $[w]_{W_p(S)}$ be the operator norm of S on $L^p(w)$.

The class $A_1(S)$ is defined as

$$A_1(S) = \{w : Sw(x) \le Cw(x) \text{ a.e.}\}$$

and $[w]_{A_1(S)}$ is the smallest C for which the inequality holds.

We have the following factorization result generalizing part (a) of Proposition 3.1.

Prop 8.1. Let S be a nonnegative sublinear operator. Assume that $|S(fg)(x)| \le \|f\|_\infty |Sg(x)|$. Let $1 \le p < p_0 < \infty$. If $w \in W_p(S)$ and $u \in A_1(S)$, then wu^{p-p_0} is in $W_{p_0}(S)$ and

$$[wu^{p-p_0}]_{W_{p_0}(S)} \le [w]_{W_p(S)}^{p/p_0} [u]_{A_1(S)}^{1-p/p_0}.$$

Proof. Define $S^\dagger f(x) = u(x)^{-1} S(fu)(x)$. For $u \in A_1(S)$, $S^\dagger$ satisfies

$$\|S^\dagger f\|_\infty \le [u]_{A_1(S)} \|f\|_\infty.$$

On the other hand, by assumption

$$\int_{\mathbf{R}^n} |S^\dagger f|^p u^p w \le [w]^p_{W_p(S)} \int_{\mathbf{R}^n} |f|^p u^p w.$$

By interpolation $S^\dagger$ is bounded on $L^{p_0}(u^p w)$, hence S is bounded on $L^{p_0}(u^{p-p_0} w)$. The estimate for the constant follows also from the interpolation theorem. □

With both ingredients, factorization and Rubio de Francia algorithm, we can obtain immediately the following generalization of the descending half of the extrapolation theorem.

Theorem 8.1. *Let S be as in Proposition 8.1. Assume that for some $1 < p_0 < \infty$ and all $w \in W_{p_0}(S)$ it holds that*

$$\|g\|_{p_0,w} \le C \|f\|_{p_0,w},$$

for some pairs of nonnegative functions (f, g) and with a constant C depending only on $[w]_{W_{p_0}(S)}$. Then for $1 \le p < p_0$ and $w \in W_p(S)$ the following inequality holds:

$$\|g\|_{p,w} \le C \|f\|_{p,w},$$

whenever the left-hand side is finite.

The other half of the factorization result and of the extrapolation theorem needs some extra assumptions in general. Indeed, take an operator S such that $W_{p_1}(S) \ne \{0\}$ and $W_{p_0}(S) = \{0\}$ for some $p_0 < p_1$. (For instance, if S is the spherical maximal operator, this holds with $p_0 < n/(n-1)$ and $p_1 > n/(n-1)$.) The hypothesis of the extrapolation theorem would be required only for $w \equiv 0$, so that we could not expect any conclusion from it. In a similar way, combining weights in $W_{p_1}(S)$ and $A_1(S)$ as in part (b) of Proposition 3.1 we cannot expect weights in $W_{p_0}(S)$.

To obtain results in the opposite sense some kind of duality property is required. Although one might assume

$$W_{p'}(S) = \{w^{1-p'} : w \in W_p(S)\},$$

the simple example of a nonnegative bounded linear operator shows that this would be natural only if S is self-adjoint, and that the duality should read as $W_{p'}(S) = \{w^{1-p'} : w \in W_p(S^*)\}$, where S^* is the adjoint of S.

In a similar way, other typical examples show that one needs to consider two operators in the duality property. We shall not pursue the details here. Related results are in[22] and[23].

9. Some results in two-weighted extrapolation

The result in the previous section extends easily to the two-weight setting. Let us denote by $W_p^*(S)$ the class of pairs (u, v) such that S is bounded from $L^p(v)$ to $L^p(u)$, for $1 \leq p < \infty$ and let $[u, v]_{W_p^*}$ be the corresponding operator norm of S. For the Hardy-Littlewood maximal operator these classes were characterized by E. Sawyer in[24], but we shall not need any characterization here.

First we state an appropriate reverse factorization.

Prop 9.1. Let S be as in Proposition 8.1. Let $1 \leq p < p_0 < \infty$. If $(u, v) \in W_p^*(S)$ and for some nonnegative locally integrable g it holds that $Sg(x)$ is finite a.e., then $(ug^{p-p_0}, v(Sg)^{p-p_0}) \in W_{p_0}^*(S)$ and

$$[(ug^{p-p_0}, v(Sg)^{p-p_0})]_{W_{p_0}^*} \leq [(u, v)]_{W_p^*}^{p/p_0}.$$

The proof is similar to that of Proposition 8.1 but now we define $S^\dagger f(x) = Sg(x)^{-1}S(fg)(x)$.

With this reverse factorization in hand we can prove the following two-weight extrapolation theorem as for the one-weight case (the proof is even simpler since we do not need the Rubio de Francia algorithm).

Theorem 9.1. *Assume that for some* $1 < p_0 < \infty$ *and all* $(u, v) \in W_{p_0}^*$ *it holds that*

$$\|g\|_{p_0,u} \leq C\|f\|_{p_0,v},$$

for some pairs of nonnegative functions (f, g) *with a constant* C *depending only on* $[u, v]_{W_{p_0}^*}$. *Then for* $1 \leq p < p_0$ *and* $(u, v) \in W_p^*$ *the following inequality holds:*

$$\|g\|_{p,u} \leq C\|f\|_{p,v},$$

whenever the left-hand side is finite.

With the same type of reverse factorization as in Proposition 9.1 we can give a simple proof of a theorem of C. Neugebauer in[25] for the extrapolation of weak type inequalities. First we need to define the two-weighted $A_{p,\mathcal{B}}$ classes corresponding to the basis $\mathcal{B}$ as in Section 2. (We use for the class the same notation as in the one-weight case.)

Definition 9.1. We say that $(u, v) \in A_{p,\mathcal{B}}$ for $1 < p < \infty$ if

$$[(u,v)]_{A_{p,\mathcal{B}}} := \left(\frac{u(B)}{|B|}\right)\left(\frac{v^{1-p'}(B)}{|B|}\right)^{p-1} < +\infty,$$

We say that (u, v) is in $A_{1,\mathcal{B}}$ if

$$M_{\mathcal{B}}u(x) \leq Cv(x) \quad \text{a.e.}$$

and we denote by $[(u,v)]_{A_{1,\mathcal{B}}}$ the smallest value of C in the right-hand side.

The following factorization is immediate from the definition of $A_{p,\mathcal{B}}$ and the property of the pairs in $A_{1,\mathcal{B}}$ similar to (2).

Prop 9.2. Let $1 \leq p < p_0 < \infty$. If $(u, v) \in A_{p,\mathcal{B}}$ and for some f it holds that $M_{\mathcal{B}}f(x)$ is finite a.e., then $(u(M_{\mathcal{B}}f)^{p-p_0}, v|f|^{p-p_0}) \in A_{p_0,\mathcal{B}}$.

Using the method of[26] (see also[27]) we obtain the extrapolation theorem of Neugebauer in the setting of Muckenhoupt bases.

Theorem 9.2. *Assume that $M_{\mathcal{B}}$ is of weak type (p,p) for the pair $(u, v) \in A_{p,\mathcal{B}}$. Assume also that for some $p_0 > p$ and all pairs $(u_0, v_0) \in A_{p_0,\mathcal{B}}$ the inequality*

$$\sup_{\lambda>0} \lambda u_0(\{x : g(x) > \lambda\})^{1/p_0} \leq C\|f\|_{p_0,v_0}$$

holds for a family of pairs of nonnegative functions (f, g) with a constant C depending on $[u, v]_{A_{p_0,\mathcal{B}}}$, but not otherwise on u and v. Then it holds that

$$\sup_{\lambda>0} \lambda u(\{x : g(x) > \lambda\})^{1/p} \leq C\|f\|_{p,v}$$

whenever the left-hand side is finite.

Proof. Let $f \in L^p(v)$. We built a weight in $A_{p_0,\mathcal{B}}$ following Proposition 9.2. We have

$$\begin{aligned} u(\{x : g(x) > \lambda\}) \leq u(\{x &: M_{\mathcal{B}}f(x) > \lambda\}) \\ &+ u(\{x : g(x) > \lambda, M_{\mathcal{B}}f(x) \leq \lambda\}). \end{aligned}$$

The first term in the right-hand side satisfies the desired inequality by the assumption on the pair (u, v). As for the second term we have

$$\begin{aligned} u(\{x : g(x) > \lambda, M_{\mathcal{B}}f(x) \leq \lambda\}) &\leq \int_{\{x:g(x)>\lambda\}} u(x)\left(\frac{\lambda}{M_{\mathcal{B}}f(x)}\right)^{p_0-p} dx \\ &\leq \frac{C}{\lambda^p}\int |f(x)|^{p_0} v(x) f(x)^{p-p_0} dx, \end{aligned}$$

as required. □

The operator $M_{\mathcal{B}}$ is of weak type (p,p) for every $(u,v) \in A_{p,\mathcal{B}}$ in many cases. For instance, it holds for the Hardy-Littlewood maximal operator for all $1 \leq p < \infty$. (The result in[25] considers only these weights and the conclusion is not given for $p = 1$.) For the strong maximal operator the assumption holds for $1 < p < \infty$.

References

1. G. H. Hardy and J. E. Littlewood, A maximal theorem with function-theoretic applications, *Acta Math.* **54**, 81 (1930).
2. N. Wiener, The ergodic theorem, *Duke Math. J.* **5**, 1 (1939).
3. J. Duoandikoetxea, *Fourier analysis*, Graduate Studies in Mathematics, Vol. 29 (American Mathematical Society, Providence, RI, 2001), Translated and revised from the 1995 Spanish original by David Cruz-Uribe.
4. J. García-Cuerva and J. L. Rubio de Francia, *Weighted norm inequalities and related topics*, North-Holland Mathematics Studies, Vol. 116 (North-Holland Publishing Co., Amsterdam, 1985), Notas de Matemática [Mathematical Notes], 104.
5. B. Muckenhoupt, Weighted norm inequalities for the Hardy maximal function, *Trans. Amer. Math. Soc.* **165**, 207 (1972).
6. J. L. Rubio de Francia, Factorization and extrapolation of weights, *Bull. Amer. Math. Soc. (N.S.)* **7**, 393 (1982).
7. J. L. Rubio de Francia, Factorization theory and A_p weights, *Amer. J. Math.* **106**, 533 (1984).
8. D. V. Cruz-Uribe, J. M. Martell and C. Pérez, *Weights, extrapolation and the theory of Rubio de Francia*, Operator Theory: Advances and Applications, Vol. 215 (Birkhäuser/Springer Basel AG, Basel, 2011).
9. B. Jawerth, Weighted inequalities for maximal operators: linearization, localization and factorization, *Amer. J. Math.* **108**, 361 (1986).
10. J. Duoandikoetxea, Extrapolation of weights revisited: new proofs and sharp bounds, *J. Funct. Anal.* **260**, 1886 (2011).
11. P. W. Jones, Factorization of A_p weights, *Ann. of Math. (2)* **111**, 511 (1980).
12. R. R. Coifman and R. Rochberg, Another characterization of BMO, *Proc. Amer. Math. Soc.* **79**, 249 (1980).
13. F. Soria, A remark on A_1-weights for the strong maximal function, *Proc. Amer. Math. Soc.* **100**, 46 (1987).
14. J. Duoandikoetxea, F. J. Martín-Reyes and S. Ombrosi, Calderón

weights as Muckenhoupt weights, *Indiana Univ. Math. J.* **62**, 891 (2013).

15. D. Cruz-Uribe and C. Pérez, Two weight extrapolation via the maximal operator, *J. Funct. Anal.* **174**, 1 (2000).
16. E. Harboure, R. A. Macías and C. Segovia, Extrapolation results for classes of weights, *Amer. J. Math.* **110**, 383 (1988).
17. L. Grafakos and J. M. Martell, Extrapolation of weighted norm inequalities for multivariable operators and applications, *J. Geom. Anal.* **14**, 19 (2004).
18. S. M. Buckley, Estimates for operator norms on weighted spaces and reverse Jensen inequalities, *Trans. Amer. Math. Soc.* **340**, 253 (1993).
19. A. K. Lerner, An elementary approach to several results on the Hardy-Littlewood maximal operator, *Proc. Amer. Math. Soc.* **136**, 2829 (2008).
20. T. P. Hytönen, The sharp weighted bound for general Calderón-Zygmund operators, *Ann. of Math. (2)* **175**, 1473 (2012).
21. O. Dragičević, L. Grafakos, M. C. Pereyra and S. Petermichl, Extrapolation and sharp norm estimates for classical operators on weighted Lebesgue spaces, *Publ. Mat.* **49**, 73 (2005).
22. E. Hernández, Factorization and extrapolation of pairs of weights, *Studia Math.* **95**, 179 (1989).
23. F. J. Ruiz and J. L. Torrea, Factorization and extrapolation of pairs of weights in two different measure spaces, *Math. Nachr.* **150**, 175 (1991).
24. E. T. Sawyer, A characterization of a two-weight norm inequality for maximal operators, *Studia Math.* **75**, 1 (1982).
25. C. J. Neugebauer, A double weight extrapolation theorem, *Proc. Amer. Math. Soc.* **93**, 451 (1985).
26. M. J. Carro, J. Soria and R. H. Torres, Rubio de Francia's extrapolation theory: estimates for the distribution function, *J. Lond. Math. Soc. (2)* **85**, 430 (2012).
27. M. J. Carro, J. Duoandikoetxea and M. Lorente, Weighted estimates in a limited range with applications to the Bochner-Riesz operators, *Indiana Univ. Math. J.* **61**, 1485 (2012).
28. L. Grafakos, *Modern Fourier analysis*, Graduate Texts in Mathematics, Vol. 250, second edn. (Springer, New York, 2009).

A note on the off-diagonal Muckenhoupt-Wheeden conjecture

David Cruz-Uribe, SFO

Department of Mathematics, Trinity College,
Hartford, CT 06106-3100, USA
E-mail: david.cruzuribe@trincoll.edu

José María Martell

Instituto de Ciencias Matemáticas CSIC-UAM-UC3M-UCM,
Consejo Superior de Investigaciones Científicas,
C/ Nicolás Cabrera, 13-15, E-28049 Madrid, Spain,
E-mail: chema.martell@icmat.es

Carlos Pérez

Departamento de Análisis Matemático,
Facultad de Matemáticas, Universidad de Sevilla,
41012 Sevilla, Spain
E-mail: carlosperez@us.es

We obtain the off-diagonal Muckenhoupt-Wheeden conjecture for Calderón-Zygmund operators. Namely, given $1 < p < q < \infty$ and a pair of weights (u, v), if the Hardy-Littlewood maximal function satisfies the following two weight inequalities:

$$M : L^p(v) \to L^q(u) \quad \text{and} \quad M : L^{q'}(u^{1-q'}) \to L^{p'}(v^{1-p'}),$$

then any Calderón-Zygmund operator T and its associated truncated maximal operator $T_\star$ are bounded from $L^p(v)$ to $L^q(u)$. Additionally, assuming only the second estimate for M then T and $T_\star$ map continuously $L^p(v)$ into $L^{q,\infty}(u)$. We also consider the case of generalized Haar shift operators and show that their off-diagonal two weight estimates are governed by the corresponding estimates for the dyadic Hardy-Littlewood maximal function.

Keywords: Haar shift operators; Calderón-Zygmund operators; two-weight inequalities; testing conditions.

The first author was supported by a grant from the Faculty Research Committee and the Stewart-Dorwart Faculty Development Fund at Trinity College; the first and third authors are supported by grant MTM2009-08934 from the Spanish Ministry of Science and Innovation; the second author is supported by grant Grant MTM2010-16518 from the same institution

2000 Mathematics Subject Classification: 42B20, 42B25.

1. Introduction and Main results

In the 1970s, Muckenhoupt and Wheeden conjectured that given p, $1 < p < \infty$, a sufficient condition for the Hilbert transform to satisfy the two weight norm inequality

$$H : L^p(v) \to L^p(u)$$

is that the Hardy-Littlewood maximal operator satisfy the pair of norm inequalities

$$M : L^p(v) \to L^p(u),$$
$$M : L^{p'}(u^{1-p'}) \to L^{p'}(v^{1-p'}).$$

Moreover, they conjectured that the Hilbert transform satisfies the weak-type inequality

$$H : L^p(v) \to L^{p,\infty}(u)$$

provided that the maximal operator satisfies the second "dual" inequality. Both of these conjectures readily extend to all Calderón-Zygmund operators (see the definition below). Very recently, both conjectures were disproved: the strong-type inequality by Reguera and Scurry[1] and the weak-type inequality by the first author, Reznikov and Volberg[2].

Remark 1.1. A special case of these conjectures, involving the A_p bump conditions, has been considered by several authors: see[2–7].

In this note we prove the somewhat surprising fact that the Muckenhoupt-Wheeden conjectures are true for off-diagonal inequalities. Our main result is Theorem 1.1 below. We also prove an analogous result for the Haar shift operators (the so-called dyadic Calderón-Zygmund operators) with the Hardy-Littlewood maximal operator replaced by the dyadic maximal operator: see Theorem 1.2 below.

To state our results we first give some preliminary definitions. By weights we will always mean non-negative, measurable functions. Given a pair of weights (u, v), hereafter we will assume that $u > 0$ on a set of positive measure and $u < \infty$ a.e., and $v > 0$ a.e. and $v < \infty$ on a set of positive measure. We will also use the standard notation $0 \cdot \infty = 0$.

Calderón-Zygmund operators

A Calderón-Zygmund operator T is a linear operator that is bounded on $L^2(\mathbb{R}^n)$ and

$$Tf(x) = \int_{\mathbb{R}^n} K(x,y)f(y)dy, \qquad f \in L_c^\infty(\mathbb{R}^n), \quad x \notin \operatorname{supp} f,$$

where the kernel K satisfies the size and smoothness estimates

$$|K(x,y)| \le \frac{C}{|x-y|^n}, \qquad x \neq y,$$

and

$$|K(x,y) - K(x',y)| + |K(y,x) - K(y,x')| \le C\frac{|x-x'|^\delta}{|x-y|^{n+\delta}},$$

for all $|x-y| > 2|x-x'|$.

Associated with T is the truncated maximal operator

$$T_\star f(x) = \sup_{0<\varepsilon<\varepsilon'<\infty} \left| \int_{\varepsilon<|x-y|<\varepsilon'} K(x,y)f(y)dy \right|.$$

Let M denote the Hardy-Littlewood maximal operator, that is,

$$Mf(x) = \sup_{Q\ni x} \fint_Q |f(y)|dy = \sup_{Q\ni x} \frac{1}{|Q|}\int_Q |f(y)|dy.$$

where the supremum is taken over all cubes in $\mathbb{R}^n$ with sides parallel to the coordinate axes.

Theorem 1.1. *Given a Calderón-Zygmund operator T, let $1 < p < q < \infty$ and let (u,v) be a pair of weights. If the maximal operator satisfies*

$$M : L^p(v) \to L^q(u) \quad \text{and} \quad M : L^{q'}(u^{1-q'}) \to L^{p'}(v^{1-p'}), \tag{1}$$

then

$$\|Tf\|_{L^q(u)} \leqslant C\|f\|_{L^p(v)} \quad \text{and} \quad \|T_\star f\|_{L^q(u)} \leqslant C\|f\|_{L^p(v)}. \tag{2}$$

Analogously, if the maximal operator satisfies

$$M : L^{q'}(u^{1-q'}) \to L^{p'}(v^{1-p'}), \tag{3}$$

then

$$\|Tf\|_{L^{q,\infty}(u)} \leqslant C\|f\|_{L^p(v)} \quad \text{and} \quad \|T_\star f\|_{L^{q,\infty}(u)} \leqslant C\|f\|_{L^p(v)}. \tag{4}$$

If the pairs of weights (u, v) satisfy any of the conditions in (1), then the weights u and $v^{1-p'}$ are locally integrable. This is a consequence of a characterization of the two weight norm inequalities for the maximal operator due to Sawyer[8]. He proved that the $L^p - L^q$ inequality holds if and only if for every cube Q,

$$\left(\int_Q M(v^{1-p'}\chi_Q)(x)^q u(x)\,dx\right)^{1/q} \leqslant C\left(\int_Q v(x)^{1-p'}\,dx\right)^{1/p} < \infty,$$

and the $L^{q'} - L^{p'}$ inequality holds if and only if

$$\left(\int_Q M(u\chi_Q)(x)^{p'} v(x)^{1-p'}\,dx\right)^{1/p'} \leqslant C\left(\int_Q u(x)\,dx\right)^{1/q'} < \infty.$$

It is straightforward to construct pairs of weights that satisfy these conditions. For instance, in $\mathbb{R}$ both of these conditions follow easily for every $1 < p \leq q < \infty$ and the pair of weights (u, v) with $u = \chi_{[0,1]}$ and $v^{-1} = \chi_{[2,3]}$ (i.e., $v = 1$ in $[2, 3]$ and $v = \infty$ elsewhere). Indeed, we only need to check Sawyer's inequalities for cubes Q that intersect both $[0, 1]$ and $[2, 3]$, in which case we have $M(\chi_{[2,3]\cap Q})(x) \leq |[2, 3] \cap Q|$ for every $x \in [0, 1] \cap Q$, and $M(\chi_{[0,1]\cap Q})(x) \leq |[0, 1] \cap Q|$ for every $x \in [2, 3] \cap Q$. These readily imply the desired estimates.

Dyadic Calderón-Zygmund operators

A generalized dyadic grid $\mathscr{D}$ in $\mathbb{R}^n$ is a set of generalized dyadic cubes with the following properties: if $Q \in \mathscr{D}$ then $\ell(Q) = 2^k$, $k \in \mathbb{Z}$; if $Q, R \in \mathscr{D}$ and $Q \cap R \neq \emptyset$ then $Q \subset R$ or $R \subset Q$; the cubes in $\mathscr{D}$ with $\ell(Q) = 2^{-k}$ form a disjoint partition of $\mathbb{R}^n$ (see[7] and[9] for more details).

We say that g_Q is a generalized Haar function associated with $Q \in \mathscr{D}$ if

(*a*) $\operatorname{supp}(g_Q) \subset Q$;

(*b*) if $Q' \in \mathscr{D}$ and $Q' \subsetneq Q$, then g_Q is constant on Q';

(*c*) $\|g_Q\|_\infty \leqslant 1$.

Given a dyadic grid $\mathscr{D}$ and a pair $(m, k) \in \mathbb{Z}_+^2$, a linear operator $\mathcal{S}$ is a generalized Haar shift operator (that is, a dyadic Calderón-Zygmund operator) of complexity type (m, k) if it is bounded on $L^2(\mathbb{R}^n)$ and

$$\mathcal{S}f(x) = \sum_{Q\in\mathscr{D}} \mathcal{S}_Q f(x) = \sum_{Q\in\mathscr{D}} \sum_{\substack{Q'\in\mathscr{D}_m(Q)\\ Q''\in\mathscr{D}_k(Q)}} \frac{\langle f, g_{Q'}^{Q''}\rangle}{|Q|} g_{Q''}^{Q'}(x),$$

where $\mathscr{D}_j(Q)$ stands for the dyadic subcubes of Q with side length $2^{-j}\ell(Q)$, $g_{Q'}^{Q''}$ is a generalized a Haar function associated with Q' and $g_{Q''}^{Q'}$ is a generalized a Haar function associated with Q''. We say that the complexity of $\mathcal{S}$ is $\kappa = \max(m,k)$. We also define the truncated Haar shift operator

$$\mathcal{S}_\star f(x) = \sup_{0<\varepsilon<\varepsilon'<\infty} |\mathcal{S}_{\varepsilon,\varepsilon'} f(x)| = \sup_{0<\varepsilon<\varepsilon'<\infty} \Big| \sum_{\substack{Q\in\mathscr{D} \\ \varepsilon\leq\ell(Q)\leq\varepsilon'}} \mathcal{S}_Q f(x) \Big|.$$

An important example of a Haar shift operator on the real line is the Haar shift (also known as the dyadic Hilbert transform) H^d, defined by

$$H^d f(x) = \sum_{I\in\Delta} \langle f, h_I\rangle \big(h_{I_-}(x) - h_{I_+}(x)\big),$$

where, given a dyadic interval I, I_+ and I_- are its right and left halves, and

$$h_I(x) = |I|^{-1/2}\big(\chi_{I_-}(x) - \chi_{I_+}(x)\big).$$

After renormalizing, h_I is a Haar function on I and one can write H^d as a generalized Haar shift operator of complexity 1. These operators have played a very important role in the proof of the A_2 conjecture: see[6,10,11] and the references they contain for more information.

Associated with the dyadic grid $\mathscr{D}$ is the dyadic maximal function

$$M_{\mathscr{D}} f(x) = \sup_{x\in Q\in\mathscr{D}} \fint_Q |f(y)|dy.$$

Note that $M_{\mathscr{D}}$ is dominated pointwise by the Hardy-Littlewood maximal operator.

We can now state our result for dyadic Calderón-Zygmund operators.

Theorem 1.2. *Let $\mathcal{S}$ be a generalized Haar shift operator of complexity κ. Given $1 < p < q < \infty$ and a pair of weights (u,v), if the dyadic maximal operator satisfies*

$$M_{\mathscr{D}} : L^p(v) \to L^q(u) \quad \text{and} \quad M_{\mathscr{D}} : L^{q'}(u^{1-q'}) \to L^{p'}(v^{1-p'}), \tag{5}$$

then

$$\|\mathcal{S}f\|_{L^q(u)} \leqslant C\kappa^2 \|f\|_{L^p(v)} \quad \text{and} \quad \|\mathcal{S}_\star f\|_{L^q(u)} \leqslant C\kappa^2 \|f\|_{L^p(v)}. \tag{6}$$

Analogously, if the dyadic maximal operator satisfies

$$M_{\mathscr{D}} : L^{q'}(u^{1-q'}) \to L^{p'}(v^{1-p'}) \tag{7}$$

then

$$\|\mathcal{S}f\|_{L^{q,\infty}(u)} \leqslant C\kappa^2 \|f\|_{L^p(v)} \quad \text{and} \quad \|S_\star f\|_{L^{q,\infty}(u)} \leqslant C\kappa^2 \|f\|_{L^p(v)}. \tag{8}$$

2. Proofs of the Main results

Proof of Theorem 1.1

We will prove our estimates for $T_\star$; the ones for T are completely analogous.

Given a dyadic grid $\mathscr{D}$ we say that $\{Q_j^k\}_{j,k}$ is a *sparse family* of dyadic cubes if for any k the cubes $\{Q_j^k\}_j$ are pairwise disjoint; if $\Omega_k := \cup_j Q_j^k$, then $\Omega_{k+1} \subset \Omega_k$; and $|\Omega_{k+1} \cap Q_{j,k}| \leq \frac{1}{2}|Q_j^k|$. Given $\mathscr{D}$ and a sparse family $\mathscr{S} = \{Q_j^k\}_{j,k} \subset \mathscr{D}$, define the positive dyadic operator $\mathscr{A}$ by

$$\mathscr{A} f(x) = \mathscr{A}_{\mathscr{D},\mathscr{S}} f(x) = \sum_{j,k} f_{Q_j^k} \chi_{Q_j^k}(x)$$

where $f_Q = \fint_Q f(y)dy$.

For our proof we will use the main result in [7,9]. Given a Banach function space X and a non-negative function f,

$$\|T_\star f\|_X \leq C(T,n) \sup_{\mathscr{D},\mathscr{S}} \|\mathscr{A}_{\mathscr{D},\mathscr{S}} f\|_X, \tag{1}$$

where the supremum is taken over all dyadic grids $\mathscr{D}$ and sparse families $\mathscr{S} \subset \mathscr{D}$. To prove Theorem 1.1 we apply this result with $X = L^q(u)$ or $X = L^{q,\infty}(u)$; it will then suffice to show that our assumptions on M guarantee that $\mathscr{A}_{\mathscr{D},\mathscr{S}}$ satisfies the corresponding two weight inequalities.

To prove this fact we will use a result by Lacey, Sawyer and Uriarte-Tuero [12]. Given a sequence of non-negative constants $\alpha = \{\alpha_Q\}_{Q\in\mathscr{D}}$, define the positive operator

$$T_\alpha f(x) = \sum_{Q\in\mathscr{D}} \alpha_Q f_Q \chi_Q(x).$$

Further, given $R \in \mathscr{D}$ we define the "outer truncated" operator

$$T_\alpha^R f(x) = \sum_{\substack{Q\in\mathscr{D}\\ Q\supset R}} \alpha_Q f_Q \chi_Q(x).$$

In [12] it was shown that for all $1 < p < q < \infty$, $T_\alpha : L^p(v) \to L^q(u)$ if and only if there exist constants C_1 and C_2 such that for every $R \in \mathscr{D}$

$$\left(\int_{\mathbb{R}^n} T_\alpha^R(v^{1-p'}\chi_R)(x)^q u(x)dx\right)^{\frac{1}{q}} \leq C_1 \left(\int_R v(x)^{1-p'}dx\right)^{\frac{1}{p}}, \tag{2}$$

and

$$\left(\int_{\mathbb{R}^n} T_\alpha^R(u\chi_R)(x)^{p'} v(x)^{1-p'}dx\right)^{\frac{1}{p'}} \leq C_2 \left(\int_R u(x)dx\right)^{\frac{1}{q'}}. \tag{3}$$

Furthermore, for $1 < p < q < \infty$, $T_\alpha : L^p(v) \to L^{q,\infty}(u)$ holds if and only if there exists a constant C_2 such that for every $R \in \mathscr{D}$, (3) holds.

We can apply these results to the operator $\mathscr{A} = \mathscr{A}_{\mathscr{D},\mathscr{S}}$ where $\mathscr{D}$ and $\mathscr{S}$ are fixed, since $\mathscr{A} = T_\alpha$ with $\alpha_Q = 1$ if $Q \in \mathscr{S}$ and $\alpha_Q = 0$ otherwise. Fix $R \in \mathscr{D}$; to estimate $\mathscr{A}^R$, take the increasing family of cubes $R = R_0 \subsetneq R_1 \subsetneq R_2 \subsetneq \ldots$ with $R_k \in \mathscr{D}$ and $\ell(R_k) = 2^k \ell(R)$. Define $R_{-1} = \emptyset$. Note that $\operatorname{supp} \mathscr{A}^R \subset \cup_{k\geq 0} R_k$. Then for every non-negative function f and for every $x \in R_k \setminus R_{k-1}$ with $k \geq 0$ we have that

$$\begin{aligned} 0 \leq \mathscr{A}^R(f\chi_R)(x) \leq \sum_{j=0}^{\infty} (f\chi_R)_{R_j} \chi_{R_j}(x) = f_R \sum_{j=k}^{\infty} 2^{-j\,n} \\ \lesssim f_R 2^{-k\,n} = (f\chi_R)_{R_k} \leq M_{\mathscr{D}}(f\chi_R)(x). \end{aligned}$$

Consequently, for every $x \in \mathbb{R}^n$,

$$0 \leq \mathscr{A}^R(f\chi_R)(x) \lesssim M_{\mathscr{D}}(f\chi_R)(x) \leqslant M(f\chi_R)(x). \tag{4}$$

Inequality (4) together with our hypothesis (1) implies (2) and (3). Therefore, we have that $\mathscr{A} : L^p(v) \to L^q(u)$ with constants depending on the dimension, p, q and the implicit constants in (1). Therefore, by Lerner's estimate (1) we get $T_\star : L^p(v) \to L^q(u)$ as desired.

For the weak-type estimates we proceed in the same manner, using the fact that (3) yields (3) and therefore $\mathscr{A} : L^p(v) \to L^{q,\infty}(u)$. This in turn implies, by Lerner's estimate (1) applied to $X = L^{q,\infty}(u)$, that $T_\star : L^p(v) \to L^{q,\infty}(u)$.

Proof of Theorem 1.2

Fix $\mathscr{D}$ and a generalized Haar shift operator of complexity κ. As before we can work with $\mathcal{S}_\star$. We can repeat the previous argument except that we want to keep the fixed dyadic structure $\mathscr{D}$. A careful examination of[7] Section 5 shows that, given a Banach function space X, we have

$$\|\mathcal{S}_\star f\|_X \leq C_n \kappa^2 \sup_{\mathscr{S}} \|\mathscr{A}_{\mathscr{D},\mathscr{S}} f\|_X, \qquad f \geq 0, \tag{5}$$

where the supremum is taken over all sparse families $\mathscr{S} \subset \mathscr{D}$. We emphasize that in[7] Section 5 there is an additional supremum over the dyadic grids $\mathscr{D}$. This is because at some places the dyadic maximal operator is majorized by the regular Hardy-Littlewood maximal operator and the latter is in turn controlled by a sum of $\mathscr{A}_{\mathscr{D}_\alpha,\mathscr{S}_\alpha}$ for 2^n dyadic grids $\mathscr{D}_\alpha$. However, keeping $M_{\mathscr{D}}$ one can easily show that (5) holds. Details are left to the interested reader.

Given (5), we fix a sparse family $\mathscr{S} \subset \mathscr{D}$ and write $\mathscr{A} = \mathscr{A}_{\mathscr{D},\mathscr{S}}$. Arguing exactly as before we obtain (4). Thus, (5) implies (2) and (3) and therefore the result from[12] yields $\mathscr{A} : L^p(v) \to L^q(u)$ with constants depending on the dimension, p, q and the implicit constants in (5). Combining this with Lerner's estimate (5) applied to $X = L^q(u)$ we conclude as desired that $\mathcal{S}_\star : L^p(v) \to L^q(u)$. We get the weak-type estimate by adapting the above proof in exactly the same way.

References

1. M. C. Reguera and J. Scurry, On joint estimates for maximal functions and singular integrals on weighted spaces, *Proc. Amer. Math. Soc.* **141**, 1705 (2013).
2. D. Cruz-Uribe, A. Reznikov and A. Volberg, Logarithmic bump conditions and the two-weight boundedness of Calderón-Zygmund operators, *Adv. Math.* **255**, 706 (2014).
3. D. Cruz-Uribe, J. M. Martell and C. Pérez, Sharp two-weight inequalities for singular integrals, with applications to the Hilbert transform and the Sarason conjecture, *Adv. Math.* **216**, 647 (2007).
4. D. Cruz-Uribe, J. M. Martell and C. Pérez, Sharp weighted estimates for approximating dyadic operators, *Electron. Res. Announc. Math. Sci.* **17**, 12 (2010).
5. D. Cruz-Uribe, J. Martell and C. Pérez, *Weights, extrapolation and the theory of Rubio de Francia*, Operator Theory: Advances and Applications, Vol. 215 (Birkhäuser/Springer Basel AG, Basel, 2011).
6. D. Cruz-Uribe, J. Martell and C. Pérez, Sharp weighted estimates for classical operators, *Adv. in Math* **229**, 408 (2012).
7. A. K. Lerner, On an estimate of Calderón-Zygmund operators by dyadic positive operators, *J. Anal. Math.* **121**, 141 (2013).
8. E. Sawyer, A characterization of a two-weight norm inequality for maximal operators, *Studia Math.* **75**, 1 (1982).
9. A. K. Lerner, A simple proof of the A_2 conjecture, *Int. Math. Res. Not. IMRN* **14**, 3159 (2013).
10. T. P. Hytönen, The sharp weighted bound for general Calderón-Zygmund operators, *Ann. of Math. (2)* **175**, 1473 (2012).
11. T. Hytönen, C. Pérez, S. Treil and A. Volberg, Sharp weighted estimates for dyadic shifts and the A_2 conjecture, *J. Reine Angew. Math.* **687**, 43 (2014).
12. M. T. Lacey, E. T. Sawyer, C.-Y. Shen and I. Uriarte-Tuero,

Two-weight inequality for the Hilbert transform: a real variable characterization, I, *Duke Math. J.* **163**, 2795 (2014).

On the interplay between nonlinear partial differential equations and game theory

Julio D. Rossi

Departamento de Análisis Matemático, Universidad de Alicante,
Ap. correos 99, 03080 Alicante, Spain
On leave from Departamento de Matemática, FCEyN UBA,
Ciudad Universitaria, Pab 1, (1428),
Buenos Aires, Argentina
E-mail: jrossi@dm.uba.ar

We review some recent results concerning game theory and their relation to some well known PDEs. In particular, we will show that solutions to certain PDEs can be obtained as limits of values of tug-of-war games when the parameter that controls the length of the possible movements goes to zero.

Keywords: Viscosity solutions; p-Laplacian; infinity Laplacian; absolutely minimizing Lipschitz extension; game theory.

1. Description of the contents

The goal of this paper is to introduce the reader (expert or not) to some techniques and results relevant for the theory of second order elliptic partial differential equations (PDEs) and their connections with game theory.

Our aim is to explain through examples a way in which PDEs arise in Probability. For instance, we show how simple is the relation between probabilistic issues on random walks and the Laplace operator and also other elliptic operators, as well as the heat equation. Next, we will enter in what is the core of these notes, the approximation by means of values of games of solutions to nonlinear problems like $p-$harmonic functions, that is, solutions to the PDE, $\text{div}(|\nabla u|^{p-2}\nabla u) = 0$ (including the nowadays popular case $p = \infty$).

We will assume that the reader is familiar with basic tools from probability theory (like conditional expectations) and with the (not so basic) concept of viscosity solutions for second order elliptic and parabolic PDEs (we refer to the book[1] for this last issue).

Partially supported by UBA X066 and CONICET, Argentina.

To keep a reasonable size of this article we do not provide full proofs of the results but include ideas and formal procedures and refer to the literature for details. The References does not escape the usual rule of being incomplete. In general, we have listed those papers which are closer to the topics discussed here (some of them are not explicitly cited in the text), but, even for those papers, the list is far from being exhaustive and we apologize for omissions.

2. Linear PDEs and probability

2.1. *The probability of hitting the exit and harmonic functions*

Let us begin by considering a bounded and smooth two-dimensional domain $\Omega \subset \mathbb{R}^2$ and assume that the boundary, $\partial\Omega$ is decomposed in two parts, Γ_1 and Γ_2 (that is, $\Gamma_1 \cup \Gamma_2 = \partial\Omega$ with $\Gamma_1 \cap \Gamma_2 = \emptyset$). We begin with a position $(x, y) \in \Omega$ and ask the following question: assume that you move completely at random beginning at (x, y) (we assume that we are in an homogeneous environment and that we do not privilege any direction. In addition, we assume that every time the particle moves independently of its past history) what is the probability $u(x, y)$ of hitting the first part of the boundary Γ_1 the first time that the particle hits the boundary ?.

We will call Γ_1 the "open part" of the boundary and think that when we hit this part we can "exit" the domain, while we will call Γ_2 the "closed part" of the boundary.

A simple way to get some insight to solve the question runs as follows: First, we simplify the problem and approximate the movement by random increments of step h in each of the axes directions, with $h > 0$ small. From (x, y) the particle can move to $(x + h, y)$, $(x - h, y)$, $(x, y + h)$, or $(x, y - h)$, each movement being chosen at random with probability $1/4$. Starting at (x, y), let $u_h(x, y)$ be the probability of hitting the exit part $\Gamma_1 + B_\delta(0)$ the first time that $\partial\Omega + B_\delta(0)$ is hitt when we move on the lattice of side h. Observe that we need to enlarge a little the boundary to capture points on the lattice of size h (that do not necessarily lie on $\partial\Omega$). Applying conditional expectations we get

$$u_h(x,y) = \frac{1}{4}u_h(x+h,y)+\frac{1}{4}u_h(x-h,y)+\frac{1}{4}u_h(x,y+h)+\frac{1}{4}u_h(x,y-h). \quad (1)$$

That is,

$$0 = \{u_h(x+h,y) - 2u_h(x,y) + u_h(x-h,y)\}$$

$$+ \{u_h(x, y+h) - 2u_h(x, y) + u_h(x, y-h)\}. \quad (2)$$

Now, assume that u_h converges as $h \to 0$ to a function u uniformly in $\overline{\Omega}$. Note that this convergence can be proved rigorously.

It can be proved, see[2] for details, that a uniform limit of the approximate values u_h, u, has the following property: *each time that a smooth function ϕ touches u from below at a point (x_0, y_0) the derivatives of ϕ must satisfy,*

$$0 \geqslant \frac{\partial^2 \phi}{\partial x^2}(x_0, y_0) + \frac{\partial^2 \phi}{\partial y^2}(x_0, y_0).$$

An analogous argument considering ψ a smooth function such that $u - \psi$ has a strict maximum at $(x_0, y_0) \in \Omega$ shows a reverse inequality. Therefore, *each time that a smooth function ψ touches u from above at a point (x_0, y_0) the derivatives of ψ must verify*

$$0 \leqslant \frac{\partial^2 \psi}{\partial x^2}(x_0, y_0) + \frac{\partial^2 \psi}{\partial y^2}(x_0, y_0).$$

But at this point we realize that this is exactly the definition of being u a **viscosity solution to the Laplace equation**

$$\Delta u = \frac{\partial^2 u}{\partial x^2} + \frac{\partial^2 u}{\partial y^2} = 0.$$

Hence, we obtain that the uniform limit of the sequence of solutions to the approximated problems u_h, u is the unique viscosity solution (that is also a classical solution in this case) to the following boundary value problem

$$\begin{cases} -\Delta u = 0 & \text{in } \Omega, \\ u = 1 & \text{on } \Gamma_1, \\ u = 0 & \text{on } \Gamma_2. \end{cases} \quad (3)$$

The boundary conditions can be easily obtained from the fact that $u_h \equiv 1$ in a neighborhood (of width h) of Γ_1 and $u_h \equiv 0$ in a neighborhood of Γ_2.

Note that we have only required *uniform* convergence to get the result, and hence no requirement is made on derivatives of the approximating sequence u_h. Moreover, we do not assume that u_h is continuous.

Now, we just notice that in higher dimensions $\Omega \subset \mathbb{R}^N$, the discretization method described above leads in the same simple way to viscosity solutions to the Laplace operator in higher dimensions and then to the fact that limits of exiting probabilities are harmonic functions.

Another way to understand this strong relation between probabilities and the Laplacian is through the *mean value property of harmonic functions.* In the same context of the problem solved above, assume that a closed ball $B_r(x_0, y_0)$ of radius r and centered at a point (x_0, y_0) is contained in Ω. Starting at (x_0, y_0), the probability density of hitting first a given point on the sphere $\partial B_r(x_0, y_0)$ is constant on the sphere, that is, it is uniformly distributed on the sphere. Therefore, the probability $u(x_0, y_0)$ of exiting through Γ_1 starting at (x_0, y_0) is the average of the exit probabilities u on the sphere, here we are using again the formula of conditional probabilities. That is, u satisfies the mean value property on spheres:

$$u(x_0, y_0) = \frac{1}{|\partial B_r(x_0, y_0)|} \int_{\partial B_r(x_0,y_0)} u(x, y)\, dS(x, y)$$

with r small enough. It is well known that this property leads to u being harmonic.

We can also say that, if the movement is completely random and equidistributed in the ball $B_h(x_0, y_0)$, then, by the same conditional expectation argument used before, we have

$$u_h(x_0, y_0) = \frac{1}{|B_h(x_0, y_0)|} \int_{B_h(x_0,y_0)} u_h(x, y)\, dx\, dy.$$

Again one can take the limit as $h \to 0$ and obtain that a uniform the limit of the u_h, u, is harmonic (in the viscosity sense).

2.2. *Counting the number of steps needed to reach the exit*

Another motivating problem is the following: with the same setting as before (Ω a bounded smooth domain in $\mathbb{R}^2$) we would like to compute the expected time, that we call T, that we have to spend starting at (x, y) before hitting the boundary $\partial\Omega$ for the first time.

We can proceed exactly as before computing the time of the random walk at the discrete level, that is, in the lattice of size h. This amounts to adding a constant (the unit of time that we spend in each movement), which depends on the step h, to the right hand side of (1). We have

$$T_h(x, y) = \frac{1}{4}T_h(x+h, y) + \frac{1}{4}T_h(x-h, y) + \frac{1}{4}T_h(x, y+h) + \frac{1}{4}T_h(x, y-h) + t(h).$$

That is,

$$0 = \{T_h(x+h,y) - 2T_h(x,y) + T_h(x-h,y)\}$$
$$+ \{T_h(x,y+h) - 2T_h(x,y) + T_h(x,y-h)\} + t(h).$$

Proceeding as we did before, and since we need to divide by h^2, a natural choice is to set that $t(h)$ is of order h^2. Choosing $t(h) = Kh^2$ and letting $h \to 0$, we conclude that a uniform limit of the approximate solutions T_h, T is the unique solution to

$$\begin{cases} -\Delta T = 4K & \text{in } \Omega, \\ T = 0 & \text{on } \partial\Omega. \end{cases} \tag{4}$$

The boundary condition is natural since if we begin on the boundary the expected time needed to reach it is zero.

From the previous probabilistic interpretations for the solutions of problems (3) and (4) as limits of random walks, one can imagine a probabilistic model for which the solution of the limit process is a solution to the general Poisson problem

$$\begin{cases} -\Delta u(x) = g(x) & \text{in } \Omega, \\ u(x) = F(x) & \text{on } \partial\Omega. \end{cases} \tag{5}$$

In this general model the functions g and F can be thought as costs that one pays, respectively, along the random movement and at the stopping time on the boundary.

2.3. *Anisotropic media*

Suppose now that the medium in which we perform our random movements is neither isotropic (that is, it is directionally dependent) nor homogeneous (that is, it differs from one point to another).

We can imagine a random discrete movement as follows. We move from a point (x, y) to four possible points at distance h located at two orthogonal axis forming a given angle α with the horizontal, and with different probabilities $q/2$ for the two points on the first axis and $(1-q)/2$ for the two points on the second axis. The angle α (that measures the orientation of the axis) and the probability q depend on the point (x, y). After the same analysis as above, we encounter now the general elliptic equation

$$\sum_{ij} a_{ij}(x,y) u_{x_i x_j}(x,y) = 0.$$

2.4. *The heat equation*

Now assume that we are on the real line $\mathbb{R}$ and consider that we move in a two dimensional lattice as follows: when we are at the point (x_0, t_0), the time increases by $\delta t := h^2$ and the spacial position moves with an increment of size $\delta x = h$ and goes to $x_0 - h$ or to $x_0 + h$ with the same probability. In this way the new points in the lattice that can be reached starting from (x_0, t_0) are $(x_0 - h, t_0 + h^2)$ or to $(x_0 + h, t_0 + h^2)$, each one with probability $1/2$. The choice $\delta t = (\delta x)^2$ is made to ensure that a certain limit as $h \to 0$ exists. Let us start at $x = 0$, $t = 0$ and let $u_h(x,t)$ be the probability that we are at x at time t (here $x = kh$ and $t = lh^2$ is a point in the two dimensional lattice). Conditional probabilities give the identity

$$u_h(x,t) = \frac{1}{2} u_h(x - h, t - h^2) + \frac{1}{2} u_h(x + h, t - h^2).$$

That is,

$$\frac{u_h(x,t) - u_h(x, t - h^2)}{h^2} = \frac{1}{h^2} \Big\{ \frac{1}{2} u_h(x - h, t - h^2) + \frac{1}{2} u_h(x + h, t - h^2) - u_h(x, t - h^2) \Big\}.$$

Now, as before, we let $h \to 0$ and, assuming uniform convergence (that can be proved !), we arrive to the fact that the limit should be a viscosity solution to

$$u_t(x,t) = \frac{1}{2} u_{xx}(x,t).$$

It is at this point where the relation $\delta t = (\delta x)^2$ is needed.

2.5. *Comments*

(*a*) We want to remark that two facts are crucial in the previous analysis. The first one is the formula for the discrete version of the problem obtained using conditional expectations, (1), and the second one is the use of the theory of viscosity solutions to perform the passage to the limit without asking for more than uniform convergence.

(*b*) One can show the required uniform convergence from the following two facts: first, the values u_h are uniformly bounded (they all lie between 0 and 1 in the case of the problem of exiting the domain through Γ_1

and by the maximum and the minimum of the datum F in the general case assuming $g \equiv 0$) and second the family u_h is equicontinuous for "not too close points" (see the following sections), this can be proved using coupling methods. In fact, one can mimic a path of the game starting at x but starting at y and when one of the two paths hits the boundary the other is at a position that is close (at a distance smaller than $|x - y|$) of the boundary. It remains to prove a uniform in h estimate that says that "given $\varepsilon > 0$ there exists $\delta > 0$ such that if one begins close to the boundary Γ_1 (with $\text{dist}(x_0, \Gamma_1) < \delta$) then the probability of hitting this part of the boundary is bounded below by $1 - \varepsilon$" (and of course an analogous statement for positions that start close to Γ_2). For this argument to work in the general case one can impose that F is uniformly continuous.

(*c*) The initial condition for the heat equation in subsection 2.4 is $u(x, 0) = \delta_0$ and the solution can be explicitly obtained as the Gaussian

$$u(x,t) = \frac{1}{\sqrt{2\pi t}} e^{-\frac{x^2}{2t}}.$$

For an initial distribution of particles given by $u(x, 0) = u_0(x)$ we get

$$u(x,t) = \frac{1}{\sqrt{2\pi t}} e^{-\frac{x^2}{2t}} * u_0(x).$$

(*d*) Also note that the identities obtained using conditional expectations are the same that correspond to the discretization of the equations using finite differences. This gives a well-known second order numerical scheme for the Laplacian.
Hence, we remark that probability theory can be used to obtain numerical schemes for PDEs.

(*e*) Finally, note that when a general Poisson problem is considered in (5) the functions g and F that appear can be thought as costs that one pays. The first one is a *running cost* that is payed at each movement, while the second one is a *final cost* that is payed when the game ends reaching the boundary.

3. Tug-of-war games and the ∞–Laplacian

In this section we will look for a probabilistic approach to approximate solutions to the ∞–Laplacian. This is the nonlinear degenerate elliptic operator, usually denoted by Δ_∞, given by,

$$\Delta_\infty u := \big(D^2u\,\nabla u\big)\cdot\nabla u = \sum_{i,j=1}^{N} \frac{\partial u}{\partial x_i}\frac{\partial u}{\partial x_j}\frac{\partial^2 u}{\partial x_i x_j},$$

and arises from taking limit as $p\to\infty$ in the p-Laplacian operator in the viscosity sense, see[3] and[4]. In fact, let us present a formal derivation. First, expand (formally) the $p-$laplacian:

$$\begin{aligned}\Delta_p u &= \mathrm{div}\left(|\nabla u|^{p-2}\nabla u\right) = |\nabla u|^{p-2}\Delta u + (p-2)|\nabla u|^{p-4}\sum_{i,j} u_{x_i}u_{x_j}u_{x_i,x_j}\\ &= (p-2)|\nabla u|^{p-4}\Big\{\frac{1}{p-2}|\nabla u|^2\Delta u + \sum_{i,j} u_{x_i}u_{x_j}u_{x_i,x_j}\Big\}\end{aligned}$$

and next, using this formal expansion, pass to the limit in the equation $\Delta_p u = 0$, to obtain

$$\Delta_\infty u = \sum_{i,j} u_{x_i}u_{x_j}u_{x_i,x_j} = Du\cdot D^2u\cdot(Du)^t = 0.$$

Note that this calculation can be made rigorous in the viscosity sense, see[4].

The ∞-laplacian operator appears naturally when one considers absolutely minimizing Lipschitz extensions of a boundary function F; see[5] and also the survey[3]. A fundamental result of Jensen[5] establishes that the Dirichlet problem for Δ_∞ is well posed in the viscosity sense. Solutions to $-\Delta_\infty u = 0$ (that are called infinity harmonic functions) are also used in several applications, for instance, in optimal transportation and image processing (see, e.g.,[6],[7] and the references therein). Also the eigenvalue problem related to the ∞-laplacian has been exhaustively studied, see[8],[9],[10],[11].

Our main aim in this section is to describe a game that approximates solutions to the problem

$$\begin{cases} -\Delta_\infty u(x) = 0 & \text{in } \Omega,\\ u(x) = F(x) & \text{on } \partial\Omega. \end{cases}$$

in the same way as problems involving the random walk described in the previous section approximate harmonic functions.

3.1. *Description of the game*

We follow[12] and[13], but we restrict ourselves to the case of a game in a bounded smooth domain $\Omega\subset\mathbb{R}^N$ (the results presented in[12] are valid in general length spaces).

A tug-of-war is a two-person, zero-sum game, that is, two players are in contest and the total earnings of one are the losses of the other. Hence, one of them, say Player I, plays trying to maximize his expected outcome, while the other, say Player II is trying to minimize Player I's outcome (or, since the game is zero-sum, to maximize his own outcome). Consider a bounded domain $\Omega \subset \mathbb{R}^N$, and take $\Gamma_D \subset \partial\Omega$ and $\Gamma_N \equiv \partial\Omega \setminus \Gamma_D$. Let $F : \Gamma_D \to \mathbb{R}$ be a Lipschitz continuous function. At an initial time, a token is placed at a point $x_0 \in \overline{\Omega} \setminus \Gamma_D$. Then, a (fair) coin is tossed and the winner of the toss is allowed to move the game position to any $x_1 \in \overline{B_\varepsilon(x_0)} \cap \overline{\Omega}$. At each turn, the coin is tossed again, and the winner chooses a new game state $x_k \in \overline{B_\varepsilon(x_{k-1})} \cap \overline{\Omega}$. Once the token has reached some $x_\tau \in \Gamma_D$, the game ends and Player I earns $F(x_\tau)$ (while Player II earns $-F(x_\tau)$). This is the reason why we will refer to F as the *final payoff function.* In more general models, it is considered also a *running payoff* $g(x)$ defined in Ω, which represents the reward (respectively, the cost) at each intermediate state x, and gives rise to nonhomogeneous problems. We will assume here that $g \equiv 0$. This procedure yields a sequence of game states $x_0, x_1, x_2, \ldots, x_\tau$, where every x_k except x_0 are random variables, depending on the coin tosses and the strategies adopted by the players.

Now we want to give a precise definition of the *value of the game.* To this end we have to introduce some notation and put the game into its normal or strategic form (see[14] and[15]). The initial state $x_0 \in \overline{\Omega}\setminus\Gamma_D$ is known to both players (public knowledge). Each player i chooses an *action* $a_0^i \in \overline{B_\varepsilon(0)}$ which is announced to the other player; this defines an action profile $a_0 = \{a_0^1, a_0^2\} \in \overline{B_\varepsilon(0)} \times \overline{B_\varepsilon(0)}$. Then, the new state $x_1 \in \overline{B_\varepsilon(x_0)}$ (namely, the current state plus the action) is selected according to a probability distribution $p(\cdot|x_0, a_0)$ in $\overline{\Omega}$ which, in our case, is given by the fair coin toss. At stage k, knowing the history $h_k = (x_0, a_0, x_1, a_1, \ldots, a_{k-1}, x_k)$, (the sequence of states and actions up to that stage), each player i chooses an action a_k^i. If the game ends at time $j < k$, we set $x_m = x_j$ and $a_m = 0$ for $j \leqslant m \leqslant k$. The current state x_k and the profile $a_k = \{a_k^1, a_k^2\}$ determine the distribution $p(\cdot|x_k, a_k)$ (again given by the fair coin toss) of the new state x_{k+1}.

Denote $H_k = (\overline{\Omega} \setminus \Gamma_D) \times \left(\overline{B_\varepsilon(0)} \times \overline{B_\varepsilon(0)} \times \overline{\Omega}\right)^k$, the set of *histories up to stage* k, and by $H = \bigcup_{k\geqslant 1} H_k$ the set of all histories. Notice that H_k, as a product space, has a measurable structure. The *complete history space* H_∞ is the set of plays defined as infinite sequences $(x_0, a_0, \ldots, a_{k-1}, x_k, \ldots)$ endowed with the product topology. Then, the final payoff for Player I, i.e. F, induces a Borel-measurable function on H_∞. A *pure strategy* $S_i = \{S_i^k\}_k$

for Player i, is a sequence of mappings from histories to actions, namely, a mapping from H to $\overline{B_\varepsilon(0)}$ such that S_i^k is a Borel-measurable mapping from H_k to $\overline{B_\varepsilon(0)}$ that maps histories ending with x_k to elements of $\overline{B_\varepsilon(0)}$ (roughly speaking, at every stage the strategy gives the next movement for the player, provided he win the coin toss, as a function of the current state and the past history). The initial state x_0 and a profile of strategies $\{S_I, S_{II}\}$ define (by Kolmogorov's extension theorem) a unique probability $\mathbb{P}^{x_0}_{S_I,S_{II}}$ on the space of plays H_∞. We denote by $\mathbb{E}^{x_0}_{S_I,S_{II}}$ the corresponding expectation.

Then, if S_I and S_{II} denote the strategies adopted by Player I and II respectively, we define the expected payoff for player I as

$$V_{x_0,I}(S_I,S_{II}) = \begin{cases} \mathbb{E}^{x_0}_{S_I,S_{II}}[F(x_\tau)], & \text{if the game terminates a.s.} \\ -\infty, & \text{otherwise.} \end{cases}$$

Analogously, we define the expected payoff for player II as

$$V_{x_0,II}(S_I,S_{II}) = \begin{cases} \mathbb{E}^{x_0}_{S_I,S_{II}}[F(x_\tau)], & \text{if the game terminates a.s.} \\ +\infty, & \text{otherwise.} \end{cases}$$

Finally, we can define the ε*-value of the game for Player I* as

$$u_I^\varepsilon(x_0) = \sup_{S_I} \inf_{S_{II}} V_{x_0,I}(S_I,S_{II}),$$

while the ε-value of the game for Player II is defined as

$$u_{II}^\varepsilon(x_0) = \inf_{S_{II}} \sup_{S_I} V_{x_0,II}(S_I,S_{II}).$$

In some sense, $u_I^\varepsilon(x_0)$ and $u_{II}^\varepsilon(x_0)$ are the least possible outcomes that each player expects to get when the ε-game starts at x_0. Notice that, as in [12], we penalize severely the games that never end. If $u_I^\varepsilon = u_{II}^\varepsilon := u_\varepsilon$, we say that *the game has a value.* In [12] it is shown that, under very general hypotheses, that are fulfilled in the present setting, the ε-tug-of-war game has a value.

All these $\varepsilon-$values are Lipschitz functions with respect to the discrete distance d^ε given by

$$d_\varepsilon(x,y) = \begin{cases} 0 & \text{if } x = y, \\ \varepsilon\left(\left[\!\left[\frac{|x-y|}{\varepsilon}\right]\!\right] + 1\right) & \text{if } x \neq y. \end{cases} \tag{1}$$

where $|.|$ is the Euclidean norm and $[\![r]\!]$ is defined for $r > 0$ by $[\![r]\!] := n$, if $n < r \leq n+1$, $n = 0,1,2,\ldots$, see [12] (but in general they are not continuous). Let us present a simple example where we can compute the value of the game explicitly.

3.2. *The* 1 − *d game*

Let us analyze in detail the one-dimensional game and its limit as $\varepsilon \to 0$. We set $\Omega = (0,1)$ and play the ε−game. To simplify we assume that $\varepsilon = 1/2^n$ and that the running payoff is zero. Concerning the final payoff, we end the game at $x = 0$ (with zero final payoff) and at $x = 1$ (with final payoff equals to one). Note that the general result from[12] applies and hence we can assert the existence of a value for this game. Nevertheless, in this simple $1 - d$ case we can obtain the existence of such value by direct computations. For the moment, let us assume that there exists a value that we call u_ε and proceed, in several steps, with the analysis of this sequence of functions u_ε for ε small. All the calculations below hold both for u_I^ε and for u_{II}^ε.

Step 1. $u_\varepsilon(0) = 0$ and $u_\varepsilon(1) = 1$. Moreover, $0 \leqslant u_\varepsilon(x) \leqslant 1$ (the value functions are uniformly bounded).

Step 2. u_ε is increasing in x and strictly positive in $(0,1]$. Indeed, if $x < y$ then for every pair of strategies S_I, S_{II} for Player I and II beginning at x we can construct strategies beginning at y in such a way that $x_{i,x} \leqslant x_{i,y}$ (here $x_{i,x}$ and $x_{i,y}$ are the positions of the game after i movements beginning at x and y respectively). In fact, just reproduce the movements shifting points by $y - x$ when possible (if not, that is, if the jump is too large and ends outside the interval, just remain at the larger interior position $x = 1$). In this way we see that the probability of reaching $x = 1$ beginning at y is bigger than the probability of reaching $x = 0$ and hence, taking expectations, infimum and supremum, it follows that $u_\varepsilon(x) \leqslant u_\varepsilon(y)$.

Now, we just observe that there is a positive probability of obtaining a sequence of $1/\varepsilon$ consecutive heads (exactly $2^{-1/\varepsilon}$), hence the probability of reaching $x = 1$ when the first player uses the strategy that points ε to the right is strictly positive. Therefore, $u_\varepsilon(x) > 0$, for every $x \neq 0$.

Step 3. In this one dimensional case it is easy to identify the optimal strategies for players I and II: to jump ε to the right for Player I and to jump ε to the left for Player II. That is, if we are at x, the optimal strategies lead to

$$x \to \min\{x + \varepsilon, 1\}$$

for Player I, and to

$$x \to \max\{x - \varepsilon, 0\}$$

for Player II. This follows from step 2, where we have proved that the function u_ε is increasing in x. As a consequence, the optimal strategies follow: for instance, Player I will choose the point where the expected payoff is maximized and this is given by $\min\{x+\varepsilon, 1\}$,

$$\sup_{z\in[x-\varepsilon,x+\varepsilon]\cap[0,1]} u_\varepsilon(z) = \max_{z\in[x-\varepsilon,x+\varepsilon]\cap[0,1]} u_\varepsilon(z) = u_\varepsilon(\min\{x+\varepsilon, 1\}),$$

since u_ε is increasing.

This is also clear from the following intuitive fact: player I wants to maximize the payoff (reaching $x = 1$) and player II wants to minimize the payoff (hence pointing to 0).

Step 4. u_ε is constant in every interval of the form $(k\varepsilon, (k+1)\varepsilon)$ for $k = 1, ..., N$ (we denote by N the total number of such intervals in $(0, 1])$. Indeed, from step 3 we know what are the optimal strategies for both players, and hence the result follows noticing that the number of steps that one has to advance to reach $x = 0$ (or $x = 1$) is the same for every point in $(k\varepsilon, (k+1)\varepsilon)$.

Remark 3.1. Note that u_ε is necessarily discontinuous at every point of the form $y_k = k\varepsilon \in (0, 1)$.

Step 5. Let us call $a_k := u_\varepsilon |_{(k\varepsilon,(k+1)\varepsilon)}$. Then we have

$$a_0 = 0, \qquad a_k = \frac{1}{2}(a_{k-1} + a_{k+1}),$$

for every $i = 2, ..., n-1$, and

$$a_n = 1.$$

Notice that these identities follow from the Dynamic Programming Principle, using that from step 3 we know the optimal strategies, that from step 4 u_ε is constant in every subinterval of the form $(k\varepsilon, (k+1)\varepsilon)$, we immediately get the conclusion.

Remark 3.2. Note the similarity with a finite difference scheme used to solve $u_{xx} = 0$ in $(0, 1)$ with boundary conditions $u(0) = 0$ and $u(1) = 1$. In fact, a discretization of this problem in a uniform mesh of size ε leads to the same formulas obtained in step 5.

Step 6. We have

$$u_\varepsilon(x) = \varepsilon k, \qquad x \in (k\varepsilon, (k+1)\varepsilon). \tag{2}$$

Indeed, the constants $a_k = \varepsilon k$ are the unique solution to the formulas obtained in step 5.

Remark 3.3. Since formula (2) is in fact valid for u_I^ε and u_{II}^ε, this proves that the game has a value.

Remark 3.4. Note that u_ε verifies that $0 \leqslant u_\varepsilon(x) - u_\varepsilon(y) \leqslant 2(x-y)$ for every $x > y$ with $x - y > \varepsilon$. This is a sort of equicontinuity valid for "far apart points".

In this one dimensional case, we can pass to the limit directly, by using the explicit formula for u_ε (see Step 7 below). However, in the N-dimensional case there is no explicit formula, and then we will need a compactness result (a sort of Arzela-Ascoli lemma), see[16].

Step 7.

$$\lim_{\varepsilon \to 0} u_\varepsilon(x) = x,$$

uniformly in $[0,1]$. Indeed, this follows from the explicit formula for u_ε in every interval of the form $(k\varepsilon, (k+1)\varepsilon)$ found in step 6 and from the monotonicity stated in step 2 (to take care of the values of u_ε at points of the form $k\varepsilon$, we have $a_{k-1} \leqslant u_\varepsilon(k\varepsilon) \leqslant a_k$).

Remark 3.5. Note that the limit function $u(x) = x$ is the unique viscosity (and classical) solution to $\Delta_\infty u(x) = (u_{xx}(u_x)^2)(x) = 0$ for $x \in (0,1)$, with boundary conditions $u(0) = 0$ and $u(1) = 1$.

3.3. *Comments*

(*a*) When we add a running cost of the form $\varepsilon^2 g(x)$ we obtain a solution to

$$\begin{cases} -\Delta_\infty u(x) = g(x) & \text{in } \Omega, \\ \dfrac{\partial u}{\partial n}(x) = 0 & \text{on } \Gamma_N, \\ u(x) = F(x) & \text{on } \Gamma_D. \end{cases}$$

(*b*) Many of the results presented here are valid in general length spaces (see[12]) and on graphs. In case of a graph the next positions that can be selected by both players are the points that are connected with the actual position of the game. This leads to consider the infinity laplacian on a graph.

(c) One can consider games in which the coin is biased, that is the probability of getting a head is p (with $p \neq 1/2$). In this case the limit as $\varepsilon \to 0$ (with $p = p(\varepsilon) \to 1/2$) was analyzed in[17] and the PDE that appears reads as $\Delta_\infty u + \beta|\nabla u| = 0$.

4. p–harmonious functions

The aim of this section is to describe games that approximate the p–Laplacian that is given by $\Delta_p u = \mathrm{div}(|\nabla u|^{p-2}\nabla u)$. We assume that $2 \leqslant p < \infty$. The case $p = \infty$ was considered in the previous section.

4.1. *Description of the game*

Consider a two-player zero-sum-game in a domain Ω described as follows: starting from a point $x_0 \in \Omega$, Players I and II play the original tug-of-war game described in[12] (see the previous section for details) with probability α, and with probability β (recall that $\alpha+\beta = 1$), a random point in $B_\varepsilon(x_0)$ is chosen. Once the game position reaches a strip near the boundary of width ε, Player II pays Player I the amount given by a pay-off function. Naturally, Player I tries to maximize and Player II to minimize the payoff. Hence, the equation

$$u_\varepsilon(x) = \frac{\alpha}{2}\left\{ \sup_{\overline{B}_\varepsilon(x)} u_\varepsilon + \inf_{\overline{B}_\varepsilon(x)} u_\varepsilon \right\} + \beta \fint_{B_\varepsilon(x)} u_\varepsilon \, dy,$$

describes the expected payoff of the above game. Intuitively, the expected payoff at the point can be calculated by summing up all the three cases: Player I moves, Player II moves, or a random point is chosen, with their corresponding probabilities.

In this variant of tug-of-war with noise the noise is distributed uniformly on $B_\varepsilon(x)$. This approach allows us to use the dynamic programming principle in the form

$$u_\varepsilon(x) = \frac{\alpha}{2}\left\{ \sup_{\overline{B}_\varepsilon(x)} u_\varepsilon + \inf_{\overline{B}_\varepsilon(x)} u_\varepsilon \right\} + \beta \fint_{B_\varepsilon(x)} u_\varepsilon \, dy, \tag{1}$$

to conclude that our game has a value and that the value is p-harmonious. There are indications, see Barles-Souganidis[18] and Oberman[19], that our results based on the mean value approach are likely to be useful in applications for example to numerical methods as well as in problems of analysis, cf. Armstrong-Smart[20]. For further results on games see[21] and[22].

We start by the statement of the *Dynamic Programming Principle* (DPP) applied to our game.

Lemma 4.1 (DPP). *The value function for Player I satisfies*

$$
\begin{aligned}
u_I^\varepsilon(x_0) &= \frac{\alpha}{2}\left\{\sup_{\overline{B}_\varepsilon(x_0)} u_I^\varepsilon + \inf_{\overline{B}_\varepsilon(x_0)} u_I^\varepsilon\right\} + \beta \fint_{B_\varepsilon(x_0)} u_I^\varepsilon\, dy, \qquad x_0 \in \Omega, \\
u_I^\varepsilon(x_0) &= F(x_0), \qquad x_0 \in \Gamma_\varepsilon.
\end{aligned} \tag{2}
$$

The value function for Player II, u_{II}^ε, satisfies the same equation.

Formulas similar to (2) can be found in Chapter 7 of[23]. A detailed proof adapted to our case can also be found in[24].

Let us explain intuitively why the DPP holds by considering the expectation of the payoff at x_0. Whenever the players get heads (probability α) in the first coin toss, they toss a fair coin and play the tug-of-war. If Player I wins the fair coin toss in the tug-of-war (probability $1/2$), she steps to a point maximizing the expectation and if Player II wins, he steps to a point minimizing the expectation. Whenever they get tails (probability β) in the first coin toss, the game state moves to a random point according to a uniform probability on $B_\varepsilon(x_0)$. The expectation at x_0 can be obtained by summing up these different alternatives.

We warn the reader that, as happens for the tug-of-war game without noise described previously, the value functions are discontinuous in general as the next example shows.

Example 4.1. Consider the case $\Omega = (0,1)$ and

$$
F(x) = u_{\mathrm{I}}^\varepsilon(x) = \begin{cases} 1, & x \geqslant 1 \\ 0, & x \leqslant 0. \end{cases}
$$

In this case the optimal strategies for both players are clear: Player I moves ε to the right and Player II moves ε to the left. Now, there is a positive probability of reaching $x \geqslant 1$ that can be uniformly bounded from below in $(0,1)$ by $C = (2/\alpha)^{-(1/\varepsilon+1)}$. This can be seen by considering the probability of Player I winning all the time until the game ends with $x \geqslant 1$. Therefore $u_{\mathrm{I}}^\varepsilon > C > 0$ in the whole $(0,1)$. This implies a discontinuity at $x = 0$ and hence a discontinuity at $x = \varepsilon$. Indeed, first, note that u_ε is nondecreasing and hence

$$
u_{\mathrm{I}}^\varepsilon(\varepsilon-) = \lim_{x \nearrow \varepsilon} \frac{\alpha}{2} \sup_{|x-y|\leqslant\varepsilon} u_{\mathrm{I}}^\varepsilon(y) + \frac{\beta}{2\varepsilon}\int_0^{2\varepsilon} u_{\mathrm{I}}^\varepsilon\, dy = \frac{\alpha}{2} u_{\mathrm{I}}^\varepsilon(2\varepsilon-) + \frac{\beta}{2\varepsilon}\int_0^{2\varepsilon} u_{\mathrm{I}}^\varepsilon\, dy,
$$

because $\sup_{|x-y|\leqslant\varepsilon} u_{\rm I}^{\varepsilon}(y) = u_{\rm I}^{\varepsilon}(x+\varepsilon)$ and $\inf_{|x-\varepsilon|\leqslant\varepsilon} u_{\rm I}^{\varepsilon}$ is zero for $x \in (0,\varepsilon)$. However,

$$u_{\rm I}^{\varepsilon}(\varepsilon+) \geqslant \frac{\alpha}{2}C + \lim_{x\searrow\varepsilon}\frac{\alpha}{2}\sup_{|x-y|\leqslant\varepsilon} u_{\rm I}^{\varepsilon}(y) + \frac{\beta}{2\varepsilon}\int_0^{2\varepsilon} u_{\rm I}^{\varepsilon}\,dy \geqslant \frac{\alpha}{2}C + u_{\rm I}^{\varepsilon}(\varepsilon-)$$

because $\sup_{|x-y|\leqslant\varepsilon} u_{\rm I}^{\varepsilon}(y) = u_{\rm I}^{\varepsilon}(x+\varepsilon) \geqslant u_{\rm I}^{\varepsilon}(2\varepsilon-)$ and $\inf_{|x-\varepsilon|\leqslant\varepsilon} u_{\rm I}^{\varepsilon} \geqslant C$ for $x > \varepsilon$.

4.2. *p−harmonious functions*

The goal of this section is to study functions that satisfy (1) with fixed $\varepsilon > 0$ and suitable nonnegative α and β, with $\alpha+\beta = 1$ (we will call such functions p−harmonious functions).

Consider a bounded domain $\Omega \subset \mathbb{R}^N$ and fix $\varepsilon > 0$. To prescribe boundary values for p−harmo-nious functions, let us denote the compact boundary strip of width ε by

$$\Gamma_\varepsilon = \{x \in \mathbb{R}^N \setminus \Omega \,:\, \operatorname{dist}(x,\partial\Omega) \leqslant \varepsilon\}.$$

Definition 4.1. The function u_ε is p-harmonious in Ω with boundary values $F : \Gamma_\varepsilon \to \mathbb{R}$ if

$$u_\varepsilon(x) = \frac{\alpha}{2}\left\{\sup_{\overline{B}_\varepsilon(x)} u_\varepsilon + \inf_{\overline{B}_\varepsilon(x)} u_\varepsilon\right\} + \beta \fint_{B_\varepsilon(x)} u_\varepsilon\,dy \qquad \text{for every} \quad x \in \Omega,$$

where α, β are defined in (9), and $u_\varepsilon(x) = F(x)$, for every $x \in \Gamma_\varepsilon$.

The reason for using the boundary strip Γ_ε instead of simply using the boundary $\partial\Omega$ is the fact that for $x \in \Omega$ the ball $\overline{B}_\varepsilon(x)$ is not necessarily contained in Ω.

Let us first explain the name p-harmonious. When u is harmonic, then it satisfies the well known mean value property

$$u(x) = \fint_{B_\varepsilon(x)} u\,dy, \tag{3}$$

that is (1) with $\alpha = 0$ and $\beta = 1$. On the other hand, functions satisfying (1) with $\alpha = 1$ and $\beta = 0$

$$u_\varepsilon(x) = \frac{1}{2}\left\{\sup_{\overline{B}_\varepsilon(x)} u_\varepsilon + \inf_{\overline{B}_\varepsilon(x)} u_\varepsilon\right\} \tag{4}$$

are called *harmonious* functions in [25] and [26] and are values of tug-of-war games like the ones described in the previous section. As we have seen, as ε goes to zero, they approximate solutions to the infinity Laplacian.

Now, recall that the p-Laplacian is given by

$$\Delta_p u = \operatorname{div}(|\nabla u|^{p-2}\nabla u) = |\nabla u|^{p-2}\left\{(p-2)\Delta_\infty u + \Delta u\right\}. \tag{5}$$

Since the p-Laplace operator can be written as a combination of the Laplace operator and the infinity Laplacian, it seems reasonable to expect that the combination (1) of the averages in (3) and (4) give an approximation to a solution to the p-Laplacian. We will show that this is indeed the case. To be more precise, in [16] it is proved that p-harmonious functions are uniquely determined by their boundary values and that they converge uniformly to the p-harmonic function with the given boundary data. Below we present an heuristic argument to support this claim and refer again to [16] for a rigorous proof.

4.3. *A heuristic argument*

It follows from expansion (5) that u is a solution to $\Delta_p u = 0$ if and only if

$$(p-2)\Delta_\infty u + \Delta u = 0, \tag{6}$$

because this equivalence can be justified in the viscosity sense even when $\nabla u = 0$ as shown in [27]. Averaging the classical Taylor expansion

$$u(y) = u(x) + \nabla u(x)\cdot(y-x) + \frac{1}{2}\langle D^2u(x)(y-x),(y-x)\rangle + O(|y-x|^3),$$

over $B_\varepsilon(x)$, we obtain

$$u(x) - \fint_{B_\varepsilon(x)} u\,dy = -\frac{\varepsilon^2}{2(n+2)}\Delta u(x) + O(\varepsilon^3), \tag{7}$$

when u is smooth. Here we used the shorthand notation

$$\fint_{B_\varepsilon(x)} u\,dy = \frac{1}{|B_\varepsilon(x)|}\int_{B_\varepsilon(x)} u\,dy.$$

Then observe that gradient direction is almost the maximizing direction. Thus, summing up the two Taylor expansions roughly gives us

$$\begin{aligned} u(x) - \frac{1}{2}&\left\{\sup_{\overline{B}_\varepsilon(x)} u + \inf_{\overline{B}_\varepsilon(x)} u\right\} \\ &\approx u(x) - \frac{1}{2}\left\{u\left(x+\varepsilon\frac{\nabla u(x)}{|\nabla u(x)|}\right) + u\left(x-\varepsilon\frac{\nabla u(x)}{|\nabla u(x)|}\right)\right\} \\ &= -\frac{\varepsilon^2}{2}\Delta_\infty u(x) + O(\varepsilon^3). \end{aligned} \tag{8}$$

Next we multiply (7) and (8) by suitable constants α and β, $\alpha+\beta=1$, and add up the formulas to obtain

$$u(x) - \frac{\alpha}{2}\left\{ \sup_{\overline{B}_\varepsilon(x)} u - \inf_{\overline{B}_\varepsilon(x)} u \right\} + \beta \fint_{B_\varepsilon(x)} u\, dy$$
$$= -\alpha\frac{\varepsilon^2}{2}\Delta_\infty u(x) - \beta\frac{\varepsilon^2}{2(n+2)}\Delta u(x) + O(\varepsilon^3).$$

Next, we choose α and β so that we have the operator in (6) on the right hand side. This process gives us the choices of the constants

$$\alpha = \frac{p-2}{p+N}, \quad \text{and} \quad \beta = \frac{2+N}{p+N}. \tag{9}$$

and we deduce

$$u(x) = \frac{\alpha}{2}\left\{ \sup_{\overline{B}_\varepsilon(x)} u + \inf_{\overline{B}_\varepsilon(x)} u \right\} + \beta \fint_{B_\varepsilon(x)} u\, dy + O(\varepsilon^3)$$

as $\varepsilon \to 0$.

4.4. *Comments*

(*a*) When we add a running cost of the form $\varepsilon^2 g(x)$ we obtain a solution to the following inhomogeneous problem that involves the 1−homogeneous p−Laplacian,

$$\begin{cases} -|\nabla u|^{2-p}\Delta_p u(x) = kg(x) & \text{in } \Omega, \\ u(x) = F(x) & \text{on } \partial\Omega. \end{cases}$$

Here k is a constant that depends only on p and N.
Note that the 1−homogeneous p−Laplacian is not variational (it is not in divergence form).

5. Spatial dependence

Following[28], our main concern in this section is to answer the following question:

What are the PDEs that can be obtained as continuous values of Tug-of-War games when we replace the ball $\overline{B_\varepsilon(x)}$ with a more general family of sets $\mathcal{A}_\varepsilon(x)$?

To answer this question we have to assume certain conditions on the family of sets $\mathcal{A}_\varepsilon(x)$ and the way that they behave as $\varepsilon \to 0$. In our case the DPP reads as

$$u_\varepsilon(x) = \frac{1}{2} \sup_{y \in \mathcal{A}_\varepsilon(x)} u_\varepsilon(y) + \frac{1}{2} \inf_{y \in \mathcal{A}_\varepsilon(x)} u_\varepsilon(y).$$

Following our previous discussion for the case of balls we can guess that the limit PDE as $\varepsilon \to 0$ will depend on the point at which a smooth function ϕ with non-zero gradient attains its maximum (and its minimum) in $\mathcal{A}_\varepsilon(x)$. The assumed conditions on the sets $\mathcal{A}_\varepsilon(x)$ are such that there is a preferred direction where the maxima and the minima of a smooth function ϕ with non-zero gradient are closely located when $\varepsilon \to 0$. This preferred direction depends on the spatial location and on the gradient of ϕ at that point. We call such direction $J_x(D\phi(x))$.

The main result in[28] reads as follows:

Under adequate conditions on the family of possible movements $\mathcal{A}_\varepsilon(x)$ and assuming that the set Ω has boundary with strictly positive curvature, the values of the tug-of-war game with the ball $\overline{B_\varepsilon(x)}$ replaced by $\mathcal{A}_\varepsilon(x)$ converge uniformly (along subsequences) to some continuous limit v that is a viscosity solution to

$$\begin{cases} -\langle D^2 v \cdot J_x(Dv); J_x(Dv)\rangle(x) = 0 & x \in \Omega, \\ v(x) = F(x) & x \in \partial\Omega. \end{cases} \tag{1}$$

Here, as we have mentioned, $J_x(Dv)$ depends on x (this dependence comes from the dependence of the sets $\mathcal{A}_\varepsilon(x)$ on x) and on $Dv(x)$. Note that in this limit equation we also have a second derivative of v but now the direction is not given by $Dv(x)$ but by the vector $J_x(Dv(x))$ that depends also on the spatial location x. In this sense we have found a natural way of introducing spatial dependence in the infinity Laplacian. Note that when one considers the Lipschitz extension problem with respect to different norms the corresponding PDE also takes the form (1), see[29] and the examples presented below.

As a first example of possible sets $\mathcal{A}_\varepsilon(x)$ we mention balls in ℓ^q with $1 < q < +\infty$, that is,

$$\mathcal{A}_\varepsilon(x) = B_\varepsilon^{\ell^q}(x) = \left\{ y \in \mathbb{R}^N \ : \ \left(\sum_{i=1}^{N} |y_i - x_i|^q \right)^{1/q} \leqslant \varepsilon \right\}. \tag{2}$$

In this case, given a direction v, the resulting $J_x(v)$ does not depend on x and reads as

$$J_x(v) = \frac{\left(|v_1|^{-\frac{q-2}{q-1}}v_1, ..., |v_N|^{-\frac{q-2}{q-1}}v_N\right)}{\left\|\left(|v_1|^{-\frac{q-2}{q-1}}v_1, ..., |v_N|^{-\frac{q-2}{q-1}}v_N\right)\right\|_{\ell^q}}.$$

Then, the limit PDE that appear in (1) is given by

$$-\langle D^2v \cdot J(Dv); J(Dv)\rangle = -\sum_{i,j} \left|\frac{\partial v}{\partial x_j}\right|^{-\frac{q-2}{q-1}} \frac{\partial v}{\partial x_j} \left|\frac{\partial v}{\partial x_i}\right|^{-\frac{q-2}{q-1}} \frac{\partial v}{\partial x_i} \frac{\partial^2 v}{\partial x_i \partial x_j} = 0.$$

This equation also appears as limit of $p-$Laplacian type operators when $p \to \infty$, see[30].

To obtain an equation with dependence on x we can just consider $q = q(x)$ in the previous example, that is, we play in balls of ℓ^q with different q at each point (we have to assume here that $q(x)$ is continuous and bounded away from one and infinity).

As another example we can also consider $\mathcal{A}_\varepsilon(x)$ as ellipses as long as the eccentricity does not degenerate.

We can also consider balls in ℓ^q with $q < 1$. In this case we obtain as the limit PDE,

$$\sum_{i \in I(Dv)} \left|\frac{\partial v}{\partial x_i}\right|^2 \frac{\partial^2 v}{\partial x_i^2} = 0,$$

where $I(Dv) = \{i \ : \ |\frac{\partial v}{\partial x_i}| = \max_j |\frac{\partial v}{\partial x_j}|\}$. This equation also appears as the limit equation of values of the game when we play in lattices, that is, when we take $\mathcal{A}_\varepsilon(x) = \{x \pm \varepsilon e_i, \ i = 1, ..., N\}$. This equation can also be obtained as limit of $p-$Laplacian type equations as $p \to \infty$ and is known as the pseudo infinity Laplacian in the literature, see[31],[32] and[33].

Let us point out that for the general case of J_x depending on x in (1) it seems difficult to obtain existence of solutions as limits of $p-$Laplacian type problems. Hence, this approach provides a new existence result for (1).

We also remark that we can consider families of sets $\mathcal{A}_\varepsilon(x)$ that are not necessarily given by balls with respect to a metric $d(x, y)$. This fact introduces new difficulties in the proofs since the positions of the game are not necessarily reversible, that is, it may happen that $x \in \mathcal{A}_\varepsilon(y)$ and $y \notin \mathcal{A}_\varepsilon(x)$. See[28] for details.

6. Parabolic nonlinear problems

Motivated by the asymptotic mean value theorems in[34], we next study the functions that approximate solutions to parabolic problems, and are obtained as value functions of a tug-war-game with noise when the number of steps is limited.

Recall that $\Omega_T \subset \mathbb{R}^{n+1}$ is an open set. To prescribe boundary values, we denote the boundary strip of width ε by

$$\Gamma_\varepsilon = \Big\{\partial\Omega^\varepsilon \times (-\frac{\varepsilon^2}{2}, T]\Big\} \cup \Big\{\Omega \times (-\frac{\varepsilon^2}{2}, 0]\Big\},$$

where $\partial\Omega^\varepsilon = \{x \in \mathbb{R}^n \setminus \Omega : \operatorname{dist}(x, \partial\Omega) \leqslant \varepsilon\}$. Below $F : \Gamma_\varepsilon \to \mathbb{R}$ denotes a bounded Borel function.

Definition 6.1. The function u_ε is (p, ε)-parabolic, $2 \leqslant p \leqslant \infty$, in Ω_T with boundary values F if

$$\begin{aligned} u_\varepsilon(x,t) =& \frac{\alpha}{2}\left\{ \sup_{y \in \overline{B}_\varepsilon(x)} u_\varepsilon\Big(y, t - \frac{\varepsilon^2}{2}\Big) + \inf_{y \in \overline{B}_\varepsilon(x)} u_\varepsilon\Big(y, t - \frac{\varepsilon^2}{2}\Big)\right\} \\ & + \beta \fint_{B_\varepsilon(x)} u_\varepsilon\Big(y, t - \frac{\varepsilon^2}{2}\Big)\, dy, \text{ for every } (x,t) \in \Omega_T, \\ u_\varepsilon(x,t) =& F(x,t), \text{ for every } (x,t) \in \Gamma_\varepsilon, \end{aligned}$$

where $\alpha = \frac{p-2}{p+n}$ and $\beta = \frac{n+2}{p+n}$.

The reason for using the boundary strip Γ_ε instead of simply using the parabolic boundary $\partial_p\Omega_T$ is the fact that $\overline{B}_\varepsilon(x) \times \{t - \frac{\varepsilon^2}{2}\}$ is not necessarily contained in Ω_T.

Next we study the tug-of-war game with noise studied in[24]. In this case, there are two key differences: the game has a preset maximum number of steps and boundary values may change with time.

To be more precise, at the beginning we fix the maximum number of steps to be N and place a token at a point $x_0 \in \Omega$. The players toss a biased coin with probabilities α and β, $\alpha + \beta = 1$. If they get heads (probability α), they play a tug-of-war game, that is, a fair coin is tossed and the winner of the toss decides a new game position $x_1 \in \overline{B}_\varepsilon(x_0)$. On the other hand, if they get tails (probability β), the game state moves according to the uniform probability density to a random point in the ball $B_\varepsilon(x_0)$. They continue playing the game until either the token hits the boundary strip $\partial\Omega^\varepsilon$ or the number of steps reaches N. We denote by $\tau_N \in \{0, 1, \ldots, N\}$ the hitting time of $\partial\Omega^\varepsilon$ or N, whichever comes first, and by $x_{\tau_N} \in \Omega \cup \partial\Omega^\varepsilon$

the end point of the game. When no confusion arises, we simply write τ. At the end of the game Player I earns $\mathcal{F}(x_{\tau_N}, \tau_N)$ while Player II earns $-\mathcal{F}(x_{\tau_N}, \tau_N)$. Here

$$\mathcal{F} : \{\partial\Omega^\varepsilon \times \{0, \dots, N\}\} \cup \{\Omega \times \{N\}\} \to \mathbb{R}$$

is a given payoff function. Denote by $H = \Omega \cup \partial\Omega^\varepsilon$. A run of the game is a sequence

$$\omega = (\omega_0, \omega_1, \dots, \omega_N) \in H^{N+1}.$$

We define random variables

$$x_k(\omega) = \omega_k, \quad x_k : H^{N+1} \to \mathbb{R}^n, \ k = 0, 1, \dots, N,$$

and

$$\tau_N(\omega) = \min\{N, \inf\{k \, : \, x_k(\omega) \in \partial\Omega^\varepsilon, k = 0, 1, \dots, N\}\}.$$

A strategy S_{I} for Player I is a function which gives the next game position

$$S_{\text{I}}(x_0, x_1, \dots, x_k) = x_{k+1} \in \overline{B}_\varepsilon(x_k)$$

if Player I wins the coin toss. Similarly, Player II plays according to a strategy S_{II}.

The fixed starting point x_0, the number of steps N, the domain Ω and the strategies S_{I} and S_{II} determine a unique probability measure $\mathbb{P}^{x_0,N}_{S_I,S_{\text{II}}}$ in H^{N+1}. This measure is built by using the initial distribution $\delta_{x_0}(A)$, and the family of transition probabilities

$$\begin{aligned}
&\pi_{S_{\text{I}},S_{\text{II}}}(x_0(\omega), \dots, x_k(\omega), A) = \pi_{S_{\text{I}},S_{\text{II}}}(\omega_0, \dots, \omega_k, A) \\
&\quad = \beta \frac{|A \cap B_\varepsilon(\omega_k)|}{|B_\varepsilon(\omega_k)|} + \frac{\alpha}{2}\delta_{S_{\text{I}}(\omega_0,\dots,\omega_k)}(A) + \frac{\alpha}{2}\delta_{S_{\text{II}}(\omega_0,\dots,\omega_k)}(A).
\end{aligned}$$

For more details, we refer to[12,16,24].

The expected payoff, when starting from x_0 with the maximum number of steps N, and using the strategies $S_{\text{I}}, S_{\text{II}}$, is

$$\mathbb{E}^{x_0,N}_{S_{\text{I}},S_{\text{II}}}[\mathcal{F}(x_{\tau_N}, \tau_N)] = \int_{H^{N+1}} \mathcal{F}(x_{\tau_N}(\omega), \tau_N(\omega)) \, d\mathbb{P}^{x_0,N}_{S_{\text{I}},S_{\text{II}}}(\omega).$$

The *value of the game for Player I* when starting at x_0 with the maximum number of steps N is given by

$$u_{\text{I}}^{\varepsilon,N}(x_0, 0) = \sup_{S_{\text{I}}} \inf_{S_{\text{II}}} \mathbb{E}^{x_0,N}_{S_{\text{I}},S_{\text{II}}}[\mathcal{F}(x_{\tau_N}, \tau_N)]$$

while the *value of the game for Player II* is given by

$$u_{\mathrm{II}}^{\varepsilon,N}(x_0,0)=\inf_{S_{\mathrm{II}}}\sup_{S_{\mathrm{I}}}\mathbb{E}_{S_{\mathrm{I}},S_{\mathrm{II}}}^{x_0,N}[\mathcal{F}(x_{\tau_N},\tau_N)].$$

More generally, we define the value of the game when starting at x and playing for $h=N-k$ steps to be

$$u_{\mathrm{I}}^{\varepsilon,N}(x,k)=\sup_{S_{\mathrm{I}}}\inf_{S_{\mathrm{II}}}\mathbb{E}_{S_{\mathrm{I}},S_{\mathrm{II}}}^{x,h}[\mathcal{F}(x_{\tau_h},k+\tau_h)]$$

while the *value of the game for Player II* is given by

$$u_{\mathrm{II}}^{\varepsilon,N}(x,k)=\inf_{S_{\mathrm{II}}}\sup_{S_{\mathrm{I}}}\mathbb{E}_{S_{\mathrm{I}},S_{\mathrm{II}}}^{x,h}[\mathcal{F}(x_{\tau_h},k+\tau_h)].$$

Here $\tau_h\in\{0,1,\ldots,h\}$ is a hitting time of the boundary. The values $u_{\mathrm{I}}^{\varepsilon,N}(x,k)$ and $u_{\mathrm{II}}^{\varepsilon,N}(x,k)$ are the best expected outcomes the each player can guarantee when the game starts at x with maximum number of steps $N-k$.

The next lemma states the *Dynamic Programming Principle* (DPP) for the tug-of-war game with a maximum number of steps. For a detailed proof in the elliptic case see[16]. The parabolic case turns out to be easier since backtracking can be directly implemented. See[23].

Lemma 6.1 (DPP). *The value function for Player I satisfies*

$$\begin{aligned}
u_I^{\varepsilon,N}(x,k)=&\frac{\alpha}{2}\left\{\sup_{\overline{B}_\varepsilon(x)}u_I^{\varepsilon,N}(y,k+1)+\inf_{\overline{B}_\varepsilon(x)}u_I^{\varepsilon,N}(y,k+1)\right\}\\
&+\beta\fint_{B_\varepsilon(x)}u_I^{\varepsilon,N}(y,k+1)\,dy,\quad \textit{if}\quad x\in\Omega \textit{ and } k<N,\\
u_I^{\varepsilon,N}(x,k)=&\,\mathcal{F}(x,k),\quad \textit{if}\quad x\in\partial\Omega^\varepsilon \textit{ or } k=N.
\end{aligned}$$

The value function for Player II, $u_{II}^{\varepsilon,N}$, satisfies the same equation.

The expectation is obtained by summing up the expectations of three possible outcomes for the next step with the corresponding probabilities, Player I chooses the next position (probability $\alpha/2$), Player II chooses (probability $\alpha/2$) and the next position is random (probability β). This is the heuristic background for the DPP.

Next we describe the change of time scale that relates values of the tug-of-war games with noise and (p,ε)-parabolic functions. The definition of (p,ε)-parabolic function u_ε, Definition 6.1, refers to a forward-in-time parabolic equation. The values $u_\varepsilon(\cdot,t)$ at time t are determined by the values $u_\varepsilon(\cdot, t - \frac{\varepsilon^2}{2})$. In contrast, in Lemma 6.1 above, the values at step k are determined by the values at step $k+1$.

Let $t_N \in (-\frac{\varepsilon^2}{2}, 0]$ and let $F : \Gamma_\varepsilon \to \mathbb{R}$ be a boundary value function. We set

$$t_k = \varepsilon^2 \frac{N-k}{2} + t_N$$

and define a payoff function $\mathcal{F}_{t_N} : \{\partial\Omega^\varepsilon \times \{0, \ldots, N\}\} \cup \{\Omega \times \{N\}\} \to \mathbb{R}$ by

$$\mathcal{F}_{t_N}(x_\tau, \tau) = F(x_\tau, \varepsilon^2(N-\tau)/2 + t_N) = F(x_\tau, t_\tau), \tag{1}$$

for every $t_N \in (-\frac{\varepsilon^2}{2}, 0]$.

It might be instructive to think of a parabolic cylinder $\Omega \times (0, t_0)$ when t_0 and ε are given thus fixing the rest of the parameters. The game begins at $k = 0$ corresponding to t_0 in the time scale. When we play one round $k \to k+1$, the clock steps $\varepsilon^2/2$ backwards, $t_{k+1} = t_k - \varepsilon^2/2$, and we play until we get outside the cylinder when $k = \tau$ corresponding to t_τ in the time scale.

Next we define

$$u_{\mathrm{I}}^\varepsilon(x, t) = u_{\mathrm{I}}^{\varepsilon, N}(x, 0) \tag{2}$$

where the function on the right hand side is defined for the boundary values $\mathcal{F}_{t_N}$ given by (1) with $N = \lceil 2t/\varepsilon^2 \rceil$ and $t_N = t - N\varepsilon^2/2$. This defines values for every instant.

For these functions, the DPP takes the form,

$$\begin{aligned} u_{\mathrm{I}}^\varepsilon(x,t) =& \frac{\alpha}{2}\left\{ \sup_{y \in \overline{B}_\varepsilon(x)} u_{\mathrm{I}}^\varepsilon\left(y, t - \frac{\varepsilon^2}{2}\right) + \inf_{y \in \overline{B}_\varepsilon(x)} u_{\mathrm{I}}^\varepsilon\left(y, t - \frac{\varepsilon^2}{2}\right) \right\} \\ &+ \beta \fint_{B_\varepsilon(x)} u_{\mathrm{I}}^\varepsilon\left(y, t - \frac{\varepsilon^2}{2}\right) dy \\ u_{\mathrm{I}}^\varepsilon(x,t) =& F(x,t), \text{ for every } (x,t) \in \Gamma_\varepsilon, \end{aligned}$$

which agrees with Definition 6.1.

Our last result reads as follows (see[34] for the proof):

Theorem 6.1. *The uniform limit* $u = \lim_{\varepsilon \to 0} u_\varepsilon$ *of* (p, ε)*-parabolic functions exists and is characterized as the unique viscosity solution to the equation* $(n+p)u_t(x,t) = |\nabla u|^{2-p} \Delta_p u(x,t)$ *with boundary values* F.

Acknowledgments: This paper reflects the contents (with more details) of the talk given by the authors at the V International School in Mathematical Analysis in Andalucía held in September 2011. He wants to thank the organizers for the invitation and specially Jose Carmona for the incredible hospitality. Many useful suggestions from Fernando Charro, Jesus Garcia Azorero, Juan J. Manfredi and Mikko Parvianen are acknowledged as well.

References

1. L. A. Caffarelli and X. Cabré, *Fully nonlinear elliptic equations*, American Mathematical Society Colloquium Publications, Vol. 43 (American Mathematical Society, Providence, RI, 1995).
2. J. D. Rossi, Tug-of-war games and PDEs, *Proc. Roy. Soc. Edinburgh Sect. A* **141**, 319 (2011).
3. G. Aronsson, M. G. Crandall and P. Juutinen, A tour of the theory of absolutely minimizing functions, *Bull. Amer. Math. Soc. (N.S.)* **41**, 439 (2004).
4. T. Bhattacharya, E. DiBenedetto and J. Manfredi, Limits as $p \to \infty$ of $\Delta_p u_p = f$ and related extremal problems, *Rend. Sem. Mat. Univ. Politec. Torino* **Special Issue**, 15 (1989), Some topics in nonlinear PDEs (Turin, 1989).
5. R. Jensen, Uniqueness of Lipschitz extensions: minimizing the sup norm of the gradient, *Arch. Rational Mech. Anal.* **123**, 51 (1993).
6. L. C. Evans and W. Gangbo, Differential equations methods for the Monge-Kantorovich mass transfer problem, *Mem. Amer. Math. Soc.* **137**, viii+66 (1999).
7. J. García-Azorero, J. J. Manfredi, I. Peral and J. D. Rossi, The Neumann problem for the ∞-Laplacian and the Monge-Kantorovich mass transfer problem, *Nonlinear Anal.* **66**, 349 (2007).
8. F. Charro and I. Peral, Limit branch of solutions as $p \to \infty$ for a family of sub-diffusive problems related to the p-Laplacian, *Comm. Partial Differential Equations* **32**, 1965 (2007).
9. P. Juutinen, Minimization problems for lipschitz functions via viscosity solutions, *Ann. Acad. Sci. Fenn. Math. Diss.* **115**, 53 (1998), Dissertation, University of Jyvskul, Jyvskul, 1998.
10. P. Juutinen and P. Lindqvist, On the higher eigenvalues for the ∞-eigenvalue problem, *Calc. Var. Partial Differential Equations* **23**, 169 (2005).

11. P. Juutinen, P. Lindqvist and J. J. Manfredi, The ∞-eigenvalue problem, *Arch. Ration. Mech. Anal.* **148**, 89 (1999).
12. Y. Peres, O. Schramm, S. Sheffield and D. B. Wilson, Tug-of-war and the infinity Laplacian, *J. Amer. Math. Soc.* **22**, 167 (2009).
13. F. Charro, J. García Azorero and J. D. Rossi, A mixed problem for the infinity Laplacian via tug-of-war games, *Calc. Var. Partial Differential Equations* **34**, 307 (2009).
14. Y. Peres and S. Sheffield, Tug-of-war with noise: a game-theoretic view of the p-Laplacian, *Duke Math. J.* **145**, 91 (2008).
15. A. Neyman and S. Sorin (eds.), *Stochastic games and applications*, NATO Science Series C: Mathematical and Physical Sciences Vol. 570, (Kluwer Academic Publishers, Dordrecht, 2003). pp. 27–36.
16. J. J. Manfredi, M. Parviainen and J. D. Rossi, On the definition and properties of p-harmonious functions, *Ann. Sc. Norm. Super. Pisa Cl. Sci. (5)* **11**, 215 (2012).
17. Y. Peres, G. Pete and S. Somersille, Biased tug-of-war, the biased infinity Laplacian, and comparison with exponential cones, *Calc. Var. Partial Differential Equations* **38**, 541 (2010).
18. G. Barles and P. E. Souganidis, Convergence of approximation schemes for fully nonlinear second order equations, *Asymptotic Anal.* **4**, 271 (1991).
19. A. M. Oberman, A convergent difference scheme for the infinity Laplacian: construction of absolutely minimizing Lipschitz extensions, *Math. Comp.* **74**, 1217 (2005).
20. S. N. Armstrong and C. K. Smart, An easy proof of Jensen's theorem on the uniqueness of infinity harmonic functions, *Calc. Var. Partial Differential Equations* **37**, 381 (2010).
21. E. N. Barron, L. C. Evans and R. Jensen, The infinity Laplacian, Aronsson's equation and their generalizations, *Trans. Amer. Math. Soc.* **360**, 77 (2008).
22. R. V. Kohn and S. Serfaty, A deterministic-control-based approach to motion by curvature, *Comm. Pure Appl. Math.* **59**, 344 (2006).
23. A. P. Maitra and W. D. Sudderth, *Discrete gambling and stochastic games*, Applications of Mathematics (New York), Vol. 32 (Springer-Verlag, New York, 1996).
24. J. J. Manfredi, M. Parviainen and J. D. Rossi, Dynamic programming principle for tug-of-war games with noise, *ESAIM Control Optim. Calc. Var.* **18**, 81 (2012).
25. E. Le Gruyer, On absolutely minimizing Lipschitz extensions and PDE

$\Delta_\infty(u) = 0$, *NoDEA Nonlinear Differential Equations Appl.* **14**, 29 (2007).
26. E. Le Gruyer and J. C. Archer, Harmonious extensions, *SIAM J. Math. Anal.* **29**, 279 (1998).
27. P. Juutinen, P. Lindqvist and J. J. Manfredi, On the equivalence of viscosity solutions and weak solutions for a quasi-linear equation, *SIAM J. Math. Anal.* **33**, 699 (2001).
28. I. Gómez and J. D. Rossi, Tug-of-war games and the infinity Laplacian with spatial dependence, *Commun. Pure Appl. Anal.* **12**, 1959 (2013).
29. M. G. Crandall, G. Gunnarsson and P. Wang, Uniqueness of ∞-harmonic functions and the eikonal equation, *Comm. Partial Differential Equations* **32**, 1587 (2007).
30. M. Belloni, B. Kawohl and P. Juutinen, The p-Laplace eigenvalue problem as $p \to \infty$ in a Finsler metric, *J. Eur. Math. Soc. (JEMS)* **8**, 123 (2006).
31. M. Belloni and B. Kawohl, The pseudo-p-Laplace eigenvalue problem and viscosity solutions as $p \to \infty$, *ESAIM Control Optim. Calc. Var.* **10**, 28 (2004).
32. T. Ishibashi and S. Koike, On fully nonlinear PDEs derived from variational problems of L^p norms, *SIAM J. Math. Anal.* **33**, 545 (2001).
33. J. D. Rossi and M. Saez, Optimal regularity for the pseudo infinity Laplacian, *ESAIM Control Optim. Calc. Var.* **13**, 294 (2007).
34. J. J. Manfredi, M. Parviainen and J. D. Rossi, An asymptotic mean value characterization for a class of nonlinear parabolic equations related to tug-of-war games, *SIAM J. Math. Anal.* **42**, 2058 (2010).
35. L. Ambrosio, *Lecture Notes on Optimal Transport Problems*, tech. rep., CVGMT preprint server.
36. G. Aronsson, Extension of functions satisfying Lipschitz conditions, *Ark. Mat.* **6**, 551 (1967).
37. M. G. Crandall, H. Ishii and P.-L. Lions, User's guide to viscosity solutions of second order partial differential equations, *Bull. Amer. Math. Soc. (N.S.)* **27**, 1 (1992).
38. K. Does, An evolution equation involving the normalized p-laplacian, PhD thesis, Ph. D. thesis.
39. W. H. Fleming and P. E. Souganidis, On the existence of value functions of two-player, zero-sum stochastic differential games, *Indiana Univ. Math. J.* **38**, 293 (1989).
40. P. Juutinen, Principal eigenvalue of a very badly degenerate operator and applications, *J. Differential Equations* **236**, 532 (2007).

41. P. Juutinen, Absolutely minimizing Lipschitz extensions on a metric space, *Ann. Acad. Sci. Fenn. Math.* **27**, 57 (2002).
42. P. Juutinen and B. Kawohl, On the evolution governed by the infinity Laplacian, *Math. Ann.* **335**, 819 (2006).
43. P. Juutinen, P. Lindqvist and J. J. Manfredi, The infinity Laplacian: examples and observations, in *Papers on analysis*, eds. J. Heinonen, T. Kilpelinen and P. Koskela, Rep. Univ. Jyväskylä Dep. Math. Stat., Vol. 83 (Univ. Jyväskylä, Jyväskylä, 2001) pp. 207–217.
44. A. Maitra and W. Sudderth, Borel stochastic games with lim sup payoff, *Ann. Probab.* **21**, 861 (1993).
45. J. J. Manfredi, M. Parviainen and J. D. Rossi, An asymptotic mean value characterization for p-harmonic functions, *Proc. Amer. Math. Soc.* **138**, 881 (2010).
46. M. Ohnuma and K. Sato, Singular degenerate parabolic equations with applications to the p-Laplace diffusion equation, *Comm. Partial Differential Equations* **22**, 381 (1997).
47. S. R. S. Varadhan, *Probability theory*, Courant Lecture Notes in Mathematics, Vol. 7 (New York University, Courant Institute of Mathematical Sciences, New York; American Mathematical Society, Providence, RI, 2001).
48. P. Wang, A formula for smooth ∞-harmonic functions, *Panamer. Math. J.* **16**, 57 (2006).

The Radon-Nikodým theorem for vector measures and integral representation of operators on Banach function spaces

E. A. Sánchez Pérez

Instituto Universitario de Matemática Pura y Aplicada,
Universitat Politècnica de València,
Camino de Vera s/n, 46022 Valencia, Spain
E-mail: easancpe@mat.upv.es

A well known vector measure version of the Radon-Nikodým theorem —comparing two vector measures— can also be interpreted in terms of factorization of an operator —the one associated to the first measure— through the one associated to the second one. Using several results published in recent years, we show the factorization version of this Radon-Nikodým theorem and its applications in the setting of the Operator Theory and the Harmonic Analysis.

Keywords: Banach function spaces; vector measures; Radon-Nikodým theorem; factorization of operators; multiplication operators; product spaces.

1. Introduction

In the theory of Banach function spaces, the classical Radon-Nikodým theorem provides a fundamental tool for representing functionals of the dual space as integrals of concrete measurable functions. In the vector valued case, vector valued versions of this result has also been a classical technique for providing integral representations of operators, and leads to the rich (geometric and topological) theory of Banach spaces with the Radon-Nikodým property. In this work we consider a slightly different approach to the problem of integral representations. Consider a couple of Banach space valued vector measures $m, n : \Sigma \to X$. In which cases it can be said that n can be written as an integral of m? In other words, we are interested in studying when there is a (scalar) m-integrable function f such that for every measurable set $A \in \Sigma$,

$$n(A) := \int_A f\, dm.$$

2000 Mathematics Subject Classification: 46E30, 47B38, 46B42.

The answer is given by a known Radon-Nikodým type theorem for couples of vector measures that will be presented in this work, as well as the corresponding operator counterparts. The way of obtaining the operator version is just to consider, for a couple of Banach valued operators $T : X(\mu) \to E$ and $S : Y(\mu) \to E$ from order continuous Banach function spaces $X(\mu)$ and $Y(\mu)$ over (Ω, Σ, μ), the corresponding vector measures m_T and m_S given by $m_T(A) = T(\chi_A)$ and $m_S(A) := S(\chi_A)$, $A \in \Sigma$. These results give interesting integral representations of operators, that provide a powerful technique for obtaining factorization of operators through Banach function spaces with applications in general Functional Analysis and Harmonic Analysis.

The paper is organized as follows. After the next preliminary section, we show in Section 3 the translation in terms of factorization of the Radon-Nikodým theorem for scalarly dominated vector measures. In Section 4 we present the general arguments that allow to prove what we call Maurey-Rosenthal type factorization of operators. Finally, in Section 5 we show how these results can be used for obtaining factorizations of general operators through a given one, which will be particularized in two subsections, showing some applications for relevant operators in Functional and Harmonic Analysis. The proof of some of the results presented here can be found in our reference[1].

2. Preliminaries

Let (Ω, Σ, μ) be a finite measure space, and consider the space $L^0(\mu)$ of all (classes of μ-a.e. equal) measurable real functions on Ω. A *Banach function space over* μ (briefly B.f.s.) $X(\mu)$ is an ideal of (classes of μ-a.e. equal) locally integrable functions with a complete lattice norm (see[2] p.28). That is, it is a a Banach space of integrable functions $X(\mu) \subset L^0(\mu)$ with norm $\|\cdot\|_X$ such that if $f \in L^0(\mu)$, $g \in X(\mu)$ and $|f| \leq |g|$ μ-a.e. then $f \in X(\mu)$ and $\|f\|_X \leq \|g\|_X$. We will write X instead of $X(\mu)$ if no explicit reference to the measure μ is needed. A B.f.s. X has the *Fatou property* if for every sequence $(f_n) \subset X$ such that $0 \leq f_n \uparrow f$ μ-a.e. and $\sup_n \|f_n\|_X < \infty$, it follows that $f \in X$ and $\|f_n\|_X \uparrow \|f\|_X$. We will say that X is *order continuous* if for every $f, f_n \in X$ such that $0 \leq f_n \uparrow f$ μ-a.e., we have that $f_n \to f$ in the norm. We will assume often that the space X is order continuous; in this case, it is well-known that the dual space X^* of X coincides with the so called Köthe dual X' of X, that is given by the functionals in X^* that allow an integral representation. Thus, we will write X' in this case instead of X^*. The notions of p-convexity for the Banach function spaces and p-concavity for operators are well-known; they can be found for example in[2,3].

Through the paper $T : X(\mu) \to E$ will be a linear and continuous operator, where E is a Banach space, and m_T the vector measure associated to T by $m_T(A) := T(\chi_A)$, $A \in \Sigma$, that is countably additive if $X(\mu)$ is order continuous.

Regarding vector measures and integration, let $m : \Sigma \to E$ be a *vector measure*, that is, a countably additive set function, where E is a real Banach space. A set $A \in \Sigma$ is *m-null* if $m(B) = 0$ for every $B \in \Sigma$ with $B \subset A$. For each x^* in the topological dual E^* of E, we denote by $|x^*m|$ the variation of the real measure x^*m given by the composition of m with x^*. There exists $x_0^* \in E^*$ such that $|x_0^*m|$ has the same null sets as m. We will call $|x_0^*m|$ a *Rybakov control measure* for m.

A measurable real function $f : \Omega \to \mathbb{R}$ is *integrable with respect to m* if

(i) $\int |f|\, d|x^*m| < \infty$ for all $x^* \in E^*$.
(ii) For each $A \in \Sigma$, there exists $x_A \in E$ such that

$$x^*(x_A) = \int_A f\, dx^*m, \quad \text{for all } x^* \in E.$$

The element x_A will be written as $\int_A f\, dm$.

We will deal with the space $L^1(m)$ of integrable functions with respect to m, where functions which are equal m-a.e. are identified. The space $L^1(m)$ is a Banach space endowed with the norm

$$\|f\|_m = \sup_{x^* \in B_{E^*}} \int |f|\, d|x^*m|.$$

Note that $L^\infty(|x_0^*m|) \subset L^1(m)$, for a Rybakov measure $|x_0^*m|$. In particular every measure of the type $|x^*m|$ is finite as $|x^*m|(\Omega) \le \|x^*\| \cdot \|\chi_\Omega\|_m$. We will write $L^\infty(m)$ for $L^\infty(|x_0^*m|)$.

Given $f \in L^1(m)$, the set function $m_f : \Sigma \to E$ given by

$$m_f(A) = \int_A f\, dm, \quad \text{for all } A \in \Sigma$$

is a vector measure. Moreover, $g \in L^1(m_f)$ if and only if $gf \in L^1(m)$ and in this case $\int g\, dm_f = \int gf\, dm$.

A useful tool for this note are the spaces of multiplication operators. Given two B.f.s.' X and Y, the space of *multiplication operators* from X to

Y is defined as

$$X^Y = \{h \in L^0(\mu) : hf \in Y \text{ for all } f \in X\}.$$

Notice that the functions are identified with the corresponding multiplication operators. A seminorm $\|\cdot\|_{X^Y}$ on X^Y can be defined as $\|h\|_{X^Y} = \sup_{f \in B_X} \|hf\|_Y$ for all $h \in X^Y$. Is a norm only in the case when X is saturated, that is, there is $g \in X$ such that $g > 0$ μ-a.e. In this case, X^Y is a B.f.s., and X^{L^1} is the Köthe dual of X (X'). For each $h \in X^Y$, we denote by $M_h \colon X \to Y$ the multiplication operator defined as $M_h(f) = hf$ for all $f \in X$ (see[4,5]). Other useful tool is given by the *product spaces*, that are in a sense the dual spaces of the ones of multiplication operators. Some recent references for these concepts are[6–10] (see also the references therein). In them the reader can find all the information on these spaces that is needed. The product space $X\pi Y$ of two B.f.s.' X and Y is the space of functions $f \in L^0(\mu)$ such that $|f| \leq \sum_{i \geq 1} |x_i y_i|$ μ-a.e. for some sequences $(x_i) \subset X$ and $(y_i) \subset Y$ satisfying $\sum_{i \geq 1} \|x_i\|_X \|y_i\|_Y < \infty$.

If $f \in X\pi Y$, denote

$$\|f\|_{X\pi Y} = \inf \Big\{ \sum_{i \geq 1} \|x_i\|_X \|y_i\|_Y \Big\},$$

where the infimum is computed over all sequences $(x_i) \subset X$ and $(y_i) \subset Y$ such that $|f| \leq \sum_{i \geq 1} |x_i y_i|$ μ-a.e. and $\sum_{i \geq 1} \|x_i\|_X \|y_i\|_Y < \infty$. If X, Y and $X^{Y'}$ are saturated then $X\pi Y$ is a saturated B.f.s. with norm $\|\cdot\|_{X\pi Y}$ (see[11,12]).

A relevant fact that is necessary for the arguments in this paper is the following (see this and related arguments in[1,8,11]). Let X and Y be two B.f.s.' containing $L^\infty(\mu)$ such that X is order continuous, the simple functions are dense in Y and $X^{Y'}$ is saturated. Then $X\pi Y$ is order continuous.

3. The first tool: vector measures and factorization of operators

As we said before, given a Banach function space $X(\mu)$ over a finite measure space (Ω, Σ, μ) and a Banach space valued operator $T : X(\mu) \to E$, it is always possible to define a set function $m_T : \Sigma \to E$ given by $m_T(A) := T(\chi_A)$, $A \in \Sigma$, that is in fact always a finitely additive vector measure. If the space $X(\mu)$ is order continuous or the operator satisfies the adequate order continuity requirements, m_T gives a (countably additive) vector measure, and consequently the following well-known factorization can be obtained for T: If $X(\mu)$ is order continuous, then T factors as

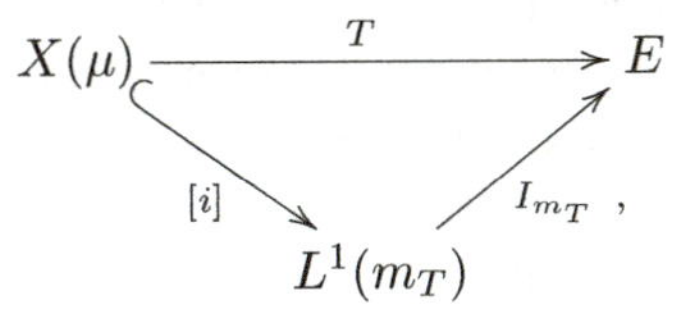

where $[i]$ is the map that send the μ equivalence class of a Σ-measurable function f to the m_T-equivalence class of f (notice that we always have that m_T is absolutely continuous with respect to μ) (see[3] Ch.3,4,5). Some new results regarding this scheme can be found in[13].

Using these ideas, we get what we call the first *General Integral Representation of Operators.* If $T : X(\mu) \to E$ is a continuous operator, then we can find a vector measure $m_T(A) := T(\chi_A)$, $A \in \Sigma$, such that

$$T(f) = \int f dm_T \quad \text{for every} \quad f \in L^1(m_T).$$

The Radon-Nikodým Theorem for vector measures (see[14]), that is closely related to our integral representation of the operator T, gives a different kind of information, since it provides a vector valued density function for the vector measure m_T. In the case that we have that m_T has an E-valued function density, we get that there is a function $\phi \in L^\infty(\mu, E)$ such that

$$T(f) = \int \phi_T \, f \, d\mu \quad \text{for every} \quad f \in L^1(m_T).$$

We are interested in exploring the possibilities of getting information on the operator T using a different kind of Radon-Nikodým theorem for vector measures. To establish it, we need the following notion. If $m, n : \Sigma \to E$ are countably additive vector measures such that there is a constant $K > 0$ such that for all $x^* \in E^*$ and for every $A \in \Sigma$,

$$|x^* n|(A) \leq K |x^* m|(A).$$

In this case, we say that n is scalarly dominated by m, and we write $n \ll m$.

Radon-Nikodým Theorem for vector measures. If $m, n : \Sigma \to E$ are countably additive vector measures such that $n \ll m$, then there is a real function $h \in L^\infty(m)$ such that

$$n(A) = \int_A h \, dm.$$

The reader can find the original proof of this result in[15]; a different proof based on the separation arguments that we explain below can be found in[1]. Related and general applications regarding this result can be found in[16,17].

A first consequence of this result is given by a direct translation of this notion in terms of operators. This provides a powerful tool for comparing operators on different Banach function spaces in terms of factorizations, whenever these spaces are defined over the same measure space. Let us use it to prove the following

Theorem 3.1. *Let (Ω, Σ, μ) a finite measure space. Consider a pair of operators $T : X(\mu) \to E$ and $S : Y(\mu) \to E$, where $X(\mu)$ and $Y(\mu)$ are order continuous Banach function spaces over μ. Suppose also that there is a constant $K > 0$ such that for each $x^* \in E^*$ and $A \in \Sigma$,*

$$\langle T(\chi_A), x^* \rangle \leq K |x^* m_S|(A).$$

Then T factors through I_{m_S} as $T = I_{m_S} \circ M_h$, where M_h is a multiplication operator associate to a function $h \in L^\infty(\mu)$, that is

$$\begin{array}{ccc} X(\mu) & \xrightarrow{\quad T \quad} & E \\ & {\scriptstyle M_h} \searrow \quad \nearrow {\scriptstyle I_{m_S}} & \\ & L^1(m_S) & \end{array} ,$$

Proof. Consider the associate vector measures m_T and m_S. Standard calculations show that the inequality in the statement of the theorem can be transferred to the corresponding variations of the scalarizations of the measures m_T and m_S, giving that for every measurable set $A \in \Sigma$,

$$|x^* m_T(A)| \leq |x^* m_S(A)|.$$

The application of the Radon-Nykodým Theorem given above provides the following integral representation for the images of the characteristic functions by T as

$$T(\chi_A) = \int_A h \, dm_S,$$

where $h \in L^\infty(m_S)$. Since the space $Y(\mu)$ is order continuous, we obtain that this integral formula can be extended to all the functions of the space, since simple functions are dense, obtaining

$$T(f) = \int_A f \, h \, dm_S, \quad f \in X(\mu).$$

Therefore, for each $f \in X(mu)$, we obtain that $T(f) = \int M_h(f)\, dm_S$, and so

$$\|T(f)\| \leq \|M_h(f)\|_{L^1(m_S)}.$$

This gives the desired factorization. □

4. The second tool: Maurey-Rosenthal type arguments for factorization of operators

Let us explain briefly which are the arguments that allow to prove a great part of the classical factorization theorems for operators on Banach function spaces though L^p-spaces, and in general, through other Banach function spaces as Orlicz spaces. The reader can find more information on these ideas in[18–21] and the references therein.

It provides results as the following one: *Let $X(\mu)$ be an order continuous p-convex Banach function space and a q-concave Banach space valued operator $T : X(\mu) \to E$. Then there is a multiplication operator $M_h : X(\mu) \to L^p(\mu)$ such that T factors as*

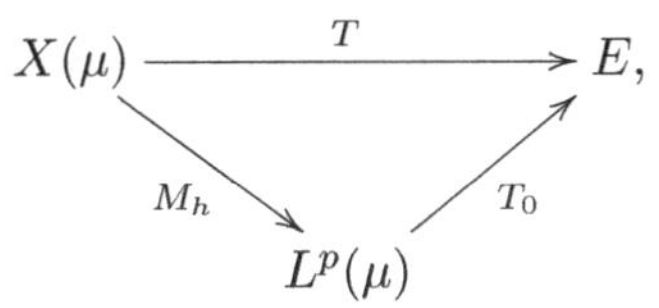

where T_0 is a continuous operator.

These kind of results are proved using a separation argument for which the main technical idea is given by the Hahn-Banach Theorem, or, alternatively by Ky Fan's Lemma. The scheme of the argument is the following.

- Take an operator as $T : X(\mu) \to E$, satisfying a lattice geometric inequality as the one given by the p-concavity of T.

$$\sum_{i=1}^{n} \|T(x_i)\|^p \leq \|\sum_{i=1}^{n} |x_i|^p\|.$$

- Since the space $X(\mu)$ is p-convex, the so called p-th power $X_p(\mu)$ of $X(\mu)$ is a Banach function space defined by the functions $f \in L^0(\mu)$ such that $f^p \in X(\mu)$ (see[3] Ch2).
- Consider the family $\mathcal{F}$ of continuous —with respect to the weak* topology— real functions $\Phi : B_{(X_p)^*} \to \mathbb{R}$ defined for $x_1, ...x_n \in X$

by

$$\Phi(x^*) := \sum_{i=1}^{n} \|T(x_i)\|^p - \sum_{i=1}^{n} \int |x_i \cdot x^*|^p \, d\mu, \quad x^* \in B_{(X_p)^*}.$$

- Hahn-Banach Theorem: For each $\Phi \in \mathcal{F}$ there exists $x^*_\Phi \in B_{(X_p)'}$, such that $\Phi(x^*_\Phi) \leq 0$.

$\Rightarrow$ CONCLUSION: There exists $x^*_0 \in B_{(X_p)^*}$ such that $\Phi(x^*_0) \leq 0$ for all $\Phi \in \mathcal{F}$.

This provides a factorization through L^p as the one given above. Although this is the better known argument for providing this kind of factorization, it can be extended to a broad class of situations, including abstract factorizations through much general Banach function spaces (see for example[19,22,23]).

This scheme, when used together with the Radon-Nikodým Theorem explained in the previous section, gives a general technique for establishing when an operator T can be factored through other given map S. We will give in the next section the main results that are nowadays known; they can be found in a series of papers recently published, mainly in[1].

5. Strong factorization of operators on Banach function spaces

Let (Ω, Σ, μ) be a fixed finite measure space and X_1, X_2, Y_1, Y_2 B.f.s.' related to μ such that $L^\infty(\mu) \subset X_1 \subset X_2$ and $L^\infty(\mu) \subset Y_2 \subset Y_1$. This guarantees that the spaces of multiplication operators $X_1^{X_2}$ and $Y_2^{Y_1}$ are B.f.s.' containing $L^\infty(\mu)$.

Consider two continuous linear operators $T : X_1 \to Y_1$ and $S : X_2 \to Y_2$. The question is: When does T *factor strongly* through S? That is, when does the following diagram commute?

$$\begin{array}{ccc} X_1 & \xrightarrow{\quad T \quad} & Y_1 \\ {\scriptstyle M_f}\downarrow & & \uparrow{\scriptstyle M_g} \\ X_2 & \xrightarrow[\quad S \quad]{} & Y_2 \end{array} \qquad (1)$$

for some $f \in X_1^{X_2}$ and $g \in Y_2^{Y_1}$?

The answer is given by the next theorems, that can be found in[1].

Theorem 5.1. *Assume that the Banach function spaces $Y_1, Y_2\pi Y_1'$ are order continuous and Y_1 has the Fatou property. The following statements are equivalent:*

(i) There exists a function $h \in X_1^{X_2}$ such that

$$\sum_{i=1}^{n} \int T(x_i)y_i'\,d\mu \le \Big\| \sum_{i=1}^{n} S(hx_i)y_i' \Big\|_{Y_2\pi Y_1'}$$

for every $x_1, ..., x_n \in X_1$ and $y_1', ..., y_n' \in Y_1'$.

(ii) There exist functions $f \in X_1^{X_2}$ and $g \in Y_2^{Y_1}$ such that $T(x) = gS(fx)$ for all $x \in X_1$, i.e. T factors strongly through S as

$$\begin{array}{ccc} X_1 & \xrightarrow{T} & Y_1 \\ {\scriptstyle M_f}\downarrow & & \uparrow{\scriptstyle M_g} \\ X_2 & \xrightarrow{S} & Y_2 \end{array} \tag{2}$$

Theorem 5.2. *Suppose that X_1, Y_1, X_2 and $X_1\pi X_2'$ are order continuous and X_2 has the Fatou property. The following statements are equivalent.*

(i) There exists a function $h \in Y_2^{Y_1}$ such that

$$\sum_{i=1}^{n} \int T(x_i)y_i'\,d\mu \le \Big\| \sum_{i=1}^{n} |S'(hy_i')x_i| \Big\|_{X_1\pi X_2'}$$

for every $x_1, ..., x_n \in X_1$ and $y_1', ..., y_n' \in Y_1'$.

(ii) There exist functions $f \in X_1^{X_2}$ and $g \in Y_2^{Y_1}$ such that $T(x) = gS(fx)$ for all $x \in X_1$, i.e. it factors as

$$\begin{array}{ccc} X_1 & \xrightarrow{T} & Y_1 \\ {\scriptstyle M_f}\downarrow & & \uparrow{\scriptstyle M_g} \\ X_2 & \xrightarrow{S} & Y_2 \end{array} \tag{3}$$

5.1. *General applications and known results*

Let us show an example of application of these results. If all X_1, X_2, Y_1, Y_2 coincide with an order continuous B.f.s. X having the Fatou property and containing $L^\infty(\mu)$, then $X\pi X' = L^1(\mu)$ (with equal norms), that is order continuous; this is a consequence of the well-known Lozanovskii's Theorem (see[11,24]). Also $X^X = L^\infty(\mu)$ (with equal norms). By the theorems given before we obtain the next

Corollary. The following statements are equivalent:

(i) T factors strongly through S, i.e.

$$\begin{array}{ccc} X & \xrightarrow{\quad T \quad} & X \\ {\scriptstyle M_f}\downarrow & & \uparrow{\scriptstyle M_g} \\ X & \xrightarrow{\quad S \quad} & X \end{array}$$

for some $f, g \in L^\infty(\mu)$.

(ii) There is a function $h \in L^\infty(\mu)$ such that

$$\sum_{i=1}^{n} \int T(x_i)x_i' \, d\mu \leq \int \Big| \sum_{i=1}^{n} S(hx_i)x_i' \Big| \, d\mu$$

for every $x_1, ..., x_n \in X$ and $x_1', ..., x_n' \in X'$.

(iii) There exists a function $h \in L^\infty(\mu)$ such that

$$\int T(x)x' \, d\mu \leq \int |S'(hx')x| \, d\mu$$

for every $x \in X$ and $x' \in X'$.

The Maurey-Rosenthal factorization explained in the previous section can also be considered as a direct consequence of these results. Let $X(\mu)$ be an order continuous p-convex Banach function space. As we explained there, an operator $T : X(\mu) \to E$ factors strongly through $L^p(\mu)$ if and only if T is p-concave, i.e.

$$\left(\sum \|T(f_i)\|^p\right)^{1/p} \leq Q \left\| \left(\sum |f_i|^p\right)^{1/p} \right\|$$

for $f_1, ..., f_n \in X(\mu)$. It can be considered as a particular case of our results. Indeed, consider the factorization diagram

$$\begin{array}{ccc} X(\mu) & \xrightarrow{\quad T \quad} & Y(\mu), \\ {\scriptstyle M_f}\downarrow & & \uparrow{\scriptstyle Id} \\ L^p(\mu) & \xrightarrow{\quad S \quad} & Y(\mu) \end{array}$$

where Id is the identity map in $Y(\mu)$. This provides a factorization as the ones in our theorems, when the operators T and S are considered.

5.2. *Kernel operators*

As a final application, let us show the characterization of the strong factorization of an operator through a fixed *kernel operator*. Consider a measurable function $K : \Omega \times \Omega \to [0, \infty)$ such that the kernel operator $S_K \colon X_2 \to Y_2$ given by

$$S_K(f)(s) = \int f(t)K(s,t)\, d\mu(t)$$

for all $f \in X_2$ and $s \in \Omega$, is well defined and so continuous. Under the requirements of the theorems, we have that there exists a function $h \in Y_2^{Y_1}$ such that

$$\sum_{i=1}^{n} \int T(x_i)y_i'\, d\mu \leq \left\| \sum_{i=1}^{n} |(S_K)'(hy_i')x_i| \right\|_{X_1 \pi X_2'}$$

for every $x_1, ..., x_n \in X_1$ and $y_1', ..., y_n' \in Y_1'$, if and only if T factors strongly through S_K, that is, there exist $f \in X_1^{X_2}$ and $g \in Y_2^{Y_1}$ such that

$$T(x)(s) = g(s) \int f(t)x(t)K(s,t)\, d\mu(t)$$

for all $x \in X_1$ and $s \in \Omega$. In this case, T is also a kernel operator with kernel $\widetilde{K}(s,t) = g(s)f(t)K(s,t)$.

Let us center our attention in a particular case. If X_1, X_2, Y_1, Y_2 all coincide with an order continuous B.f.s. X having the Fatou property and containing $L^\infty(\mu)$, by the corollary, we obtain that

- T factors strongly through S_K if and only if
- there exists $h \in L^\infty(\mu)$ such that

$$\int T(x)(t)x'(t)\, d\mu(t) \leq \int \left| x(t) \int h(s)x'(s)K(s,t)\, d\mu(s) \right| d\mu(t) \tag{4}$$

  for all $x \in X$ and $x' \in X'$.

Indeed, $S_K' \colon X' \to X'$ satisfies that

$$\begin{aligned}\langle S_K'(x'), x\rangle = \langle x', S_K(x)\rangle &= \int x'(s) \int x(t)K(s,t)\, d\mu(t)\ d\mu(s)\\ &= \int x(t) \int x'(s)K(s,t)\, d\mu(s)\ d\mu(t)\\ &= \left\langle \int x'(s)K(s,\cdot)\, d\mu(s), x \right\rangle\end{aligned}$$

for all $x' \in X'$ and $x \in X$.

Let us finish by showing a concrete example.

Example 5.1. The **Hardy operator** is given by the kernel

$$K(s,t) = \frac{1}{s}\chi_{[0,s]}(t)$$

for all $s, t \in [0,1]$. Suppose that $S_K : X \to X$ is well defined and continuous (e.g. $X = L^p[0,1]$ with $1 < p < \infty$). As a consequence of our theorems, we get that the following assertions are equivalent.

(i) There exist $g, f \in L^\infty[0,1]$ such that

$$T(x)(s) = \frac{g(s)}{s} \int_0^s f(t)x(t)\,dt \quad \text{a.e.}(s)$$

for all $x \in X$.

(ii) There exists $h \in L^\infty[0,1]$ such that

$$\int_0^1 T(x)(t)x'(t)\,dt \leq \int_0^1 \left| x(t) \int_t^1 \frac{h(s)x'(s)}{s}\,ds \right| dt$$

for all $x \in X$ and $x' \in X'$.

(iii) There exists $h \in L^\infty[0,1]$ such that

$$|T'(x')(t)| \leq \left| \int_t^1 \frac{h(s)x'(s)}{s}\,ds \right| \quad \text{a.e.}(t)$$

for all $x' \in X'$.

Acknowledgment

The author gratefully acknowledges the support of the Ministerio de Economía y Competitividad (project #MTM2009-14483-C02-02) (Spain).

References

1. O. Delgado and E. A. Sánchez Pérez, Strong factorizations between couples of operators on Banach function spaces, *J. Convex Anal.* **20**, 599 (2013).
2. J. Lindenstrauss and L. Tzafriri, *Classical Banach spaces. II*, Ergebnisse der Mathematik und ihrer Grenzgebiete [Results in Mathematics and Related Areas], Vol. 97 (Springer-Verlag, Berlin-New York, 1979), Function spaces.
3. S. Okada, W. J. Ricker and E. A. Sánchez Pérez, *Optimal domain and integral extension of operators*, Operator Theory: Advances and Applications, Vol. 180 (Birkhäuser Verlag, Basel, 2008), Acting in function spaces.
4. J. M. Calabuig, O. Delgado and E. A. Sánchez Pérez, Generalized perfect spaces, *Indag. Math. (N.S.)* **19**, 359 (2008).

5. L. Maligranda and L. E. Persson, Generalized duality of some Banach function spaces, *Nederl. Akad. Wetensch. Indag. Math.* **51**, 323 (1989).
6. P. Kolwicz, K. Leśnik and L. Maligranda, Pointwise products of some Banach function spaces and factorization, *J. Funct. Anal.* **266**, 616 (2014).
7. P. Kolwicz, K. Leśnik and L. Maligranda, Pointwise multipliers of Calderón-Lozanovskiĭ spaces, *Math. Nachr.* **286**, 876 (2013).
8. E. A. Sánchez Pérez, Factorization theorems for multiplication operators on Banach function spaces, *Integral Equations Operator Theory* **80**, 117 (2014).
9. E. A. Sánchez Pérez, Product spaces generated by bilinear maps and duality, To appear in Czec. Math. J.
10. F. Sukochev and A. Tomskova, (E, F)-Schur multipliers and applications, *Studia Math.* **216**, 111 (2013).
11. O. Delgado and E. A. Sánchez Pérez, Summability properties for multiplication operators on Banach function spaces, *Integral Equations Operator Theory* **66**, 197 (2010).
12. A. R. Schep, Products and factors of Banach function spaces, *Positivity* **14**, 301 (2010).
13. J. M. Calabuig, E. Jiménez, M. A. Juan and E. A. Sánchez Pérez, Optimal extensions of compactness properties for operators on banach function spaces, To appear in Topology Appl.
14. J. Diestel and J. J. Uhl, Jr., *Vector measures* (American Mathematical Society, Providence, R.I., 1977), With a foreword by B. J. Pettis, Mathematical Surveys, No. 15.
15. K. Musia$\mathcal{L}$, A Radon-Nikodým theorem for the Bartle-Dunford-Schwartz integral, *Atti Sem. Mat. Fis. Univ. Modena* **41**, 227 (1993).
16. J. M. Calabuig, O. Delgado and E. A. Sánchez Pérez, Factorizing operators on Banach function spaces through spaces of multiplication operators, *J. Math. Anal. Appl.* **364**, 88 (2010).
17. J. M. Calabuig, P. Gregori and E. A. Sánchez Pérez, Radon-Nikodým derivatives for vector measures belonging to Köthe function spaces, *J. Math. Anal. Appl.* **348**, 469 (2008).
18. A. Defant, Variants of the Maurey-Rosenthal theorem for quasi Köthe function spaces, *Positivity* **5**, 153 (2001).
19. A. Defant and E. A. Sánchez Pérez, Maurey-Rosenthal factorization of positive operators and convexity, *J. Math. Anal. Appl.* **297**, 771 (2004), Special issue dedicated to John Horváth.
20. A. Fernández, F. Mayoral, F. Naranjo, C. Sáez and E. A. Sánchez

Pérez, Vector measure Maurey-Rosenthal-type factorizations and l-sums of L^1-spaces, *J. Funct. Anal.* **220**, 460 (2005).
21. M. Masty$\mathcal{L}$o and E. A. Sánchez Pérez, Maurey-rosenthal factorization of operators through orlicz spaces, Preprint.
22. A. Defant and E. A. Sánchez Pérez, Domination of operators on function spaces, *Math. Proc. Cambridge Philos. Soc.* **146**, 57 (2009).
23. M. A. Juan and E. A. Sánchez Pérez, Maurey-Rosenthal domination for abstract Banach lattices, *J. Inequal. Appl.* **2013:213**, 1 (2013).
24. G. Y. Lozanovskii, On some banach lattices, *Siberian Math. J.* **10**, 419 (1969).

The Orlicz-Pettis theorem for multiplier convergent series

Charles Swartz

Mathematics Department, New Mexico State University,
Las Cruces, NM 88003, USA
E-mail: cswartz@nmsu.edu

1. Introduction

The Orlicz-Pettis Theorem was among the first abstract results in the early history of functional analysis. The version of the theorem for normed spaces asserts that a series in a normed space is subseries convergent in the weak topology is subseries convergent in the norm topology (a series $\sum_j z_j$ in a topological vector space Z is subseries convergent if the subseries $\sum_{j=1}^{\infty} z_{n_j}$ converges in Z for every subsequence $\{n_j\}$). The theorem was first established by Orlicz in[1] for weakly sequentially complete normed spaces but the general version was known to the Polish mathematicians and appears in Banach's book[2]. The first version available in English is due to Pettis and appears in[3]. It is not clear what motivated Orlicz to consider such a result but Pettis used the result to show that the integral which he defined in[3], and which now bears his name, was countably additive in the norm topology. It is interesting to observe that both proofs by Orlicz and Pettis used the equivalence of weak and norm convergence of sequences in l^1. For the history of the Orlicz-Pettis Theorem, see[4–6]. The theorem has been generalized in many directions. It was extended to locally convex spaces by McArthur[7] and Grothendieck[8], to a non-locally convex setting by Stiles[9], and even to topological groups by Kalton[10]. In this exposition we describe another generalization to multiplier convergent series.

Note that a series $\sum_j z_j$ in a topological vector space Z is subseries convergent iff the series $\sum_{j=1}^{\infty} \chi_\sigma(j) z_j$ converges in Z for every $\sigma \subset \mathbb{N}$, where χ_σ is the characteristic function of σ. Thus, if $m_0 = \text{span}\{\chi_\sigma : \sigma \subset \mathbb{N}\}$ is the space of all scalar sequences with finite range, a series $\sum_j z_j$ in Z is subseries convergent iff the series $\sum_{j=1}^{\infty} t_j z_j$ converges in Z for every $t = \{t_j\} \in m_0$. The Orlicz-Pettis Theorem can then be rephrased to state:

- If the series $\sum_{j=1}^{\infty} t_j z_j$ in the normed space Z converges in the weak topology for every $t = \{t_j\} \in m_0$, then the series $\sum_{j=1}^{\infty} t_j z_j$ converges in the norm topology for every $t = \{t_j\} \in m_0$.

This form of the theorem suggests the following question:

Can the sequence space m_0 be replaced in the theorem by other sequence spaces?

This is the question which we now address.

Let λ be a vector space of scalar sequences which contains the space c_{00} of sequences which are eventually 0.

Definition 1.1. A series $\sum_j z_j$ in a topological vector space Z is λ multiplier convergent if the series $\sum_{j=1}^{\infty} t_j z_j$ converges in Z for every $t = \{t_j\} \in \lambda$. The elements of λ are called multipliers.

Thus, a series is subseries convergent iff the series is m_0 multiplier convergent. A series which is l^{∞} multiplier convergent is often called bounded multiplier convergent.

We refer to a result which asserts that a series which is λ multiplier convergent in some weak topology is λ multiplier convergent in some stronger topology as an *Orlicz-Pettis Theorem.*

We first give an example which shows that an Orlicz-Pettis Theorem for a general space of multipliers is not possible.

Example 1.1. Let μ be scalar sequence space which consists of the sequences which are eventually constant. Then a series $\sum_j z_j$ in a topological vector space Z is μ multiplier convergent in Z iff the series converges in Z. Take a sequence $\{x_j\}$ in a normed space X which converges weakly to 0 but is not norm convergent (e.g., $\{e^j\}$ in c_0, where e^j is the sequence with 1 in the jth coordinate and 0 in the other coordinates). Then the series $\sum_j (x_{j+1} - x_j)$ is μ multiplier convergent in the weak topology but not μ multiplier convergent in the norm topology.

This example shows that to obtain an Orlicz-Pettis Theorem for multiplier convergent series, the multiplier space λ must satisfy some condition.

We now describe the setting which we will consider. Let X, X' be a pair of vector spaces in duality with respect to the bilinear pairing $\langle , \rangle$. Let $\sigma(X, X')$ be the weak topology on X from this pairing. Let $\mathcal{A}$ be a family of $\sigma(X', X)$ bounded sets which contains the finite sets and whose union is all of X' and $\tau_{\mathcal{A}}$ the polar topology of uniform convergence on the members of $\mathcal{A}$ which is generated by the semi-norms $p_A(x) = \sup\{|\langle x', x\rangle| : x' \in A\}$, $A \in \mathcal{A}$ (see[11], [12], [13] for polar topologies). The strong topology $\beta(X, X')$ [$\gamma(X, X')$, $\lambda(X, X')$, $\tau(X, X')$, respectively] is the topology $\tau_{\mathcal{A}}$ where $\mathcal{A}$ is the family of all $\sigma(X', X)$ bounded subsets of X' [$\mathcal{A}$ is the family of conditionally $\sigma(X', X)$ sequentially compact subsets (a subset $K \subset X'$ is conditionally $\sigma(X', X)$ sequentially compact if each sequence $\{x'_j\} \subset K$ has a subsequence $\{x'_{n_j}\}$ which is $\sigma(X', X)$ Cauchy, i.e., $\lim_j \left\langle x'_{n_j}, x\right\rangle$ exists for each $x \in X$), $\mathcal{A}$ is the family of all $\sigma(X', X)$ compact subsets, $\mathcal{A}$ is the family of all $\sigma(X', X)$ convex, compact subsets ($\tau(X, X')$ is the Mackey topology), respectively]. Any Banach (metrizable) space carries the strong topology (Mackey topology).

Let $w(X, X')$ be a polar topology defined for all dual pairs X, X'. We have the following useful notion introduced by Wilansky[13].

Definition 1.2. The topology $w(\cdot, \cdot)$ is a Hellinger-Toeplitz topology if whenever

$$T : (X, \sigma(X, X')) \to (Y, \sigma(Y, Y'))$$

is linear and continuous, then

$$T : (X, w(X, X')) \to (Y, w(Y, Y'))$$

is continuous.

For example, all of the polar topologies defined above are Hellinger-Toeplitz topologies.

The β-dual of a sequence space λ is defined to be

$$\lambda^\beta = \left\{ s = \{s_j\} : \sum_{j=1}^{\infty} s_j t_j = \{s_j\} \cdot \{t_j\} \text{ converges for every } t = \{t_j\} \in \lambda \right\}.$$

Then λ, λ^β form a dual pair under the bilinear map $s \cdot t = \{s_j\} \cdot \{t_j\}$ (since $c_{00} \subset \lambda$).

Definition 1.3. A sequence space λ with a Hausdorff vector topology is an AK-space if for every $t = \{t_j\} \in \lambda$, $t = \sum_{j=1}^{\infty} t_j e^j$, where the series converges in λ.

For example, c_0, l^p $(0 < p < \infty)$, and cs are AK-spaces under their natural topologies; $(m_0, \|\cdot\|_\infty)$, $(l^\infty, \|\cdot\|_\infty)$ are not AK-spaces.

2. The results

We have a necessary and sufficient condition for an Orlicz-Pettis Theorem for multiplier convergent series.

Theorem 2.1 ([14–16]). *Let w be a Hellinger-Toeplitz topology for dual pairs. The following are equivalent:*

(i) For every dual pair X, X' a series which is λ multiplier convergent for the weak topology $\sigma(X, X')$ is λ multiplier convergent with respect to $w(X, X')$.
(ii) $(\lambda, w(\lambda, \lambda^\beta))$ is an AK-space.

Proof. Assume (i). Then $\sum_j e^j$ is λ multiplier convergent with respect to $\sigma(\lambda, \lambda^\beta)$ so by (i), $\sum_j e^j$ is λ multiplier convergent with respect to $w(\lambda, \lambda^\beta)$. But, this means that if $t \in \lambda$, then $t = \sum_{j=1}^\infty t_j e^j$, where the series is $w(\lambda, \lambda^\beta)$ convergent so (ii) holds.

Assume (ii). Let $\sum_j x_j$ be λ multiplier convergent with respect to $\sigma(X, X')$. Consider the summing operator $S : \lambda \to X$,

$$St = \sum_{j=1}^{\infty} t_j x_j \qquad [\ \sigma(X, X') \text{ limit }].$$

Let $x' \in X'$, $t = \{t_j\} \in \lambda$. Then $\langle x', St\rangle = \sum_{j=1}^\infty t_j \langle x', x_j\rangle = \{\langle x', x_j\rangle\} \cdot t$, since $\{\langle x', x_j\rangle\} \in \lambda^\beta$ by the convergence of the series. This implies S is $\sigma(\lambda, \lambda^\beta) - \sigma(X, X')$ continuous and, therefore, $w(\lambda, \lambda^\beta) - w(X, X')$ continuous. If $t = \{t_j\} \in \lambda$, then $t = w(\lambda, \lambda^\beta) - \lim \sum_{j=1}^n t_j e^j$ so

$$St = w(X, X') - \lim \sum_{j=1}^{n} t_j x_j = \sum_{j=1}^{\infty} t_j x_j.$$

Hence, (i) holds. □

Condition (i) is, of course, just the conclusion of the Orlicz-Pettis Theorem for the Hellinger Toeplitz topology $w(X, X')$. Thus, in order to establish an Orlicz-Pettis Theorem for a Hellinger-Toeplitz topology, it suffices to check the AK-property for the topology $w(\lambda, \lambda^\beta)$. We now give several examples where this is the case. Theorem 2.1 applies to the spaces c_0 and l^p $(1 \leqslant p < \infty)$ since they are AK-spaces under their natural norm topologies. The spaces $(m_0, \|\cdot\|_\infty)$ and $(l^\infty, \|\cdot\|_\infty)$ are not AK-spaces but we will show that they are AK-spaces under different topologies.

Corollary 2.1. *Let λ be a barrelled AK-space. If $\sum_j x_j$ is λ multiplier convergent with respect to the weak topology $\sigma(X,X')$, then $\sum_j x_j$ is λ multiplier with respect to the strong topology $\beta(X,X')$.*

Proof. A barrelled space always carries the strong topology ([11] 21.2(2), [12] 24.7) so the result follows from Theorem 2.1 since the strong topology is a Hellinger-Toeplitz topology. □

Corollary 2.1 can be misleading. In general the Orlicz-Pettis Theorem does not hold for the strong topology as the following example shows.

Example 2.1. The series $\sum_j e^j$ is subseries convergent in l^∞ with respect to the weak topology $\sigma(l^\infty, l^1) = \sigma(l^\infty, (l^\infty)^\beta)$ but is not subseries convergent with respect to the strong topology $\beta(l^\infty, l^1) = \|\cdot\|_\infty$.

We obtain Orlicz-Pettis Theorems for the strong topology later by imposing strong conditions on the multiplier space λ.

We next establish an AK theorem for a general class of sequence spaces. The space of multipliers which we consider satisfy a gliding hump condition. An interval in $\mathbb{N}$ is a set of the form $[m,n] = \{j \in \mathbb{N} : m \leqslant j \leqslant n\}$ and a sequence of intervals $\{I_j\}$ is increasing if $\max I_j < \min I_{j+1}$. If s and t are scalar sequences, st will denote their coordinatewise product.

Definition 2.1. A sequence space λ has the signed weak gliding hump property (signed-WGHP) if whenever $t \in \lambda$ and $\{I_j\}$ is an increasing sequence of intervals, then there exist a sequence of signs $\{s_j\}$ and a subsequence $\{n_j\}$ such that the coordinate sum $\sum_{j=1}^\infty s_j \chi_{I_{n_j}} t \in \lambda$; if the signs can all be chosen to be equal to 1, then λ is said to have the weak gliding hump property (WGHP).

For example, any monotone space such as c_0, l^p $(0 < p \leqslant \infty)$, m_0 has WGHP (a sequence space λ is monotone if $\chi_\sigma t \in \lambda$ whenever $t \in \lambda$); the non-monotone space cs of convergent series has WGHP while the space bs of bounded series has signed-WGHP but not WGHP [17–19].

We require two additional results for our main theorem.

Definition 2.2. Let X be a vector space and σ and τ two vector topologies on X. We say that τ is linked to σ if τ has a neighborhood base at 0 consisting of σ closed sets.
[The terminology is that of Wilansky ([13] 6.1.9).]

For example, the polar topologies $\beta(X,X')$, $\tau(X,X')$, $\gamma(X,X')$ and $\lambda(X,X')$ are linked to the weak topology $\sigma(X,X')$.

Lemma 2.1. *Let X be a vector space and σ and τ two vector topologies on X such that τ is linked to σ.*

(i) If $\{x_j\} \subset X$ is τ Cauchy and if σ-$\lim x_j = x$, then τ-$\lim x_j = x$.
(ii) If (X, σ) is sequentially complete and $\sigma \subset \tau$, then (X, τ) is sequentially complete.

Proof. (i): Let U be a τ neighborhood of 0 which is σ closed. There exists N such that $j, k \geqslant N$ implies $x_j - x_k \in U$. Since U is σ closed, $x_j - x \in U$ for $j \geqslant N$.

(ii) follows from (i). □

The second result needed is a generalization of a matrix theorem of Antosik and Mikusinski due to Stuart [17–19].

Theorem 2.2 (Antosik-Mikusinski, Stuart). *Let Z be a topological vector space and $z_{ij} \in Z$ for $i, j \in \mathbb{N}$. Assume*

(a) $\lim_i z_{ij} = z_j$ exists for each j and
(b) For each increasing sequence of positive integers $\{n_j\}$ there is a subsequence $\{m_j\}$ and a sequence of signs $\{s_j\}$ such that the series $\sum_{j=1}^{\infty} s_j z_{im_j}$ converges for each i and $\lim_i \sum_{j=1}^{\infty} z_{im_j}$ exists.

Then $\lim_i z_{ii} = 0$.

For a proof of a more general result, see [17–19]. The theorem without the statement concerning the signs is due to Antosik and Mikusinski; the signed version is due to Stuart. A matrix satisfying conditions (1) and (2) is called a signed $\mathcal{K}$ matrix; if the signs $\{s_j\}$ can all be chosen equal to 1, the matrix is called a $\mathcal{K}$ matrix.

Theorem 2.3. *Assume that λ has signed-WGHP. Then*

(i) $(\lambda, \gamma(\lambda, \lambda^\beta))$ is an AK-space,
(ii) $(\lambda, \lambda(\lambda, \lambda^\beta))$ is an AK-space.

Proof. (i): Since $\gamma(\lambda, \lambda^\beta)$ is linked to $\sigma(\lambda, \lambda^\beta)$, it suffices to show that for every $t \in \lambda$ the series $\sum_{j=1}^{\infty} t_j e^j$ is $\gamma(\lambda, \lambda^\beta)$ Cauchy since the series is $\sigma(\lambda, \lambda^\beta)$ convergent to t (Lemma 2.1). Suppose that there exist $\varepsilon > 0$, $K \subset \lambda^\beta$ which is conditionally $\sigma(\lambda^\beta, \lambda)$ sequentially compact and increasing intervals $\{I_j\}$ such that

$$\sup_{u \in K} \left| u \cdot \sum_{j \in I_k} t_j e^j \right| > \varepsilon.$$

For each k pick $u^k \in K$ such that

$$\left| u^k \cdot \sum_{j \in I_k} t_j e^j \right| > \varepsilon. \tag{1}$$

There exists an increasing sequence $\{n_k\}$ such that $\{u^{n_k}\}$ is $\sigma(\lambda^\beta, \lambda)$ Cauchy. Define the matrix

$$M = [m_{ij}] = \left[u^{n_i} \cdot \sum_{l \in I_{n_j}} t_l e^l \right].$$

We show that M is a signed $\mathcal{K}$ matrix (Theorem 2.2). First the columns of M converge since $\{u^{n_i}\}$ is $\sigma(\lambda^\beta, \lambda)$ Cauchy. Next, if $\{p_j\}$ is an increasing sequence, there is a subsequence $\{q_j\}$ of $\{p_j\}$ and a sequence of signs $\{s_j\}$ such that $v = \sum_{j=1}^\infty s_j \sum_{l \in I_{n_{q_j}}} t_l e^l \in \lambda$. Then

$$\sum_{j=1}^\infty s_j m_{iq_j} = u^{n_i} \cdot v$$

so $\lim_i \sum_{j=1}^\infty s_j m_{iq_j}$ exists since $\{u^{n_i}\}$ is $\sigma(\lambda^\beta, \lambda)$ Cauchy. Hence, M is a signed $\mathcal{K}$ matrix and by the signed version of the Antosik-Mikusinski Matrix Theorem 2.2, the diagonal of M converges to 0. But, this contradicts (1) and establishes (i).

(ii): Consider λ with the Mackey topology $\tau(\lambda, \lambda^\beta)$ so the dual of $(\lambda, \tau(\lambda, \lambda^\beta))$ is λ^β. We claim that $(\lambda, \tau(\lambda, \lambda^\beta))$ is $\tau(\lambda, \lambda^\beta)$ separable. This follows since $(\lambda, \sigma(\lambda, \lambda^\beta))$ is an AK-space so the $\sigma(\lambda, \lambda^\beta)$ closure of $S = \text{span}\{e^k : k \in \mathbb{N}\}$ is $\sigma(\lambda, \lambda^\beta)$ dense in λ. But, S has the same closure in $\sigma(\lambda, \lambda^\beta)$ and $\tau(\lambda, \lambda^\beta)$ so S is $\tau(\lambda, \lambda^\beta)$ dense in λ and $(\lambda, \tau(\lambda, \lambda^\beta))$ is $\tau(\lambda, \lambda^\beta)$ separable. This implies that $\sigma(\lambda^\beta, \lambda)$ compact sets are sequentially compact ([13] 9.5.3). Now, the proof of part (i) may be repeated using a $\sigma(\lambda^\beta, \lambda)$ compact (sequentially compact) set $K \subset \lambda^\beta$. □

From Theorems 2.1 and 2.3 we obtain an Orlicz-Pettis Theorem for λ multiplier convergent series.

Corollary 2.2. *Assume that λ has signed-WGHP and let X be a locally convex space (LCTVS). If $\sum_j x_j$ is λ multiplier convergent with respect to the weak topology $\sigma(X, X')$, then $\sum_j x_j$ is λ multiplier convergent with respect to the topologies $\gamma(X, X')$ and $\lambda(X, X')$. In particular, if $\sum_j x_j$ is λ multiplier convergent with respect to $\sigma(X, X')$, then $\sum_j x_j$ is λ multiplier convergent with respect to the Mackey topology $\tau(X, X')$.*

The space m_0 has signed-WGHP (being monotone and having WGHP) so $(m_0, \tau(m_0, l^1))$ is an AK-space and Corollary 2.2 applies to subseries convergent series. We give a formal statement of the subseries result.

Corollary 2.3. *Let X be a LCTVS. If $\sum_j x_j$ is subseries convergent with respect to the weak topology $\sigma(X, X')$, then $\sum_j x_j$ is subseries convergent with respect to the topologies $\gamma(X, X')$ and $\lambda(X, X')$. In particular, $\sum_j x_j$ is subseries convergent with respect to the Mackey topology $\tau(X, X')$.*

The usual version of the Orlicz-Pettis Theorem for the Mackey topology is due to McArthur[7] and Grothendieck[8]. The version of the Orlicz-Pettis Theorem for $\gamma(X, X')$ is due to Dierolf[20] and the version for $\lambda(X, X')$ is due to Bennett and Kalton[21].

Dierolf has also shown that there is a strongest polar topology for which the conclusion of the Orlicz-Pettis Theorem holds[20]; a multiplier version is given in[19]. Tweddle has shown that there is a strongest locally convex topology for which the conclusion of the Orlicz-Pettis topology holds[22]; a multiplier version is given in[19].

Example 2.1 shows that in general the Orlicz-Pettis Theorem does not hold for the strong topology even when the multiplier space λ satisfies WGHP (is monotone). We show, however, that if the multiplier space satisfies stronger gliding hump conditions, then an Orlicz-Pettis Theorem does hold for the strong topology.

Definition 2.3. The sequence space λ has the infinite gliding hump property ($\infty - GHP$) if whenever $t \in \lambda$ and $\{I_j\}$ is an increasing sequence of intervals, there exist a subsequence $\{n_j\}$ and $a_{n_j} > 0$, $a_{n_j} \to \infty$ such that every subsequence of $\{n_j\}$ has a further subsequence $\{p_j\}$ such that the coordinate sum of the series $\sum_{j=1}^{\infty} a_{p_j} \chi_{I_{p_j}} t \in \lambda$.

The term "infinite" gliding hump property is used to suggest that the humps $\{\chi_{I_j} t\}$ are multiplied by a sequence of scalars which tend to infinity.

For example, the spaces c_0, l^p $(0 < p < \infty)$, cs have $\infty - GHP$; the spaces m_0, l^∞, bs do not have the property. Further examples are given in Appendix B of[19].

We use Theorem 2.1 to establish an Orlicz-Pettis Theorem for spaces with $\infty - GHP$.

Theorem 2.4. *Let λ have $\infty - GHP$ and let X be a LCTVS. Then $(\lambda, \beta(\lambda, \lambda^\beta))$ is an AK-space.*

Proof. Let $t = \{t_j\} \in \lambda$ and note the series $\sum_{j=1}^{\infty} t_j e^j$ converges to t with respect to $\sigma(\lambda, \lambda^\beta)$. Since $\beta(\lambda, \lambda^\beta)$ is linked to $\sigma(\lambda, \lambda^\beta)$, it suffices to show that for every $t \in \lambda$ the series $\sum_j t_j e^j$ is $\beta(\lambda, \lambda^\beta)$ Cauchy (Lemma 2.1). If this fails, there exist $\varepsilon > 0$, a $\sigma(\lambda^\beta, \lambda)$ bounded subset $B \subset \lambda^\beta$ and an increasing sequence of intervals $\{I_k\}$ such that

$$\sup \left\{ \left| u \cdot \sum_{j \in I_k} t_j e^j \right| : u \in B \right\} > \varepsilon.$$

For each k pick $u^k \in B$ such that

$$\left| u^k \cdot \sum_{j \in I_k} t_j e^j \right| > \varepsilon. \tag{2}$$

By $\infty - GHP$, there exist $\{p_j\}$, $a_{p_j} \to \infty$, $a_{p_j} > 0$ such that every subsequence of $\{p_j\}$ has a further subsequence $\{q_j\}$ such that $\sum_{j=1}^{\infty} a_{q_j} \chi_{I_{q_j}} t \in \lambda$. Define an infinite matrix

$$M = [m_{ij}] = [a_{p_j} \sum_{l \in I_{p_j}} t_l \frac{u_l^{p_i}}{a_{p_i}} e^l] = [a_{p_j} \frac{u^{p_i}}{a_{p_i}} \cdot \chi_{I_{p_j}} t].$$

We claim M is a $\mathcal{K}$ matrix. First, the columns of M converge to 0 since $\{u^{p_i}\}$ is $\sigma(\lambda^\beta, \lambda)$ bounded and $1/a_{p_i} \to 0$. Next, given any subsequence there is a further subsequence $\{q_j\}$ such that $v = \sum_{j=1}^{\infty} a_{q_j} \chi_{I_{q_j}} t \in \lambda$. Then

$$\sum_{j=1}^{\infty} m_{iq_j} = u^{p_i}/a_{p_i} \cdot v \to 0.$$

By Theorem 2.2 the diagonal of M converges to 0, but this contradicts (2). □

We now consider other conditions on the multiplier space which yield an Orlicz-Pettis Theorem with respect to the strong topology.

Definition 2.4. The space λ is c_0-factorable (called c_0-invariant by Garling[23] and c_0-composite by Li/Wang[24]) if $t \in \lambda$ implies that there exist $s \in c_0$, $u \in \lambda$ with $t = su$ [coordinate product].

For example, c_0, l^p $(0 < p < \infty)$ are c_0-factorable; other examples are given in Appendix B of[19].

We require a lemma of Li/Wang. We state a scalar version of their result. A far more general version for convex sets in vector spaces is given in[24].

Lemma 2.2. *Let* $\{x_{1,\ldots,}x_n\} \subset \mathbb{R}$, $\varepsilon > 0$, $M > 0$. *If*

$$M\left|\sum_{i\in\Delta} x_i\right| < \varepsilon \quad \textit{for } \Delta \subset \{x_{1,\ldots,}x_n\},$$

then

$$\left|\sum_{i=1}^{n} s_i x_i\right| < \varepsilon \quad \textit{for } 0 \leqslant s_i \leqslant M.$$

Theorem 2.5. *Assume* λ *is* c_0*-factorable and monotone. Then* $(\lambda, \beta(\lambda, \lambda^\beta))$ *is an AK-space.*

Proof. Let the notation be as in Theorem 2.4 up to condition (2).

Since λ is c_0-factorable, $t = sv$, where $s \in c_0$, $v \in \lambda$, and since λ is monotone, we may assume $s \geqslant 0$. Then $\left|\sum_{l\in I_k} u_l^k s_l v_l\right| > \varepsilon$. Put $r_k = \sup\{s_l : l \in I_k\}$. By Lemma 2.2 there exists $\Delta_k \subset I_k$ such that

$$\left|r_k \sum_{l\in\Delta_k} u_l^k v_l\right| = \left|r_k u^k \cdot \chi_{\Delta_k} v\right| > \varepsilon. \tag{3}$$

Define an infinite matrix

$$M = [m_{ij}] = [r_i u^i \cdot \chi_{\Delta_j} v].$$

We claim M is a $\mathcal{K}$ matrix. First, the columns of M converge to 0 since $r_i \to 0$ and $\{u^k\}$ is $\sigma(\lambda^\beta, \lambda)$ bounded. Next, given any subsequence $\{n_j\}$, $w = \sum_{j=1}^\infty \chi_{\Delta_{n_j}} v \in \lambda$ since λ is monotone. Then

$$\sum_{j=1}^{\infty} m_{in_j} = r_i u^i \cdot w \to 0.$$

By Theorem 2.2 the diagonal of M converges to 0, but this contradicts (3). □

From Theorems 2.1, 2.4 and 2.5, we have an Orlicz-Pettis Theorem for the strong topology.

Corollary 2.4. *Assume that* λ *has* $\infty - GHP$ *or is monotone and* c_0*-factorable and* X *is a locally convex space. If* $\sum_j x_j$ *is* $\sigma(X, X')$ λ *multiplier convergent, then* $\sum_j x_j$ *is* $\beta(X, X')$ λ *multiplier convergent.*

The version of the Orlicz-Pettis Theorem when λ has $\infty - GHP$ is established in[25] and when λ is monotone and is c_0-factorable is established in[26] (actually more general results for spaces of continuous linear operators are given in[26] and[25]).

Chen and Li have given examples of a space which has $\infty - GHP$ but is not monotone and c_0-factorable and a space which is monotone and c_0-factorable but fails $\infty - GHP$ so the results in Theorems 2.4 and 2.5 are independent.

References

1. W. Orlicz, Beitrage zur theorie der orthogonalent wichlungen ii, *Studia Math.* **1**, 241 (1929).
2. S. Banach, *Théorie des opérations linéaires* (Warsaw, 1932).
3. B. J. Pettis, On integration in vector spaces, *Trans. Amer. Math. Soc.* **44**, 277 (1938).
4. J. Diestel and J. J. Uhl, Jr., *Vector measures*, Mathematical Surveys, Vol. 15 (American Mathematical Society, Providence, R.I., 1977).
5. W. Filter and I. Labuda, Essays on the Orlicz-Pettis theorem. I. The two theorems, *Real Anal. Exchange* **16**, 393 (1990/91).
6. N. J. Kalton, The Orlicz-Pettis theorem, in *Proceedings of the Conference on Integration, Topology, and Geometry in Linear Spaces (Univ. North Carolina, Chapel Hill, N.C., 1979)*, ed. W. H. Graves, Contemp. Math., Vol. 2 (Amer. Math. Soc., Providence, R.I., 1980) pp. 91–100.
7. C. W. McArthur, On a theorem of Orlicz and Pettis, *Pacific J. Math.* **22**, 297 (1967).
8. A. Grothendieck, *Produits Tensoriels Topologiques et Espaces Nucleaires* (AMS, Providence, 1966).
9. W. J. Stiles, On subseries convergence in F-spaces, *Israel J. Math.* **8**, 53 (1970).
10. N. J. Kalton, Subseries convergence in topological groups and vector spaces, *Israel J. Math.* **10**, 402 (1971).
11. G. Köthe, *Topological vector spaces* (Springer-Verlag New York Inc., New York, 1969).
12. C. Swartz, *An introduction to functional analysis* (Marcel Dekker, Inc., New York, 1992).
13. A. Wilansky, *Modern methods in topological vector spaces* (McGraw-Hill International Book Co., New York, 1978).

14. C. Swartz and C. Stuart, Orlicz-Pettis theorems for multiplier convergent series, *Z. Anal. Anwendungen* **17**, 805 (1998).
15. C. Stuart and C. Swartz, Generalizations of the Orlicz-Pettis theorem, *Proy. J. Math.* **24**, 37 (2005).
16. Y. Tao and R. Li, Orlicz-Pettis theorem for λ-multiplier convergent operator series, *Bull. Austral. Math. Soc.* **75**, 247 (2007).
17. C. E. Stuart, *Weak sequential completeness in sequence spaces* (ProQuest LLC, Ann Arbor, MI, 1993), Thesis (Ph.D.)–New Mexico State University.
18. C. E. Stuart, Weak sequential completeness of β-duals, *Rocky Mountain J. Math.* **26**, 1559 (1996).
19. C. Swartz, *Multiplier convergent series* (World Scientific Publishing Co. Pte. Ltd., Hackensack, NJ, 2009).
20. P. Dierolf, Theorems of the Orlicz-Pettis-type for locally convex spaces, *Manuscripta Math.* **20**, 73 (1977).
21. G. Bennett and N. J. Kalton, FK-spaces containing c_0, *Duke Math. J.* **39**, 561 (1972).
22. I. Tweddle, Unconditional convergence and vector-valued measures, *J. London Math. Soc.* **2**, 603 (1970).
23. D. J. H. Garling, The β- and γ-duality of sequence spaces, *Proc. Cambridge Philos. Soc.* **63**, 963 (1967).
24. R. Li and J. Wang, Invariants in abstract mapping pairs, *J. Aust. Math. Soc.* **76**, 369 (2004).
25. C. Swartz, Orlicz-Pettis theorems for multiplier convergent operator valued series, *Proy. J. Math.* **23**, 61 (2004).
26. A. Chen and R. Li, A version of Orlicz-Pettis theorem for quasi-homogeneous operator space, *J. Math. Anal. Appl.* **373**, 127 (2011).
27. D. Noll, Sequential completeness and spaces with the gliding humps property, *Manuscripta Math.* **66**, 237 (1990).